Globalization and Environment

Globalization and Environment

Discourse, Policies and Practices

Edited by

Manish K. Verma

RAWAT PUBLICATIONS

Jaipur • New Delhi • Bangalore • Guwahati • Kolkata

ISBN 978-81-316-0709-1

Published by
Prem Rawat for **Rawat Publications**
Satyam Apts, Sector 3, Jawahar Nagar, Jaipur 302 004 (India)
Phone: 0141 265 1748 / 7006 Fax: 0141 265 1748
E-mail: info@rawatbooks.com
Website: www.rawatbooks.com

New Delhi Office
4858/24, Ansari Road, Daryaganj, New Delhi 110 002
Phone: 011 2326 3290

Also at *Bangalore, Guwahati* and *Kolkata*

Typeset by Rawat Computers, Jaipur
Printed at Nice Printing Press, New Delhi

Contents

Preface

It is quite evident to see that all over the world, globalization has mesmerized one and all in the last few decades. The process of globalization leading to social, economic, political and ecological changes and transformation has attracted the attention of scholars, policy makers, media and the people worldwide and thus remained the centre of debates and discussions. However, it is an important issue of concern whether to consider it benign and a blessing for the overall benefit of the humankind or a curse, as it is causing environmental degradation and giving rise to upsurge of poverty, unemployment and discrimination.

Contemporary academicians and intelligentsia, all over the globe, are in intense debate. There are scholars who perceive globalization as historically rooted and originated, in some tacit form, during the early modern period. This perspective finds support in the writings of various classical thinkers such as August Comte, Hegel, Karl Marx, Max Weber, Emile Durkheim etc. who, in some form or other, discussed in detail the process of transformation of the traditional society into a modern industrial society and associated trends of rising complexity, rationalities and unemployment and, thus, various characteristics of globalization in their writings. However, there is another school of thought which considers globalization as a completely new and the latest phenomenon the world has come across. According to them, it originated on the global level in the post-liberal era with liberalization, privatization and globalization (LPG) format in the early 1990s and provided a platform for the extraordinary growth of information and communication technologies (ICT) as well as played a significant role in terms of unification of the world in general and market in particular. Scholars like Joseph Stiglitz, Ronald Robertson, Anthony Giddens, Kaplan etc. provided substantial evidences through their writings in support of this viewpoint. However, it would be worth mentioning here that the idea and process of globalization gathered momentum in actual sense due to three historical happenings in the 1980s. They were the collapse of the erstwhile Soviet Union, the demolition of the Berlin Wall and the advent of the world wide web (www).

The most important point of debate among scholars is whether globalization is just an extension of the modernization process, or we can call it as a new model of development. There are scholars who use different concepts to understand the phenomenon of globalization as – 'knowledge society', 'information society', 'reflex society', 'global village' etc. Contrarily, there is another perspective which calls the globalization a process sponsored by the western hegemonic drive wherein the developed, powerful and technological giant western countries are using globalization as a tool to augment their control over not only the distribution of scarce resources and goods all over the world but also the ecology, environment, economy, society and governance of the developing and underdeveloped countries. Thus, in a way, the western countries are not only continuing with their age-old colonial policy of drain of wealth but to extend it further and making the globalization project more sharper, they are by and large intending to control knowledge, information, technology, economy, society and ultimately environment. The view of globalization as a project of western hegemony finds expression in the writings of scholars of 'world system theory', 'theory of development of underdevelopment' and 'centre-periphery theory'.

However, the way globalization process has operated during the last couple of decades, it has become clear that it is not a fixed but an evolving reality. Its contours have changed significantly in due course of its evolution and diffusion at global level. Launched by the developed countries as a hegemonic process of unilateral flow of trade and commerce over the underdeveloped countries in the 1980s, globalization has now converted into a multilateral flow of investment and services in the backward direction also.

Thus, it seems that there are as many interpretations of globalization, as the number of its interpreters and this should be no surprise, given its multifaceted and dynamic character. There should not be any hesitation among scholars in admitting that the process of globalization cannot operate in a vacuum. It requires certain medium to lay its stronghold and this platform is provided by the very environment itself – both biotic and abiotic part of it. The process of globalization is taking a ride over environment for its quickest expansion. Environment is also using the globalization process for its own benefits. It may be held that globalization is directly affecting environment and at the same time getting affected by it. The relationship is symmetrical. Globalization is affecting environment and society and at the same time its nature and direction are getting determined by the environment and society.

Against this backdrop, the effort is to debunk the enigma associated with the process of globalization, through the edited volume, and to understand comprehensively the inter-linkages between globalization, environment and society. The essays of this edited volume are crafted with a purpose to analyse and understand the discourse, policies and practices associated with the process of 'Globalization' which has caught the attention of the whole world. For the convenience of understanding, the book is broadly divided into four sections apart from one introductory chapter in the beginning. Each section is focused on one important facet of globalization, viz. 'emerging discourse', 'ecological dimensions', 'social dimensions' and 'economic dimensions'. First three units of the volume comprise four articles each, whereas, in the fourth and last unit, five articles are there. All the articles of the book are organized as per the four subthemes and are written by renowned academicians and experts of the subject area. The book is of important significance for academicians, post-graduate and graduate level students of social sciences and development studies, policy makers and the NGOs working in the area of globalization, environment and development.

In the last, it is time for showing my gratitude and acknowledgement to all those who stood with me for the completion of the work of edited volume. At the outset, I owe a deep sense of gratitude to all the contributors of the volume. Without their support and contribution, the edited volume on *Globalization and Environment: Discourse, Policies and Practices* would have been a distant dream. They have extended their full support, cooperation and forbearance at every stage of the completion of the book and shown intense willingness to help me in the editing work to make the volume a reality.

I take this opportunity to convey my gratitude to Prof. R.C. Sobti, the Hon'ble Vice-Chancellor, Babasaheb Bhimrao Ambedkar University, for his inspiring words which strengthened me at every moment of time to work diligently for the completion of the book. I also express my deep sense of gratitude to Prof. N.M.P. Verma, my senior in the university, whose support and guidance proved extremely valuable for the endeavour. The inspiration of Prof. Kameshwar Choudhary needs special mention here. He not only prepared me mentally to work for this project but also provided all kinds of support and assistance in formulation and sketching of the volume. The discussions held with him to streamline the format of the book proved extremely helpful. I thank him for giving motivation, support and encouragement. I also express my sincere thanks to my friends Dr Sanatan Nayak and Dr Venkatesh Dutta for their constant help, advice and support provided during the project. I am equally thankful to my friends Prof. R.B. Ram, Sri D. Jayabalachandran, Dr Kamal Jaishwal, Prof. Vipin Saxena, Dr D.R. Modi, Dr B.N. Dubey, Dr Brajesh Kumar, Dr Gopal Singh, Dr Govind Ji Pandey, Dr Jaya Srivastava and Mrs Rajshree. I take this opportunity to thank all my friends, well-wishers and colleagues at the university for their inspiring words. Their list is so long that it would be difficult to name them individually.

It is high time that I must show my sincere gratitude towards my teachers. It is their teaching and blessings which strengthened me at every moment of my life and empowered me to complete the work. Prof. Yogendra Singh, Prof. K.L. Sharma, Prof. Nandu Ram, Prof. Anand Kumar and Prof. Ehsanul Haq are foremost among such notable teachers.

It will be amiss if I do not mention the contribution of my family members whose encouragement, cooperation and patience enormously helped me to complete the volume. The cooperation of my wife Runu and the patience of my beloved son Aradhya highly encouraged me in the completion of this endeavour. Without their help, cooperation and forbearance, this work would not have been possible. It is the inspiration and blessings of my Baba and Dadi that I am in this position to fulfil all their dreams. Even from heavenly abode, their blessings are encouraging and empowering me to do well in my life. Lastly, I take this opportunity to thank my parents, brother and sisters, in-laws etc. for their love and cooperation extended to me at every moment of my life.

Finally, I owe a deep sense of gratitude to Mr Pranit Rawat and his entire team at Rawat Publications, Jaipur for taking personal interest in getting this volume out of the press with high level of efficiency and least amount of aberrations and errors.

Finally, I must admit that I was extremely benefitted from the suggestions of many colleagues in the process of editing this volume. However, the entire responsibility about the content and form of the edited volume is mine, and if there are shortcomings I alone am to be blamed.

Manish K. Verma

Contributors

Arvinder A. Ansari is Associate Professor, Department of Sociology, Jamia Millia Islamia, New Delhi.

Saket Bihari is a faculty member at the Indian Institute of Public Administration, New Delhi.

Venkatesh Dutta is currently teaching courses in Statistics, EIA, Environmental Economics and Management at the School of Environmental Sciences, BBA Central University, Lucknow.

Chamira Gamage hails from the country town of Armidale, New South Wales, Australia where he is engaged in research and writing activities in the field of International Relations and Development.

Ramanuj Ganguly is Head and Associate Professor in Department of Sociology at West Bengal State University, Barasat, Kolkata.

Biswajit Ghosh is Professor of Sociology and Director of UGC Academic Staff College at The University of Burdwan, West Bengal.

Habibul Haque Khondker is Professor, Department of Humanities and Social Sciences, Zayed University, United Arab Emirates.

Tapan R. Mohanty is Chairperson, Centre for Socio-Legal Studies and Associate Professor, National Law University, Bhopal.

Pampa Mukherjee is Professor and Chairperson in the Department of Political Science, Panjab University, Chandigarh.

Siddhartha Mukerji is working as Assistant Professor in the Department of Political Science, Babasaheb Bhimrao Ambedkar Central University, Lucknow.

Sanatan Nayak is Associate Professor at the Department of Economics, Babasaheb Bhimrao Ambedkar University, Lucknow.

Manisha Tripathy Pandey teaches at the Department of Sociology, Jamia Millia Islamia, New Delhi as Associate Professor.

Shweta Prasad teaches sociology at Banaras Hindu University, Varanasi.

Antara Ray is Assistant Professor, Department of Sociology, West Bengal State University, Barasat, Kolkata.

Ashish Saxena presently teaches at Department of Sociology as Associate Professor, Central University of Allahabad, Allahabad.

Vijaylaxmi Saxena, an independent researcher, is teaching sociology at undergraduate level at VMLG (PG) College, Ghaziabad and Calliope Law College, University of Jammu.

Rashi Krishna Sinha is Associate Professor, Department of Economics, Dr Shakuntala Misra National Rehabilitation University, Lucknow

Manish K. Verma is Associate Professor and Head, Department of Sociology, Babasaheb Bhimrao Ambedkar (Central) University, Lucknow.

N.M.P. Verma is Professor of Economics, Babasaheb Bhimrao Ambedkar University, Lucknow.

1

Introduction

Manish K. Verma

Globalization has ushered a new vista of life for the mankind living on this planet. The rapid pace of globalization has augmented a process of swift industrialization and urbanization which has far-reaching effect on various facets of human environment. By bringing people of different cultures together, it has contributed immensely for creation of a world culture and homogeneous lifestyle for people living at diverse places of the globe. As a result, the whole world has turned into a 'global village' (McLuhan 1960) diminishing the boundaries of time and space – a notion that implied a sort of compression of the globe and intensification of the consciousness of the world as a single entity. Consequently, its ramification is experienced in both material and non-material parts of society. It has led to improved communication, innovation and transportation, faster access to technology apart from bringing economic prosperity and opening up of a vast channel of development activities. It has intensified 'consciousness of the world as a whole' (Robertson 1992: 8) and 'local happenings are now shaped by events occurring miles away and vice-versa' (Giddens 1990: 64).

Undoubtedly, due to some very significant global processes, the economic, social, cultural, political and, most importantly, ecological and environmental aspects of the society are under transformation and these have immensely affected almost everyone living on this planet. Despite their manifold and multifarious positive impacts on the society, these have an ingrained predicament, as the process of globalization heavily impinges and makes a dent on the serene environment – be it social, economic, political or ecological per se. 'For most of its proponents, it is an irresistible and desirable force sweeping away frontiers, liberating individuals and enriching all it touches, whereas for a large number of its opponents, it is no less an irresistible force, but absolutely undesirable' (Ghuman et al. 2010: 1). 'Accordingly, the scholarship on globalization is divided: at one extreme are its proponents who regard it as coterminous with progress and the most effective route to end poverty; at the other end are those vilifying it as the source of all contemporary social, political, economic and cultural ills, not only in the developing world but also in the developed world. In between these two extremes are scholars who view globalization as a process, which, though a historical inevitable, can be managed for alleviating the vulnerabilities resulting from it, if not for reaping its promised benefits' (Jayaram and Rajasekhar 2012: 1–2).

Due to its all pervasive impacts on the society, perhaps no other term has received as much scholarly and media attention in the last couple of decades as globalization. Thus, Giddens has rightly called it as an inevitable force, transforming all aspects of contemporary society, politics and economy (Giddens 1999). While drawing attention towards the inherent discontent of globalization in terms of its asymmetrical, hegemonic and devastating effects on society, Joseph Stiglitz warns against the extreme reform strategies adopted by governments which he calls, 'shock therapy' that have failed miserably. He opines, 'government needs to – and can – adopt policies that help countries grow but also ensure that growth is shared more equitably'. He further asserts that, 'there are, of course, no magic solutions. But there are a multitude of changes to be made – in policies, in institutions, in the rules of the game, and in mindsets – that hold out the promise of helping make globalization work better ...' (Stiglitz 2012: ix-xvi). Similarly, much before Stiglitz writes, Wallerstein also opined in the similar fashion by considering 'globalization no more than a buzzword that describes the current situation in which the end of the "cold war" has led to the emergence of a unipolar world with an unassailable supremacy of the USA both in global economy and geopolitics'. To him, globalization is meaningless as an analytical concept. The concept only serves primarily as a political prescription for the developing world to opt for all kinds of short-cuts to integration with the capitalist global economy' (Dhanagare 2012: 33).

Before delving deep into the discourse on globalization, it would be better to trace its origin by peeping into the history.

History of Globalization

While tracing the origin of the most discussed and deliberated epitome, we find that 'although the word global is over 400 years old, the common usage of such words as globalization, globalize and globalizing did not begin until about 1960. In 1961, Webster became the first major dictionary to offer the definitions of globalism and globalization' (Waters 2001: 2). History is witness to the fact that globalization is not a new phenomenon and its evidences are found in the writings of several classical thinkers, scholars and academicians. 'Saint Simon noticed that industrialization was inducing commonalities of practice across the disparate cultures of Europe. Seeking to hasten the process, he argued for a utopian internationalism that include a pan-European government and a new and universalizing humanistic philosophy. Saint Simon's ideas found their ways through Comte to Durkheim. Durkheim's genuine legacy to globalization is his theories of differentiation and culture. He opined that industrialization tends to weaken collective commitments and to open the way for dismantling the boundaries between societies' (Waters 2001: 8). In a similar fashion where Durkheim identified structural specialization, Weber finds rationalization as the globalizing solvent among all the western cultures so as to set up an 'iron cage' for all moderns. Rationalization, to him, implies that all the cultures will characterize by the depersonalization of social relationships, the refinement of techniques of calculations, the enhancement of the importance of specialized knowledge, and the extension of technically rational control over both natural and social processes.

Karl Marx is probably the first classical theorist who most explicitly discussed about the impact of globalization while discussing capitalism. In the opinion of Marx, globalization caused an enormous increase in the power of the capitalist class because it opened up new markets for it. In so doing he identified an economic dimension of the process. As early as in the mid-19th century, Karl Marx had observed that capitalism as an economic system is by definition and necessarily an international economic system. In the 1960s, Andre Gunder Frank analysed the implication of the phenomena when advancing his thesis on 'capitalist accumulation' and the 'development of underdevelopment'. And, before the word 'globalization' had acquired currency, Immanuel Wallerstein (1979) spoke of 'the capitalist world economy' and the 'world system theory'. Of course, it is not just the term that is new; what is new is the pervasiveness of the phenomena and the speed at which it has engulfed all parts of the world (Jayaram and Rajasekhar 2012: 2). To the question of whether globalization is truly metamorphosing economy in a fundamental way, John Kenneth Galbraith, an eminent economist replied that 'such a reference to a new economy is just short of fraudulent'. What was once termed capitalism is now being called a market economy or corporate system, i.e. profit maximization through monopolies at any cost. But then the new expression 'market system' is preferred because it is seen as a more bland, ideologically neutral, and yet sophisticated way of describing the gory side of capitalism. Globalization is, thus, yet another way of putting the fact that capitalist regime gets extended beyond the political boundaries of a nation state (Dhanagare 2012: 33–34).

Definition and Meaning of Globalization

By defining the word 'globalization', Friedman and Kaplan state, 'the integration of everything with everything else. It is the process of integration of markets, finance and technology, especially information technology in a way that shrinks the world to a miniature in size and enables people from different parts of the globe to reach around the world faster, deeper and cheaper' (Friedman and Kaplan 2002: 64). By and large, the definition given by Friedman focuses on the economic aspect and characterizes globalization by confining it to market and economy alone. A more comprehensive definition of globalization is found in the words of E. Aninat which illuminate globalization as 'the process through which an increasingly free flow of ideas, people, goods and capital leads to the integration of economies and societies' (Aninat 2002: 4). He further argued that 'such global integration has boosted growth and incomes and has thereby helped to raise general standard of living in less developed and underprivileged parts of the world. With globalization not only sophisticated technologies have become available to developing countries, but also marked improvements in average literacy, quality of health and general life expectancy and, above all, greater integration have taken place and these have promoted human freedom. This has been possible because information sharing and market expansion now provide increasing choices to people' (Aninat 2002: 4).

In a monumental work on modernity, Giddens deliberated upon globalization. According to him, 'Globalization can be defined as the intensification of the world-wide social relations which link distant localities in such a way that local happenings are shaped by events occurring many miles away and vice-versa. This is a dialectical process because such local happenings may move in an obverse direction from the very distanciated relations that shape them. Local transformation is as much a part of globalization as the lateral extension of social connections across time and space' (Giddens 1990: 64). In his definition, Giddens keeps at centre stage the explicit notion of time and space. He emphasizes on locality and territoriality by stressing that the process of globalization is not merely about grand events but also about the autonomization of local lifeworlds. Globalization thus, symbolizes for localization also. 'Localization implies a reflexive reconstruction of community in the face of the dehumanizing implications of rationalizing and commodifying' (Waters 2001: 5). 'While considering globalization coterminous with modernization, Giddens attributed four dimensions of it: the nation-state system, the world capitalist economy, the world military order and the international division of labour. Each of these four, according to him, is an extension of societal modernity' (Dhanagare 2012: 30).

In a similar fashion, Robertson holds that 'globalization as a concept refers both to the compression of the world and the intensification of the consciousness of the world as a whole ... both concrete global interdependence and the consciousness of the global whole' (Robertson 1992: 8). A more broader definition of globalization is given by Waters in the words, 'A social process in which the constraint of geography on economic, political, social and cultural arrangements recede, in which people become increasingly aware that they are receding and in which people act accordingly' (Waters 2001: 5). Thus, it can be affirmed that 'globalization has changed the face of the world irrevocably: time has been compressed, communication revolutionized, the concept of self and community metamorphosed, and the idea of the state and its boundaries changed' (Jayaram and Rajasekhar 2012: 2). It has all pervasive impact on the lives and livelihood of the global citizens; at times it is positive and at the other it proves to be detrimental.

Further, 'the discovery that globalization has a political dimension first surfaced in the work of Burton, Keohane and Nye and Rosenau. Each of these political scientists noticed that political action was decreasingly confined to the sphere of the nation-state and that an elaborated web of trans-national connections was emerging alongside it. In general, contemporary political scientists continue to insist on this duality in the globalization process – the state is argued to retain sovereignty even while losing some of its effectivity. Rosenau's later work theorizes the existence of two 'worlds' on the planet, a state centric and a multi-centric world, the interaction between which creates turbulence' (Waters 2001: 12).

On the basis of various perspectives, definitions and meanings of globalization discussed above, its important characteristics churn out: material exchanges universalize; political exchanges internationalize; social networks transnationalize; and information and communication technology globalize in the process of globalization. A close scrutiny of the studies on globalization shows five main analytical frameworks on globalization in a broader perspective. These can be:

Globalization as Internationalization: It describes the growth of international exchange and interdependence between all the countries of the world. Hence, the word 'global' in globalization symbolizes for an adjective used for cross-border relationship between various countries of the globe.

Globalization as Liberalization: The perspective emerged with the evolution of World Trade Organization (WTO) which works for removing the trade barriers on the free flow of goods across borders. WTO has diminished the government and country specific barriers and restrictions on the flow of capital and goods between countries and set forth a universal liberal trade practice.

Globalization as Universalization: Globalization is viewed here as diminishing of geographical distance between countries and nations in terms of flow of idea, information and culture with the advent of sophisticated information and communication technologies. From this perspective, the world has become 'global village' (McLuhan 1960) having 'reflex' (Giddens 1990) characteristics. There is the spread of ideas, experiences and aspirations to people in all nooks and crannies of the world in order to evolve uniform and harmonized level of culture and awareness.

Globalization as Westernization or Modernization: A very strong perspective emerges in the analysis of some noted scholars who consider globalization nothing more than a word used alternatively for the process of westernization and modernization. From this perspective, the social structures of modernity (capitalism, industrialism etc.) and the nuances of westernization have hegemonized the whole world and thereby disintegrated the indigenous culture and the capacity of local self-determination of most of the vulnerable Third World countries.

Globalization as Deterritorialization: It is indicative of diminishing time and space barriers across nations and continents. It expresses the process of the reconfiguration of geography, so that social space is no longer wholly mapped in terms of territorial places, territorial distances and territorial borders. It became possible due to the emergence of highly sophisticated transportation and communication channels.

Thus, it is evident from the above discussion that the process of globalization has largely enveloped all aspects of human society and played a pivotal role in compression of the world into a single entity in economic, political, social, cultural and technological terms and thus unifying the world. Now trickling down to the question of globalization and environment linkages, and the theme of the book under consideration, the following paragraphs will be providing an analytical discussion.

Globalization and Environment Linkages

Since last couple of decades, the academicians, planners and administrators are focusing their prime attention to understand and record the changes brought about by the global processes on the environment and society and are trying to solve multifarious riddles attached with such a rapid process of social transformation. Yet, there is no agreement amongst scholars on exactly how to define 'globalization', nor on exactly which parts of it are good or bad, and for whom. But despite this, undoubtedly, it is crystal clear that there are two-way linkages between globalization and environment. From that perspective, globalization and environment are interdependent: not only the process of globalization can impact the environment, but the dynamics of the environment can also influence and shape the nature of globalization. On a much higher level, rather than pondering into the nature of interdependence between globalization and environment, in present context, the truly relevant policy question would be about who benefits and who does not; how the benefits and the cost of these processes can be shared fairly between all the citizenry of the world; how the opportunities can be maximized by all; and lastly and most importantly, how the risks can be minimized so as to safeguard the earth from the discontents of the most glorified, discussed and debated phenomena in the 21st century – globalization.

While addressing these issues, it is understandable that globalization is quite a complex phenomena and offers not only many opportunities for the betterment of the humankind, but also involves significant number of potential perils. In its present form, contemporary globalization has manifold manifestations but three of them are particularly relevant to planners, academicians and intellectuals. They are more important because of having significant environmental opportunities and risks as well at the global level. According to Najam et al. (2008: 5–9), these are:

1. *Globalization of the Economy:* The world economy globalizes as national economies integrate into the international economy through trade; foreign direct investment; short-term capital flows; international movement of workers and people in general; and flows of technology. This has created new opportunities for many; but not for all. It has also placed pressure on the global environment and on natural resources, straining the capacity of the environment to sustain itself and exposing human dependence on our environment. A globalized economy can also produce globalized externalities and enhanced global inequalities. Local environment and economic decisions can contribute to the global solutions and prosperity, but the environmental costs, as well as the economic ramifications of our actions, can be externalized to places and people who are so far away as to seem invisible.
2. *Globalization of Knowledge:* As economies open up, more people become involved in the process of knowledge integration and the deepening of non-market connections, including flows of information, culture, ideology and technology. New technologies can solve old problems, but they can also create new ones. Technologies of environmental care can move across boundaries quicker, but so can technologies of environmental extractions. Information flows can connect workers and citizens across boundaries and oceans, but they can also threaten social and economic networks at the local level. Environmentalism as a norm has become truly global, but so has mass consumerism.
3. *Globalization of Governance:* Globalization places great stress on existing patterns of global governance with the shrinking of both time and space; the expanding role of non-state actors; and the increasing complex interstate interactions. The global nature of the environment demands global environmental governance, and indeed, a worldwide infrastructure of international agreements and institutions has emerged and continues to grow. But many of today's global environmental problems have outgrown the governance system designed to solve them. Many of these institutions, however, struggle as they have to respond to an ever-increasing set of global challenges while remaining constrained by institutional design principles inherited from an earlier, more state-centric world.

In addition to Najam's Trinitarian model, there are two other important aspects which need focused attention through the subsequent points mentioned in 4 and 5:

4. *Globalization of Ecological Resources:* The advent of capitalism saw a new era for the procreation of trade across the globe and a large segment of export and import comprising of such entrepreneurial activities was in terms of transport of goods and natural resources. In the beginning, the motto behind such trades was profit enhancement as well as fulfilment of the needs of the society. As long as it was within the ambit of ecological balance, there was no problem with the practice. But the mad race of profit enhancement as well as destruction of natural resources in order to attain various goals started proving disastrous for the ecological and environmental sustenance. Global warming, deforestation, soil erosion, air pollution, water depletion, floods, earthquakes and many more catastrophes are a direct outcome of that. If globalization has to flourish, it has to judiciously harness the environmental resources so that ecological balance can be maintained.
5. *Globalization of Culture and Society:* Even though the process of globalization started with the economic reforms, trade practices and liberal transactions but, at the same time it compressed the world into an entity with the rise of sophisticated means of transportation and communication. It resulted in increased movements of people across the globe, more societal interaction and ultimately diffusion of culture. The world has become 'reflex' and cultures and customs were no more restricted to any specific region. Such large scale diffusion of culture is helpful for a homogeneous world, but, at the same time, there is inherent danger in it. There is an invariable apprehension that the hegemonic relationship between nations may prove detrimental for cultural domination and subordination. Thus, the danger of loss of culture followed by people's identity shall also be at risk. It may ultimately result in loss of civilization.

Thus, it would be apt to state that the process of globalization has stirred the economic prospects of the world by interconnecting various economies across the globe by diminishing regional and country-specific, government-created barriers. Further, accelerated economic interactions led to the spurt in the dissemination of information and augmented a way for the creation of universal understanding and awareness. Both the practices led to the initiation of global governance, wherein many a decisions related to the human beings are taken at the stage of world bodies and its impact is experienced at the level of common man due to strengthening of global processes. Policies of liberalization and privatization followed in India, since 1991, have enhanced India's linkages globally. This process of globalization has important ecological, social and economic dimensions. The rate of economic growth has substantially increased due to policy changes in the country. But there are problems with the existing regime in the way to promote sustainable development rather than just economic growth.

The Present Volume

The whole world has been swept under the currents of globalization since last more than two decades. Every individual, group, society and nation, in contemporary period, aspires to be a part of the globalization process and thereby further accelerating it. A country's progress is supposed to be closely associated with the globalization process. At the present juncture, it is hard to believe that the process of globalization may be an impediment. Thus, it would be wise that rather than discarding it, we should try to harness maximum benefit from it. Hence, the volume under consideration goes with the assumption that globalization is essentially rooted in the environment and the effort should be to harness maximum benefit out of it.

Thus, the edited volume focuses on a more pragmatic level by establishing the inter-linkages between environment, economy, society and polity in the globalized world from the sustainable development perspective. It is held here that the fruits of globalization can only be achieved if the process keeps in tandem social equity, economic efficiency, ecological resilience and democratic governance. Only a balance between these four will be able to convert the globalization project a success for the human existence on the earth. The perspective can be illustrated with the diagram as shown in Figure 1.

As we all know the process of globalization started with the economic reforms in the early 1990s to enhance 'economic efficiency' and for attaining

Figure 1

Sustainable Development Paradigm

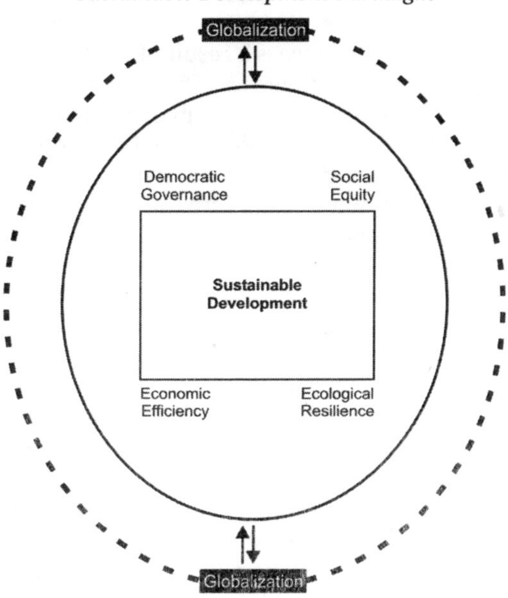

high economic growth and development of the people. While talking about 'economic efficiency', it is essential that there should be a balance between marginal cost (MC) of a product and marginal benefit (MB) inherent in it. It will be a 'win-win' situation for both market and society and will pave the way for sustainable development. However, if MC is greater than MB, there will be inefficient outcome and both economy and society will lose. On the other hand, if MC is lesser than MB, society will gain at the cost of market. The situation will be good for the society but not for the market wherein market will collapse in the long run. Thus, only a balance between MC and MB can provide a balanced 'economic environment'.

The next item under consideration in the 'Sustainable Development Paradigm' is 'Social Equity' which connects access, participation and inclusion of the masses. The allocation of resources of globalization should be accessible to all, and everyone should be equal participants in the decision-making process which will ultimately decide their fate. The mantra of attaining success in the realm of 'social environment' would be inclusion of everyone in the societal affairs and even distribution of economic resources and status as well.

Due to spurt in massive developmental activities in the present era, as a result of acceleration in urban and industrial enterprises, the ecological and environmental balance of the society is under a dreadful threat. Thus, rather than participating in the mad race of developmental activities by altering the balance of nature, the nations of the globe should adopt the policy of 'resource conservation', through which the scarce resources of nature can be utilized judiciously keeping in view the rights of the forthcoming generations on the earth. It must be noted that environmental pollution and ecological imbalance is a result of the failure of both market and society, wherein both of them are showing short-sightedness and forgetting their responsibility towards the planet earth.

All the three above-mentioned goals of sustainable development model can effectively be realized if there exists a 'democratic governance'. The notion of democratic governance starts with the practice of political participation of the masses and common people in the decision-making processes. Here, the role of civil society becomes very important to keep checks and balances between the state and the society. Failing which, the dream of egalitarian society cannot be achieved and there will be as many power centres in the society as there are groups having conflicting interests. The equilibrium of the state and society, thus, will be in disarray.

In this background, the effort is being made to keep alive the debate of concurrence between globalization, environment and society and to understand the issues comprehensively. In order to address the debate and reach a conclusive stage, from where we can foresee the future of the world, the edited volume on the theme, *Globalization and Environment: Discourse, Policies and Practices,* has been crafted by incorporating various interconnected themes having convergence with globalization and environment. For ease of understanding, the book is broadly divided into four sections apart from one introductory chapter in the beginning. Each section is focused on one important facet of globalization, viz. 'emerging discourse', 'ecological dimension', 'social dimension' and 'economic dimension'. First three units of the volume comprise four articles each, whereas, in the fourth and last unit five articles are being given. All the articles of the volume are concerned with the sub-theme and are written by renowned academicians and experts of the subject area. It is worth mentioning at the outset that though the social aspects of globalization and environment linkages are discussed in the volume, but the cultural aspects are not covered.

I

The first unit of the volume, 'Emerging Discourse on Globalization and Environment' is having four chapters in all, focusing on various emergent discourses related with globalization and environment.

The first article in this section is contributed by Saket Bihari on the theme, 'Dilemma of Globalization'. The paper traces the genesis and recapitulates the multiple interpretation of globalization as a term. Globalization is explained in various deliberations as speeding up of the flow of capital, worldwide mode of transport and communication, movements of people, goods and services, and ideas across the world; thus pointing to the general increase in the pace of global interactions and processes. It also intensifies the links, modes of interaction and flows that interconnect the world ties across borders that are not sporadic or haphazard but somewhat regularized. The trajectory also enables people to control social, cultural, political and economic practices even from distant places. The decision made in one part of the globe influences the other part. Even the ideas, happenings, decisions and practices created on the front line of the globe are having consequences on the communities and cultures of remotely located places. Thus, it also enables a reciprocal relation between local and global. The impact on local space may also have global repercussions and vice versa. The idea of cultural specificity is a core issue on which the paper focuses its attention and substantiates the propositions with the empirical understanding. Human trafficking, the growth of satellite communication technologies, consumption and consumerism, cosmopolitan lifestyle, world tourism, ecological crisis, human rights, and interchange of world religions have been considered some of the relevant issues to be discussed in this paper. It also embarks upon the emerging global experiences and traces the synergy with the Indian one in its local context. The posited idea of this paper may be useful in redrafting the programmes and policies for the development of shining India.

Pampa Mukherjee's article on 'The World Beyond State and Market: Locating Elinor Ostrom in India' is the second contribution of the unit. It begins with the opinion that in recent years the common resources have gained attention, particularly in countries like India, where the rural poor are dependent on such resources for their survival. Due to its fast depletion all over the world, the key concern among policy analysts and scholars are regarding its governance and in identifying appropriate institutions to manage them. There has been intense debate on this aspect amongst the social scientists. A group of them recommend that 'the state' is the best possible alternative when market arrangements fail, while opposed to this viewpoint, other schools see privatization as the only solution. In this context Nobel laureate, Elinor Ostrom's work on governance of common pool resources is significant as it goes beyond the dominant debate between the state and market and strongly recommends institutional diversity. Such theoretical formulations are in fact extremely relevant in the Indian context, where there exists numerous community-based institutions managing diverse resources ranging from water to forests. Taking recourse to Ostrom's work on Common-Pool Resource (CPR), this paper highlights on the debates on institutional arrangements in governing these resources. Another important thrust of the paper is to examine into the relevance of Ostrom's thesis in the Indian context.

The third submission of this unit hails from Rashi Krishna Sinha titled, 'Socio-economic Dimensions of Globalization in India: Challenges and Prospects'. Sinha opines that human history is replete with instances where people from across the globe interacted with one another at different levels through different means in order to attain divergent objectives. Globalization might not have been envisaged then as it is in the contemporary era. Integration of the economy with the rest of the world has its pros and cons – opportunities and challenges. It has within its ambit cultural, social, political, legal, economic, governance and other related aspects which directly or indirectly, constructively or obstructively affect one and all. To be out of its ambit neither was nor is an option. Knowingly or unknowingly, willingly or unwillingly we all are an integral part of it. The crux of the matter is how we are preparing ourselves to meet the challenges and how well equipped we are to avail the opportunities that are there. Are we preparing our people to get attuned to the changes we confront? Are the benefits/fruits of this confluence trickling down to the masses? These are few but obvious anxieties one comes across when we still debate over the right ways to gauge the exact numbers suffering from destitute, poverty and hunger. The author opines that welfare state does everything else than the welfare. Eradication of poor rather than poverty seems to have become the hidden agenda nowadays. In a country like India, where bottom quintile of the population spends around 67 per cent of their total expenditure on food and notwithstanding the increase in aggregate agriculture production per capita availability of cereals has gone down, it becomes pertinent to look at the state of food and nutritional security of the vulnerable people – the consumption pattern of food, sources of income generation and how that is spent so as to somehow have an idea of the state of their well-being. Vulnerable sections, groups or classes, in a society are found in developed countries but are prominent in developing and the least developed nations. In India, norms and cultural practices are rooted in a highly patriarchal social order where women are expected to adhere to strict gender roles and they are vulnerable to much more problems as compared to men. The prevalence of poverty and economic dependence among women, gender bias in the society at large, the limited power and their lack of influence in decision-making are social realities that have an adverse impact on their physical, psychological and cognitive well-being. They have little control on the resources and on important decisions related to their lives. In the Indian context, caste may be considered broadly as a proxy for socio-economic status and poverty. In this background, the paper is divided into five sections. Section I puts forth the idea of globalization and few pertinent issues pertaining to India in brief. Section II dwells upon the conceptual framework of hunger, food and nutritional security. Section III analytically examines the state of food security at the global level in brief, and the condition of food and nutritional security in India in the post-reform period. It also undertakes trend analysis of cereal consumption pattern across expenditure groups, changes in consumption pattern in rural and urban areas, examines the source of income of farmer's households. The penultimate section dwells upon the state of women's welfare and empowerment with respect to certain selected indicators. Lastly, section V is based on drawing inference based on the aforestated sections.

The fourth and last article in the section is submitted by Habibul Haque Khondker titled, 'From the Silent Spring to the Globalization of the Environmental Movement'. The paper traces the history of environmental movement and the growing awareness about sustainable living. Taking *The Silent Spring* of Rachel Spring (published in 1962) as a point of departure, this paper navigates through the interplay of consciousness raising, knowledge production, institution building, environmental movement and its globalization part. The paper examines the lack of development of a holistic understanding of environmental challenges of the world in the face of issue-based social movements. It presents the transformation of ideas of development from a narrow economic development as measured in income growth to a more inclusive sustainable development where the focus is quality of life, well-being and happiness. It also provides a brief history of the environmental activism prompted by the growing awareness about environmental and ecological sustainability to the globalization of the environmental movement. In this process, environmental activism launched by the civil society, the state and interstate institutions played a role. Movements for political rights and economic entitlements need to include environmental rights as an integral part of social movements. The findings of the paper suggest that in many parts of the Global South, as well as the Global North such an integrated, thus holistic movement is still at a rudimentary stage.

II

The second unit of the volume on Ecological Dimensions highlights the inter-linkages between globalization and environment vis-à-vis ecology.

The chapter on, 'A Review of the Climate Change Debate, its Causes and Necessary Action in the Context of Developed and Developing Countries' by Chamira Gamage is the first article of the section. This chapter explores the nature of global debate on climate change and its causes giving consideration to overpopulation and overconsumption as contributory factors. It looks at the role played by developed and developing countries in atmospheric pollution and the necessary action to be taken to address the problem on the basis of a review of relevant literature. Why it is necessary to discuss this topic is also explained while concluding that the consumption habits of middle to higher income demographics, coupled with population growth, are significant contributors to greenhouse gas emissions and change in climate. The paper concludes that the spread of consumerist lifestyle and the rise in use of manufactured products of developed countries by developing countries are considered as disturbing and complicating factors in the debate on climate change.

The second contribution of the section is made by Venkatesh Dutta on, 'India's Climate Policy and Barriers to Low-Carbon Inclusive Growth'. This paper examines India's climate policy and the bottlenecks in transition to low-carbon inclusive growth strategies through an expert survey and review of policy documents. The author states, on the basis of expert interviews, that it is a common consensus among experts that, while India has initiated several programmes for ensuring development of its economy along low-carbon pathways, the targets are unlikely to be achieved without significant new resources. The higher cost of cleaner and newer technology was reported as the most important barrier to low-carbon growth in India, followed by split incentives – cost/benefit mismatch, and limited access to private capital. The other important barrier relates to lack of institutional capacity and regulatory uncertainty. The paper concludes that climate crises will further expose the weaknesses of central, state and local governance systems and institutions, thereby simultaneously aggravating the crisis, however, offering a chance for retrofitting and renewal.

Ashish Saxena and Vijaylaxmi Saxena's article on 'Developmentality, Multimedia Technologies and Growing Ecological Concerns in Global Era: A Critical Appraisal from Rural India' is the third article of the unit. The crux of the paper goes with the assumption that people all over the world have high hopes that the new technologies will lead to better quality of life in terms of greater social freedom, increased knowledge and livelihoods that are more productive. Putting the discussion in the framework of developing country like India, one finds that transforming rural India is a big challenge and that should be the focus of attention of the best minds of Indian intellectuals and planners. At the onslaught of liberalization and globalization, the critical analysis of developmental model and the level of inequitable inclusion of the rural population and the marginals are urgently needed. The author suggests that there is an urgent need of multi-specialists' task force having sensitivity towards the critical issues and with enough courage to evolve alternative rural development policies beyond elitist paradigm.

Tapan R. Mohanty's article on, 'Perils of a Fragile Planet: Paradigms and Possibilities' follows in the last part of the unit. The chapter begins with the opinion that in the era of globalization and rapidly transforming socio-economic institutions, the politico-legal institution of state has been under severe challenge. The economics of globalization has taken away the absoluteness of sovereignty in terms of management of finances while 'Macdonaldization' and 'Cocacolanization' has ensured a greater transfer of 'consumerism' as a value, at least among the youth, resulting in continued 'civilizational clash' between the neo-rich and an austere traditional elite. The author vies that tradionally the vast rural and tribal areas were out of the orbit of such consumption pattern, thus, not posing a serious threat to ecology. The anti-hedonistic and esoteric values of Hindu social order and its strict adherence among the clergy and vast sections of peasantry helped in maintaining the balance between the 'nature' and 'culture'. However, with the modernization of tradition and globalization of consumer products, coupled with scientific and technological innovation, the ecological sensitivity has become a serious casualty. Indeed, the harmonious construction of nature and culture, viz. the mythic *Ardhanarishwara* has followed the path of dualism, thus, dichotomizing both, posing them mutually exclusive, antagonistic and opposed to each other. The result has been catastrophic in terms of environmental disasters frequented by Asia in general, and South Asia in particular, during the last decade.

The paper progresses with the view that the reaction to environmental problems at the global level, at best, can be explained as palliative rather than curative. Under the auspice of UN system, a series of initiatives have been taken including the apt quoted 'Millennium Development Goals' but the distance between the aspirations and achievements has been far too much. But then, global problems require global solution and hence global support. There may be numerous criticisms of UN systems either as a 'talking shop' or as a 'stooge of super powers' but at least we have a structure and global institution to provide us logistical and technical support in dealing with global issues. In fact, globalization as process of societal transformation and only after industrialization and colonialism that has a universal effect has benefited us in creating 'a global village'. Rise, growth and development of multilateral agencies between and among nation states have been the order of the day and powerful mechanism to deal with common issues.

There have been a number of environmental covenants, conventions and treaties between parties to deal with these international issues. In the pervasive structure of public international law, enforcement has been its soft underbelly, yet the rise and growth of international environmental jurisprudence is an offshoot of UN. Further, law needs not to be viewed essentially as an instrument of enforcement and coercion, as Hart and Fuller better know, adversaries tend to speak for once in unison. However, in the absence of enforcement, law tends to veer more towards morality and ethics and hence, ratification thus done brings the element of enforceability through the instrumentality of the state.

While coming to the macro level, India as an agrarian country producing agricultural goods, heavily depends on the management of natural resources such as soil, water, vegetation and the climate. These elements are already beginning to suffer from considerable strains, which are having a negative effect on the quality and quantity of agricultural products. The relation between growth, equity and environmental sustainability are extremely complex. On one hand, the transformation of natural resources into goods is essential for growth and the raising of living standards. Thus, for example, the expansion of the water supply channels is what makes possible the growth of human settlements. On the other hand, however, there is the danger of development processes affecting the quality of the environment. For example, air and water pollution due to industrialization and concomitant social and economic changes reduce the capacity of ecosystems to provide the community with vital goods and services.

It is in this context that the author intends to examine the contours and capacity of international environmental law in creating the structure of multilateral agency for South Asia to deal with the issues of ecology and environment, which is both sensitive and sacrosanct to all inhabitants. It is also common and endemic to the world-view and life world of a huge cache of human population diverse in culture, consumption and commitment to life, yet united in their values and love for the nature and 'mother earth'.

III

The third and penultimate unit on 'Social Dimension' highlights the nuances of globalization and environment with respect to its inter-linkages with society.

The unit begins with the contribution of Shweta Prasad paper on, 'Impact of Globalization on Women'. The paper finds that the contemporary period of globalization, as opposed to the global integration of the world economies in the late 19th and the early 20th centuries, is characterized by a never-ending controversy. On one hand, there are pro-globalizers who celebrate the virtues of the phenomenon and have faith in its inevitability, while anti-globalizers, on the other hand, highlight its vices and limitations. Among the anti-globalizers, there are some women's groups who are strongly sceptical about the consequences of globalization on women, who have termed it as 'anti-women'. These organizations have severely criticized various international trade and financial institutions for ignoring the well-being of women and not recognizing their competing demands. They are of the opinion that the international trade agreements violate the rights of the women workers which results in their exploitation. The macro-economic policy of the WTO, which fails to take into account the gender issues and micro-level impact of those policies on women, suffers from the same malice. The 'conditionalities' imposed by the international financial institutions like the IMF and the World Bank goes against the interests of the women. Besides, globalization has led to an increase in trafficking and prostitution among women and girls, especially in countries facing financial crises. Does it mean that globalization has only negatively influenced women? Or, does it have some positive impacts as well? The paper raises a debate on the utility of globalization from women's perspective and seeks answers to some most pertinent questions raised above, which will ultimately prove decisive for the fate of women.

The second contribution of the section comes from Arvinder A. Ansari on the theme, 'Understanding Gender, Water and Development: A Critical Appraisal'. While citing the importance of water in today's world, Arvinder says, it goes without saying that what water is to the 21st century, oil was to the 20th century. While access to water forms an important element of development, the gender component and relations associated form a significant subject in water collection and management. Traditionally, women in Indian context have been domestic water managers, purely due to gendered socialization and value orientation, while men have mostly attached themselves with agriculture and irrigation labour, hence sharing different water relations. This conceptual paper critically presents an overview of the social triangle of gender, water and development to explore the relationship among the three and elucidates how procuring, managing and using water, shapes and affects the gender relations in the contemporary times. The main aim of the paper is to examine how water use, water collection or water management and labour associated with it shapes up gender relations and differentiates between men and women reflecting inequality, labour, power relations and development. The paper mainly uses secondary sources besides observation and discussions/interactions for locating gender between water and development today.

Manish K. Verma's article on, 'Development-Induced Displacement: Policy, Information, Awareness and Tribulations in a Globalized World' is the third contribution in the section. The author says, while acclimatizing with the pace of global challenges, the gigantic urge of development activities saw man reaching to such a threshold where their endless craving results in the exclusion and underdevelopment of their fellow brothers. Most ironically, at such a moment, when the contemporary global society has entered into 'information age' and known as 'knowledge society' having 'reflex characteristics' in terms of transparency, dissemination of information and awareness, a large number of socially marginalized and economically deprived populace, involuntarily displaced worldwide due to development projects. They are forced to live in a communication vacuum having no information at all about their present and future destiny and thus get exposed to greater 'impoverishment risks.' Worldwide experience with resettlement has shown that people who are displaced do not easily recover, much less improve their previous standard of living. Resettlement studies have vividly documented the devastating consequences of failed resettlement projects, which create new pockets of poverty where none existed before. The plethora of situation is proving true in the instances of most of the development activities engineered to attain the developmental goals due to illiteracy, ignorance and unawareness on the part of the host communities residing at remote locations in most of the cases.

In this background, the paper attempts sociological interpretation of involuntary displacement which awfully disrupts the entire environment of the affected community. It tries to trace the global linkage on the nature and magnitude of development-induced displacement and its aftermath on the affected population between the countries across the globe and Singrauli region of India. By delineating into theoretical acumen, it introspects various policies concerned with the issue and critically evaluates its adequacies. Effort is made to ascertain the level of awareness of the host communities about relocation caused due to development projects and tribulations encountered by them due to lack of awareness. In addition to it, the discrepancy between planning and its real execution is also ascertained in the paper. Last endeavour of the paper is to comprehend people's perception, reactions and expectations from the entire process. All these investigations have been made in the domain of role played by rehabilitation agencies and its impact on Project Affected Persons (PAPs).

For the convenience of analysis, the author divided the paper into five parts. The first section provides an overall glimpse of the development-induced displacement scenario across the globe as well as in the Singrauli region of India. The second part, discusses in length the origin and evolution of various resettlement and rehabilitation laws in India and World Bank with its highlighted significance. The third section critically analyses the effectiveness and usefulness of the resettlement and rehabilitation policies discussed in the previous section at the level of information and awareness amongst the land oustees and the tribulations encountered. The fourth section focuses on various loopholes concerned with the issue. The fifth and last part of the paper focuses on aspects like how the general masses conceive these policies which become culprit for bringing turmoil in their serene lives. Coupled with the kind of reactions and resentment that the locals demonstrate alongwith their expectations from various rehabilitation agencies are noted in this section.

The last paper of the section by Antara Ray and Ramanuj Ganguly on, 'Changing Experience of Dalit Youth in Contemporary India' follows in the line. This paper attempts to focus on the changing Dalit youth culture in India. While doing so, it traces the socio-historical factors which facilitated the development of the contemporary youth identity among Dalits and critically evaluates the ways in which sociology can theorize the relationship between young people and culture vis-à-vis old and new India. It also tries to focus on some diverse kinds of motivations and ideologies which are now shaping and directing the youth of the contemporary society. The paper also tries to draw attention towards the fact that youth of India are taking prominent collective action which is guided and provoked by the changing social institutions (which are more politically active and mobile now) that are creating contradictions and new form of discriminations/exclusivist trends in the Indian society at large. The paper attempts to narrate how youth of contemporary India is getting involved in the new age caste-discriminations based on politics by using their cultural baggage. The rationale provided to the youth, either Dalit or non-Dalit, is reasoned on the basis of cultural tools in order to gain a leading role in Indian political system. The paper attempts to portray this dichotomy, contradictions and the logic of Dalit politics and the continuous involvement of youth in it.

Author finds that, media, rather than focusing on the identity of a youth as simply 'youth', is viewing and constructing their identity with their *jati* affiliations. The reality is continuously being constructed, deconstructed and reconstructed and is never being analysed without any power relations. The conclusion drawn suggests that, by giving thrust on these multiple realities which is creating multiple ideologies, political motives and youth culture on the basis of time, space and political inclinations, the youth of India is spontaneously confronting with the problem into everyday reality.

IV

The last section on 'Economic Dimensions' comprises five articles written by various subject experts concerned with the nuances of globalization and environment in the economic domain.

The section begins with the article of Siddhartha Mukerji. The paper on, 'Industrial Policy Reforms since 1991: The Contexts of Globalization and Domestic Politics in India' traces and analyses various economic factors in the aftermath of industrial policy reforms. Globalization provided an impetus for economic integration of nations. The process was initiated in the developed world and gradually, several post-colonial nations were dragged in. However, the process of integration with the global market took place in different phases. While East Asian countries made best use of the benefits accruing from global trade, others followed suit in the 1970s and 1980s. India was amongst the last few countries to initiate liberalization and adopted export-oriented industrialization. The process was started in the mid-1980s but assumed greater salience after the adoption of 'Structural Adjustment Programme' in 1991. Although the global economy had become highly competitive by this time, it was difficult to remain insulated from it as many economists felt that this would further stagnate the economy that was already trapped in a serious macro-economic crisis. The author opines that the year 1991, therefore, marks a watershed in the economic history of India as the government made fundamental amendments in its economic policies like industrial licensing, taxation, trade laws, and foreign direct investment (FDI). This was all done in a transformative political environment that was all-consensual and supportive of these new changes. However, the consistency in the process of liberalization favouring global integration was hindered intermittently by the political resistance from both extreme Right and Left organizations.

In this backdrop, the paper is an attempt to account for the structural changes that were brought in to make the policy environment conducive for private investment and global integration of the Indian economy. It then looks into the paradoxes and contradictions inherent in the political environment that resisted global-friendly policies like FDI, foreign institutional investment (FII) and export-promotion. The findings suggest that the overall scenario of global environment is found to be conducive for integration. Nevertheless, the intermittent turbulences desire a balanced policy framework and greater innovation at all levels of industrial enterprise.

Manisha Tripathy Pandey's article on 'New Capitalism and Violence: The Case of Special Economic Zones in India' comes next in the unit. The article begins with a critical note that ethical assessment of the means and ends of development is important. The paper deals with the means and ends of development with respect to the postmodern capitalism or new capitalism. India, through its five-year plans, has followed development as a philosophy and ideology of the state. But with neo-liberalism, there has been a shift in the means of development from the state to the market in many vital areas. This paradigm shift from growth with social justice and empowerment to neo-liberal development has led to systemic and structural violence. This has been made possible through Public-Private Partnerships (PPP) serving the interests of crony capitalists. The paper explores how the practice of crony capitalism in free market economy in India, as in the case of Special Economic Zones (SEZs), increases the 'crisis of modernity' leading to exploitation and social inequality. The proposed universal values of development like life-sustenance, equity, security, identity, capability expansion and 'well-being' are definitely undermined. The new capitalism and the market promote exclusion and isolation and the state is not being accessible, responsive and accountable. The author warns that this kind of development is certainly not democratic and participatory and thus violent.

The third contribution in the section is done by Biswajit Ghosh on, 'Does Excess Manpower Continue to be a Liability for India? Some Introspections in the Context of Globalization.' The paper introspects into the utility and relevance of population explosion in India in the present era, especially in the light of the revival of old debate about the role of population explosion in countries like India in the context of globalization. While globalization has enhanced the possibility of mobility and migration of excess labour power from the developing countries, there has also been talks about the shifting of manufacturing industry from the advanced industrialized countries to those where labour is cheap. The author argues, following these arguments and keeping in mind the emerging 'youth power' of the developing nations in years to come, that we need to review the old logic about population control. The question that needs to be addressed is whether excess manpower continues to be a liability to development in countries like India. This paper, based on an analysis of contemporary issues, tries to introspect into population-environment-development debate with particular reference to India in the context of globalization.

The second last article of this section and the volume is by Sanatan Nayak on, 'Participatory Approach in Management of Natural Resources: A Reflection of Global Practices at Empirical Level'. This paper examines, on the basis of a field study, the process of participatory management of natural resources, i.e. management of sodic soil in various regions of Uttar Pradesh, India. The author begins with the view that, developing the ideas of individual choice within collective action principle of management of natural resources depends on various attributes, i.e. knowledge on the attributes of goods (resources), knowledge on attributes of user group and knowledge on attributes of institutional arrangements. While establishing so, the author tries to assess the fundamental query regarding the knowledge of various stakeholders on the above-mentioned three attributes. On the basis of the analysis of 480 samples drawn from twelve selected villages, the author concludes that the knowledge on participatory management and technology is on higher side amongst the locals. However, the farmers had very dismal knowledge about the local bodies who are involved in management of the quality of the soil. The conclusion drawn in the paper suggests that, the poverty eradication programmes run through participatory soil management could not succeed as per anticipated pace due to poor institutional involvement.

The volume is concluded with the article on, 'Global Growth and Persistence of Poverty: Some Analytical Arguments' by N.M.P. Verma. In the article, the author is focusing his attention on the analysis of poverty in the aftermath of globalization. He perceives that there has been continuous expansion of market after the initiation of globalization process, but its growth is not constant in every part of the economy. As a consequence, global growth varies across economies. The higher growth is visible in South Asian developing economies. Capital flow has increased from industrialized economies to developing economies. Even labour movement has also gone up in view of capital growth. As a consequence, production activities have gone up. But the share of wage in output has not increased to the desired extent. As a result, distribution of income has become more skewed. That is why poverty has gone up. In this backdrop, the paper by Verma dwells into the issues of growth rates, malnutrition, inequalities and poverty in developing economies and the deduction of arguments have been made based on the secondary data of wealth, income and poverty during the economic reform and globalization. The paper concludes that although growth rates are high in the post-globalization period but poverty is also highly correlated. It implies that the benefits of growth hardly reach the poor. Poverty reduction is possible through enhancing the policy of empowerment vis-à-vis skill formation and entrepreneurship development. Here comes the role of good governance. The governance can control social and public corruption which can further enhance public accountability and efficiency. Public policy should be intensified and honestly implemented. Once accountability and efficiency in developing nations are raised, it will give a further spurt to growth rates and less skewed income distribution. Proper employability and empowerment policy may be helpful in correcting dull trends.

References

Aninat, E. 2002. 'Surmounting the Challenges of Globalization', *Finance and Development*, 39 (1): 4–7.

Dhanagare, D.N. 2012. 'Globalization', in N. Jayaram and D. Rajasekhar (eds), *Vulnerability and Globalization: Perspectives and Analyses from India*, Jaipur: Rawat Publications.

Friedman, T. and R. Kaplan. 2002. 'States of Discord: A Debate', *Foreign Policy*, 129, March–April: 64–70.

Ghuman, R.S., Surjit Singh and Jaswinder Singh Brar. 2010. *Globalization and Change*, Jaipur: Rawat Publications.

Giddens, Anthony. 1990. *The Consequence of Modernity*, Cambridge: Polity Press.

——. 1999. *The Third Way: The Renewal of Social Democracy*, Malden, Massachusetts: Polity Press.

Jayaram, N. and D. Rajasekhar (eds) 2012. *Vulnerability and Globalization: Perspectives and Analyses from India*, Jaipur: Rawat Publications.

McLuhan, M. 1960. 'Exploration in Communication', in E. Carpenter and M. McLuhan (eds), *Explorations in Communication*, Boston: Beacon Press.

Najam, Adil, David Runnalls and Mark Halle. 2008. *Environment and Globalization: Five Propositions*, Denmark: Ministry of Foreign Affairs, IISD.

Robertson, R. 1992. *Globalization: Social Theory and Global Culture*, London: Sage.

Stiglitz, Joseph. 2012. *Globalization and Its Discontents*, New Delhi: Penguin Books India.

Wallerstein, Immanuel. 1979. *The Capitalist World Economy,* London: Cambridge University Press.

Waters, Malcolm. 2001. *Globalization*, London: Routledge.

Part I

Emerging Discourse on Globalization and Environment

2

Dilemma of Globalization

Saket Bihari

Globalization as a process has been considered and reconsidered by almost all intellectuals of leading disciplines. Its significance has specifically been highlighted while planners and policy makers have initiated innovatively, experienced and people friendly ideas to be incorporated in the rubric of governance so that it may gain the mandate of majority. Indeed, this deserves importance because it unfolds new realities which a country or society is not necessarily exposed to. The experience to new realities comparatively becomes difficult because it carries a lower degree of consensus with the existing value and belief systems. The consensus fluctuates, as the intensity of development does not go on equal lines. Thus, it holistically sounds uneven and unequal considering the economic and social configuration in the Indian parlance, which is why the appreciation and disapproval both become inherent with this process in the course of its influence and expedition. Some people belonging to a specific class cherish the fruits of it, whereas some others perish because of its inaccessibility of minimum resources to decipher the favourable meanings to be successfully replicated to make life more appealing and shining. In the former, it has been argued that the process has opened up many opportunities for employment and advancement for unemployed sections of society while the latter curses it for its inherent requirement of orchestrated skills on the tune of market forces. It is also seen as the metaphor of cultural invasion and cultural growth by the diffusionist trends. The dilemma involved in globalization pertaining to loss and profits, advance and backward cannot be handled in a singular direction. Therefore, it is imperative to conceptualize the term so that its merits and vicious circles could be mapped out.

I

It is commonly believed that 'globalization is brought about by variety of social and cultural developments: the existence of world satellite information system; the emergence of global pattern of consumption and consumerism, the cultivation of cosmopolitan lifestyles; the emergence of global sports such as Olympic games, world football competition, the international tennis matches; the spread of world tourism; the decline of the sovereignty of the nation-state and the growth of global military system; recognition of the worldwide ecological crisis; the development of worldwide health problems such as AIDS; the emergence of world political systems such as the League of Nations and the United Nations; the creation of global political movements such as Marxism; extension of the concept of human rights; and the complex interchange between world religions'.[1] Thus, the gloss of globalization is not singularly noticed but it endows multiple currents and traditions which come in its own way. Its influence is wider and imperceptible which continually evolves and experiments with itself. The resilient spaces for one family where the members of the globe seem to become the coparceners are claiming more for their rights than the duties especially in environmental concerns, as only the well-equipped people exploit the natural resources and others are left to mould their activities coterminous with the sustainable and ecologically sensitive development. It is precisely because common environment is exploited by the communities, cutting across inherent consideration of common loss depending upon the accessibility to neo-technological experiences. In fact, the satellite information system is accessed by those who have say in the global resources and only after satiation they are ready to move into the post-industrial age. The quest for identity in the goods and services are only available to those who are guided by having impulses, leisure time and disposable incomes which are owned by the new middle class/neo-rich. The middle class follows the cosmopolitan lifestyle which allures others. In the late capitalism, the idea and understanding of identity crisis gains reasonable grounds because the conventional matrix of social structure is redefined in terms of emergent one, where the affluent man and disposable income regulates the recognizable intensity of identity. It creates a cleavage between aspiration and attainment because of unequal accessibility to the resources. Nevertheless, the following ideas synoptically describes globalization process sociologically:

(a) Globalization implies a speeding up of the flows of capital, worldwide modes of transport and communication, movements of people, goods and services, and ideas across the world, thus pointing to the general increase in the pace of global interactions and processes.
(b) It intensifies the links, modes of interaction, and flows that interconnect the world ties across borders that are not sporadic or haphazard but somewhat regularized.
(c) Globalization also enables people to control social, cultural, political and economic practices from distance. The decision made in the one

part of the globe influences the other part. Even the ideas, happenings, decisions and practices created on the front line of the globe are having consequences on the communities and cultures of remotely located places of the globe.

(d) To some extent, it also enables a reciprocal impact between local and global. The impact on local space may also have global repercussions and vice versa.

II

'Cultural aspect of globalization is associated with the symbolic exchanges by means of oral communication, publication, performance, teaching, oratory, ritual, display, entertainment, propaganda, advertisement, public demonstration, data accumulation and transfer, exhibition and spectacle.'[2] Symbolic exchanges liberate relationships from spatial referents. Symbols can be produced anywhere and at any time and there are relatively few resources constraints on their production and reproduction. Moreover, they are easily transportable and communicable. It is precisely because they frequently seek to appeal human fundamentals, they can often claim universal significance. These influences are very much guiding forces to human mind and cognition. And therefore, to live in the age of ever expanding flow of mass communication is to be influenced by globalization. The process has given rise to miniaturization,[3] personalization,[4] integration,[5] diffusion,[6] and autonomization.[7] It thus, gives the feeling of a world in motion, to an interconnected world, to a shrinking world'.[8] Globalization is illuminated as action at a distance (Giddens 1990), time space compression (Harvey 1989), accelerating interdependence (Ohmae 1990), networking (Castells 1996: 8), interconnectedness in the complex web of social relations and so on. This is to say that the multifaceted process has far-reaching consequences for the lives of all women and men, imposing constraints and opening opportunities for individual and collectivity. Indeed, homogenization, hetrogenization, glocalization and migration are the basic processes of globalization which is governed by electronic media. Globalization at the level of ideas exists in the pretext of new sources of mass communication.

Sklair[9] has explained globalization by globalizing tendencies of the trans-national practices which become vivid all the way through economic, political and cultural domains. Sklair defines transnational practices as those that cross state boundaries but do not necessarily originate at the level of the state. The transnational corporations, the transnational capitalist class and the culture ideology of consumerism that together constitute the transnational practices are the dominant institutions found in the economic, cultural and political domains respectively. Globalization has pushed the social transformation through digital TV, mobile phone and internet. These are those specific sources of communication which can easily reproduce images and symbols. The reproduced images and symbols carry subjective meaning proposed by the subject itself with a capacity to influence others. Indeed, the influence is exercised from the far distance.

III

As per the TAM Annual Universe Update – 2010, India has over 134 million households with television sets, of which 103 million have access to cable TV or satellite TV, including 20 million households which are DTH subscribers. In urban India, 85 per cent of all households have a TV and over 70 per cent of all households have access to satellite, cable or DTH services. TV owning households have been growing at between 8–10 per cent, while growth in satellite/cable homes exceeded 15 per cent and DTH subscribers grew 28 per cent over 2009. It is expected that India has 500 TV channels covering all the main languages spoken in the nation. The ministry of Information and Broadcasting issued a notification on 11 November 2011, setting 31 March 2015 as the deadline for complete shift from analog to digital systems. 'It has been estimated that there are 105 million internet users in urban India while the rural areas contribute 45 million subscribers. As of June 2012, there are 137 million claimed internet users. Out of 137 million claimed users, there are 111 million active users; the number of active users is growing at the rate of 19 per cent, faster than the claimed users. We confront a huge spike in the frequency of internet usage from a mere 28 per cent of the users in 2011 accessing internet daily to 54 per cent users now.'[10] 'Indian telecom industry underwent a high pace of market liberalization and growth since 1990s and now has become the world's most competitive and one of the fastest growing telecom markets. The industry has grown over twenty times in just ten years, from under 37 million subscribers in the year 2001 to over 846 million subscribers in the year 2011. India has the world's second largest mobile phone user base with over 929.37 million users as of May 2012. It has the world's third largest internet user base with over 137 million as of June 2012.'[11] Globalization process enables us to penetrate beyond appearance to the reality, if a systematic academic effort is applied upon. The problems emanated from globalization needs to be identified. We find it in the form of dilemma because as the intention of globalization is inextricably intermingled with positive coat with negative implications. Some of the pathologies attached to this process have been identified as:

IV

Human Trafficking

For instance, the illegal human trafficking is becoming a reality in the developing and developed parts of globe. The illegal trafficking ranges from bonded labour to prostitution and to trade of the rare human organs.

In fact, 'the United States (US) has placed India on the Tier-2 Watch List for human trafficking for the 5th consecutive year as India has failed to take effective measure in combating it. According to a report, India is a source, destination, and transit country for men, women, and children trafficked for the purposes of forced labour and commercial sexual exploitation. The numbers of persons affected could be anywhere between 20 to 65 million. According to some estimates, the expected annual turnover of human trafficking in India is around ₹ 20 billion. What is distressing is that out of the total number of persons affected by human trafficking, as many as 80 per cent are women and 50 per cent are children. The benefits of economic development have not trickled down to the marginalized sections of the society and millions of people still live below the poverty line. The poverty and hunger makes children and women belonging to the poor sections of the society highly vulnerable to human trafficking. In case of India, social and religious practices too have been a big cause of trafficking in India. Article 23 under Part 3 (Fundamental Rights) of the Indian Constitution prohibits trafficking of human beings in the territory of India. There are also more than 20 provisions in the Indian Penal Code, 1860 which deal with various aspects of human trafficking. But despite all this, there is an inexplicable apathy in the approach of law enforcement agencies when it comes to dealing with human trafficking. The Government of India has undertaken several measures in the past to combat this menace. The Ministry of Women and Child Development was made the nodal agency by the government to deal with human trafficking in India. A nodal cell against human trafficking has been constituted in the Ministry of Home Affairs (MHA). The National Human Rights Commission has formulated an integrated plan of action to prevent and combat human trafficking with special focus on women and children. However there is still a lack of clarity in government policies with regard to human trafficking. The existing laws have not been properly defined and there are several loopholes in them due to which the perpetrators of human trafficking escape from being punished. To combat human trafficking, several short-term and long-term measures are needed to be taken up at all levels. There is an urgent need to create awareness among the public about human trafficking. Media can play a very effective role in it. Poverty alleviation measures too will help in combating it in the long run. Since India is also a transit point for human trafficking, the government needs to take speedy measures to secure India's borders by completing its fencing and ensuring strict vigil.

V

There is a need to develop an institutionalized system of co-ordination between the law enforcement agencies and non-governmental organizations that sometimes prove to be more effective than government agencies in exposing human trafficking networks. There is a need to have greater co-ordination between different states in India as trafficking has a long trail from the source point to the destination with several transit points in between. Investigation in the cases involving human trafficking should be carried out with the aim to destroy this long trail. Increased co-ordination between government departments like police, public welfare, health, woman and child is required to ensure an effective response. Government and NGOs should work together to ensure post-rescue rehabilitation of the victims in terms of providing them healthcare, education and other employment opportunities'.[12] Human rights issues are being raised by developed countries as the means to safeguard largely the capitalist led criminal activities. Even the flip side development is also not being looked into. The reason is affluent have lesser concern to those who are deprived. Jajmani System[13] is no longer an issue to be taken an insight of, which has a glorious history as means to help poor. Therefore, we find a clear mismatch between the structural requirement and functional fulfilment of the system. This mismatch brings the system on the cross road where social change and its challenges become equivocal and blurred. This is one of the major challenges posed by globalization. This is not to say that these did not exist in the pre-1990s but the fact of the matter is it has grown alarmingly in the post-1990s which is considered to be the beginning of globalization process in India.

VI

World Satellite Information System

World satellite information system is an offshoot of globalization process. The seismic changes pertaining to information brought about on the globe are largely led by satellite information system. It accelerates the momentum of information in such a way that we forget the sense of time and space. Anybody existing on the globe may easily relate oneself to others standing distantly. It is capable of creating a virtual community. The world satellite information system has brought the experiences closer to one another. The close proximity apparently seems to be a symbolic representation of the value system where Indian values are contested and contrasted with the West and the rest. The contestations of values have tremendous potential to embark upon the domestic feelings and imaginations. The ideas and images seem to be imbricated with real activities. The imbricated ideas sound explicitly draconian and alien, as it fails to address the people's orientation and aspirations. This all happens in the faster, cheaper and deeper momentum of world satellite information system. But the operation and accessibility of the system essentially requires a minimum degree of skill and resources to handle. Therefore, the satellite information system may benefit a limited number of people and cannot democratize either the skill or the resources, as they are limited and regulated by the intensity of the social capital. However, it has largely influenced the structural components of the market. Those who are blessed get the pleasure and others are reduced to suffer in the pain.

VII

Consumption and Consumerism

Consumption and consumerism are also important byproducts of the globalization process. At first sight the link between globalization and food intake appears to be rather remote. However, its influence can be substantially high, though it works mostly through employment, incomes, prices and the market influences on food preferences. In recent years, the consumption pattern has shifted away from cereals to more expensive protective food items. One may notice the preferential choice towards processed foods. The transnational corporation has promoted fast food culture. This is how one may notice that the changes in the dietary pattern, in turn, influence the nutritional status of the population. Some effects such as fewer cereals and more protective foods in the diet can improve nutritional status, while a shift towards high-fat, high-sugar snacks foods may lead to obesity and chronic diseases. It is also experienced that lower affordability and reduced calorie intake leads to growth disorders such as stunting.

Imports of food such as cereals and cereal products, cashew nuts and edible oils have increased since globalization began. The imports of many food items were liberalized by putting them on open general licensing which does not necessarily require permission from the government. Import duties have been reduced to zero or negligible levels on many food items. 'The food related imports of cereals and products surged from 3,08,000 tonnes in 1990 to 16,20,000 tonnes in 1999–2000.'[14] The concept of consumerism is also seen in terms of consumer's rights and the Ministry of Consumer Affairs has designed consumer sensitization programme to protect the consumers from any illusion. However, consumerism also aims at finding one's location in terms of goods and brands which one consumes. It is often intended to delimit the routinized monotonous existence of life. It is also explained as the texture threaded through modernity.

VIII

Cosmopolitan Lifestyles

Culture and globalization are interconnected to each other via movie, television, business, tourism and internet today. Assimilation of culture and globalization is a way of communication that symbolizes one voice throughout the world. Culture is established by a group of people who live together and adhere to some principles in their society. In a broader sense, it is also a phenomenon of inheritance, unconsciously adopted by a person who is born within that group. When a culture is newly introduced to globalization, people find it detrimental to their way of life. Such as the fear that some inconsiderable concepts of other culture might take the place of their culture. Globalization with its universalistic propositions seeks to give a similar lifestyle throughout the globe. One may see the unique existence of jeans and T-shirt, symbolic expression of uniform lifestyle with the help of fashion shows etc. The lifestyle also proposes Christianity as a value to be imbibed. The Protestant ethics engaged in capital generation is gradually becoming the mantra of economic reconstruction and development. At this time every economic enterprise is suggested to require sustainability aiming at capital generation and profit accumulation. The implications of financial forces influence the lifestyle which separates people from coral reefs of social structure. 'This is the reason why the dawn of 21st century has witnessed rise of the most serious crisis in the form of several anomalies irrespective of concentration and decentralization trends of power structure.'[15]

IX

World Tourism

In recent years, world tourism has also become very significant. The geographical widening of linkages between places has become a major factor in the development of international tourism leading to the internationalization of tourism, leisure culture, more global flows of tourists, and increased competition. Though the process of intensification, the long established trickles of tourists to the most distant corners of the globe have been transformed into large scale tourist flows. It is also believed that one of the most powerful exemplars of globalization as the geographical scale of travel and trade has become greater than it used to be, with spatial interactions taking place over greater distances and ever wider spatial interactions taking place over greater distances and ever wider circles. Thus, the globalization phenomenon has contributed significantly to the spread of tourists to the furthest reaches of the planet, including Africa, leading to the shrinking of the pleasure periphery as it trends, in most innocuous sense, towards a single world society.

International trade, technological improvements, regionalization, role of multinational corporations and country images are the basic proponents of globalization influencing tourism.

It has also been noticed that the globalization of tourism often leads to the price increases of basic commodities as well as high land values. As a consequence, the availability of basic commodities has declined at the expense of international products, thus forcing local residents to travel long distances to buy basic commodities. Due to high land values, some local business people have been driven further away from the tourism core areas, a process referred to as the displacement effects. The globalization of tourism industry has tended to be characterized by expatriate labour. Generally, jobs in high positions are often reserved for expatriate employees while locals are employed in lowly paying and seasonal jobs.

Thus, the displacement of the people has been echoed by pull and push factors prepared by the processes of globalization.

X

Ecological Crisis

In fact, the technological and economic impulses of globalization furthers the capitalization of nature in the true sense of the term. It gives rise to particular resource governance regimes, the scale and scope of which are, to a great extent, determined by the logics of capital and technology. In the realm of environment, biodiversity represents one of the newest frontiers in the advance of a global political ecology, committed to nature's transformation from a phenomenal order to a value vector that meets the needs and interests of technological civilization. A symbolic conquest of nature, represents the use of scientific knowledge to bring nature into the semiotic realm of capital and, thereby, to prepare it for economic valuation. Unlike earlier use of natural resources that, when processed by human beings, are converted into commodity forms, the scientific symbolization of nature enables a new valuation process. It is also estimated that symbolization and capitalization of nature reflect the emergence of new governance regime for common resources in which nature's enduring character itself is valorized. In fact, the contemporary environmental problems represent not only a major economic crisis of supply, but also a new crisis of legitimacy for the market system. Anthropogenic climate change, desertification, and biodiversity loss, signal social interference, not only in the viability of particular eco systems, but also in the ecological process themselves. This eco-destructive possibility raises serious concerns about contradictory contemporary relations between nature and culture, and agency and structure. Here, globalization seems to have enhanced integration and integration of ideas, culture, people and place that had previously been isolated independently of each other. It also deals with the enhanced integration of the people and commodities into national and global markets. It also integrates of what had been local commons managed by informal, traditional systems into international and global economics and governance.

Thus, it shapes nature-society relations congruent with commodified and capitalized forms. The homogeneous lifestyle proposed by globalization sparks about the significance of the losses of biodiversity from which the habitat loss alone has projected to cause a species decline by 10–15 per cent. Pollution, climate change, invasive species and other environmental degradation causes further decline. In global parlance, an estimated 35 per cent of mangroves and 20 per cent of coral reefs have been lost. Activities for improving biodiversity include native planning programmes, green roof promotion and encouragement to use regional and organic products free of genetic modification. Promoting local organizations and businesses with good practices and educating citizens, especially youth, is also part of biodiversity management. Within local and regional administrations, setting dates and clear targets for biodiversity action with good assessment tools is important.

'The unemployment position has also been seen in the three categories namely number of persons usually unemployed based on usual status approach, number of persons unemployed on an average in a week based on the weekly status, and number of person days unemployed on an average during the reference period of seven days.'[16] 'Despite the positive implications, it should also be taken into the consideration that the labour force also grew faster in the second period. However, the pace of growth in labour force in the second period at 2.84 per cent per annum, so that the unemployment rate increased from 7.3 per cent in 1999–2000 to 8.3 per cent in 2004–05.'[17]

On the other hand, the notion of brand consciousness has also grown in the Indian middle class with the advent of consumerism which stands for the consumption of brands rather than concentrating on the use value of it.

XI

Health Problems

In the wake of globalization, certain diseases are also becoming universal. It has been found that AIDS, STDs are the major diseases which threaten the human health and hygiene. In fact, the channels through which globalization may affect health outcomes are multiple. The world health problem has been characterized as one of fighting the double burden of disease. 'The increased life expectancy recorded in recent decades, together with changes in the life style stemming from socioeconomic development, have increased the importance of non-communicable diseases and injuries while at the same time as many as one billion people in the third world still suffer from infectious disease, under nutrition and complication of child birth, conditions not seen among the non-poor and that are largely avoidable because inexpensive and effective tools exist to deal with much of it.'[18] Different income growth and levels among countries not only define distinct health problems, they also lead to the confirmation of diversified markets for health services. The varied ways in which the double burden may appear in a society, force difficult decisions about the allocation of scarce resources, and are a source of distributive conflict across rich and poor households in those countries. With increase in income in developing countries, the demand for private health services, including health insurance, goes up. This demand interacts with constraints on the supply of public health services in those countries, where governments face higher demands on limited resources, due to population increases and other causes. The public sector has to attend to public health such as immunization, control of infectious diseases and vectors, health education, water and food safety, and basic health services with those limited funds, leaving mostly unattended the demand for the higher levels of individual medical care. The unmet private demand eventually creates a market for private health services leading to the development of a dual market structure and to the escalating costs, all of which may affect the poor negatively.

The tension between public and private health services may exist irrespective of whether the system is closed to, or allows presence of foreign providers of health services. Therefore, the dynamics of dual market structures and of escalating costs may happen, whatever the nature of the firms involved in the private health system, to the extent that this tension mostly depends on the nature of the double burden created by the epidemiological transition and on the profits of income growth and distribution in those countries. The best human, financial, and technological resources may end up being absorbed by the high-end segment catering to a healthier and most affluent clientele, while the poor and the greater health risks may be excluded. The public sector may be burdened with the most difficult cases, in terms of health and incomes, further straining public budgets that still have to attend to nation-wide health problems and reinforcing the image of low quality public services, eroding support for the public health system.

However, increase in international travel, tourism, and food trade mean that toxic products, both legal and illegal, reach wider markets and that new and resurgent disease producing organisms can be transported rapidly from one continent to another. During the 1990s, emerging and re-emerging infectious diseases have become a major public health concern. Some 30 new and highly infectious diseases have been recorded in the last 20 years. Though contact in airports and air travel, which has skyrocketed in the last 40 years, from two million passengers a year in 1950 to cover 1.4 billion today, airborne disease such as pneumonic plague, influenza, and TB can easily be spread. HIV/AIDS has also spread through sexual tourism and, in sub-Saharan Africa, through migrant workers and truck drivers. Due to important migration flows caused by wars and civil strife, the number of refugees and displaced people has increased ninefold over the past two decades. In 1996, as many as 50 million people worldwide, or one per cent of the world's population, had been uprooted from their homes. Refugees and displaced persons living in overcrowded, unsanitary conditions are at risk of outbreaks of cholera and other waterborne diseases. Insects and other animal vectors can also move globally, carried by trade in goods by the wind, by birds, or by ocean currents. The efforts to control infectious diseases are undermined by the growing resistance of microbes to drugs that once were highly effective against infections. At the same time, there is a decrease in the speed of new drugs' development, in part due to the cost of their development and a decline in resources available to fund disease surveillance, diagnosis, and control system, based on the perception prevalent in the 1970s that communicable diseases had been controlled.

XII

Human Rights

The international bill of human rights holds states accountable for realization of human rights. But in the contemporary era it is the private global players that are frequently the most egregious violators of rights, and so far as the idea of accountability goes, they are accountable to none. Strikes and demonstrations protesting worsening labour condition have become widespread, including in communally-based societies where individual political action has been rare. Political authorities often react to this strife by increasing restrictions on civil and political rights and, at times, grossly violating basic rights, such as the right to life, in an effort to control the labour force. Human rights NGOs have extensively documented the increase in labour activism, particularly in such low wage areas as Malaysia, Indonesia, and even China, and the concomitant increase in state repression. Human rights violation as a consequence of destructive social change resulting from globalization might result, at least in some instances, in radical shift in a society's cultural values and norms that, in turn, may lead to a reconfiguration of the substance of traditional or historic notions of human rights. The outcome of this search for a revitalizing identity and meaning is unpredictable. There may be reinforcement of an exclusive communalism with little personal autonomy or there may be loosening of communal ties and an expression of individual demands based on class. The evidence points in both the directions. Clearly globalization has had a deleterious effect on the entire complex of human rights, resulting in significant transformations in the behaviour and values of masses of humanity across the globe. The negative shock of globalization has, it should be noted, affected not only non-western societies, but those of advanced, industrialized Western Europe, the United States, and Japan. Increased trade, often aids developing countries and thus contributes to the mitigation of poverty; increased communication permits countries to learn from each other. In the sphere of human rights, communication via e-mail has permitted human advocates in their locality and to communicate with other human rights advocates throughout the world.

However, there are other less beneficent effects on human rights arising from globalization. The emphasis on competitiveness and economic development has had especially negative effects on such vulnerable groups as migrant workers, indigenous peoples and migrant women. Globalization has been cited as contributing factor in violations of the right to life, the right to protection of health, the right to safe and healthy working conditions and freedom of association in many countries.

XIII

Interchange Between World Religions

In globalization, one may also come across the heavy encroaching pressure of religion for the conversion. Globalization has resulted in the intensification of ethnic and religious conflict and violence. Globalization with its attendant social change has precipitated fragmentation at the local level. On the one hand, the resurgence of ethnic and religious strife can be understood as an effort to assert identities and to ascribe meaning in the face of global forces over which one has no control.

There is a way in which globalization flattens our cultural differences, erodes local customs and beliefs, and spreads a secular, capitalist way of life that is at odds with religions of all sorts. At the same time, there is a way in which religion serves as the source of globalization's greatest resistance and as the haven for those standing in opposition to its ubiquitous yet often subtle power. Religion and globalization can also be seen as partners in historical change. In times past, religion, in various manifestations, has been a carrier of globalizing tendencies in the world. The history of Christianity, of course can be understood in part as an early effort to create a global network of believers. Its extraordinary growth and influence as a world religion is a result of a link between its own global ambitions and the expansion of various political and economic regimes. It succeeded as a globalizing force long before there was a phenomenon called globalization. Elements of this historical pattern can be found in Buddhism, Islam, and other faiths as well. Religion is hardly epiphenomenal to the processes of globalization in our own day. It continues to be a player in intricate and even contradictory ways. To be sure, it was once thought that secularization was the inevitable outcome of the process we call modernity. Clearly, this has been the case. Religious faith persists in a complex interaction with the structures and processes of the modern world and that complexity has only intensified under the conditions of contemporary globalization. Constant through it all, is the capacity of religious faith to generate commitments in ways of seeing and being in the world of extraordinary depth and passion and irreducible difference. Needless to say, all these differences are among the most contemptuous and volatile of all differences. Humans are becoming more and more closely connected on this earth and that this connection breeds tensions, misunderstandings, and violence suggests that we need to think hard about the ways we can live with our differences in peace. Agreement may not be an achievable or desirable goal, but finding possibilities for coexistence in the midst of disagreement must be.

XIV

Squaring Off

Globalization has been defined in many ways by the scholars, depending on the kind of discipline it is interpreted in. It has provided opportunity to those who can match with the requirement of its demand. It has abated mobility for those who cannot afford to provide the skills in the framework of its requirements. The tempest of hunger on the globe is gradually becoming a noticeable phenomenon. Indeed, the world's hunger is getting ridiculous. There is more fruit in a rich man's shampoo than in the poor man's plate.

The cultural part of globalization is very significant as the technological changes are making the world society more and more interconnected. This interconnectedness has many facets to be discovered. It influences the Weltanschuung. Therefore, the dynamic force of globalization continues to change the people's perception, as it reshapes lives. It seeks to find the answer with alternative universals for all, equally. But only economically competent class enjoys the benefit of it which is not in majority. For most of the people, it seems to be an image-driven proposition. The economic nationalist approach of globalization explains that it undermines the sovereignty of the nation and makes the country more beholden to transnational corporations. It is undemocratic in the sense that it increases inequality which impoverishes the people. Thus, it gives many dilemmas especially the existence of world satellite information system; the emergence of global pattern of consumption and consumerism, the cultivation of cosmopolitan lifestyles; the emergence of global sports such as Olympic Games, world football competition, the international tennis matches; the spread of world tourism; the decline of the sovereignty of the nation-state the growth of global military system; recognition of the worldwide ecological crisis; the development of worldwide health problems such as AIDS; the emergence of world political systems such as the League of Nations and the United Nations; the creation of global political movements such as Marxism; extension of the concept of human rights; and the complex interchange between world religions. We need to hang on tight because there is rough ride ahead of us.

Notes

1. G. Marshall (ed.) 1994. *A Dictionary of Sociology*, New York: Oxford University Press, 258.
2. M. Waters. 1995. *Globalisation*, London and New York: Routledge, 8–9.
3. Miniaturization is related to technologies which are reduced in size. It affects transmission as well as reception.

4. Personalization has historically been a general reduction of the scope of the audience for electronic mass communications. The music hall or the cinema could entertain several hundred, the television, a family, but the PC is literally a personal computer, although it is not actually quite as personal as a palm-top computer. The Sony Walkman and its mimics represent the ultimate in personalized consumption.
5. Integration deals with the various technologies of text, sound, visuals, and response via keyboard or microphone are progressively becoming integrated with one another. This centres on the technology of the microchip which organizes computers. The microchip provides an enormous capacity to process information.
6. Diffusion is access to technologies of mass media, and is becoming more widespread in terms of both reception and transmission. It further implies that there is increasing access to mass communications. It is not possible to maintain the national sovereignty in mass communication so long as the members of the society have access to satellite-based means of communication.
7. Consumers have their increasing potential choice of accessing a particular stuff. They have increased access to production facilities via home recording equipment and community studios; and the viewers control the timing and content with the help of compact discs, cassettes and video cassettes.
8. Z. Jonathan (ed.) 2002. *The Anthropology of Globalisation: A Reader*, USA: Blackwell Publishers Ltd., 9.
9. L. Sklair. 1991. *Sociology of the Global System*, Hemel Hempstead: Harvester Wheatsheaf.
10. *The Economic Times*, 5 September 2012.
11. *The Times of India*, 5 July 2012.
12. http://www.policyproposalsforindia.com/article.php?article_id=203&languageid=1, accessed on 29 April 2013.
13. Hindu Jajmani system emphasizes solidarity, but does not represent symmetrical inter-relationship for all members. It is communistic and conservative, built on inequality rather than equality. It represents a high degree of economic and social organization and control. Its weakness lies in its lack of individual freedom, its limited altruism, and its inequality of membership, its static nature and economic inelasticity. It has also been mentioned that the art of peace and contentment in the social and economic relationships of community are attainable only by integrating one's own aims and purpose with those of the local groups. Absolute individualism cannot lead to social and economic satisfaction and stability. It leads to privileged group license, to unbridled self-interest in business practices, to an insatiate desire for profit and the accumulation of wealth, to emphasis on competition rather than co-operation, and to social and

economic disorganization. It is an implicit lesson from Jajmani system that strong are under an ethical obligation to bear the burdens of weak. Each must work for common good. For more details please see, W. Wiser. 1954. *The Hindu Jajmani System*, Lucknow: Lucknow Publishing House, 190–192.

14. *Government of India Report*, Planning Commission, 2002.
15. S. Bihari. 2013. 'Globalization and Security Concerns: Paradoxes and Possibilities', *The Indian Police Journal*, IX (2): 155–162.
16. Eleventh Five-Year Plan, Planning Commission, Government of India, Page No. 64 (NSSO Report No. 409, Employment and Unemployment in India 1993–94, NSS 50th Round, Chapter 7).
17. Eleventh Five-Year Plan, Planning Commission, Government of India, Page No. 65.
18. E. Diaz-Bonilla and others. 2002. *Opportunities and Risks for the Poor in Developing Countries*, New Delhi: Indian Council for Research on International Economic Relations.

References

Alagh, Y.K. 2003. *Globalization and Agricultural Crisis in India*, New Delhi: Deep and Deep Publications.

Albrow, M. 1996. *The Global Age*, Cambridge: Polity Press.

Allen, G. 1996. *Kinship and Friendship in Modern Britain*, Oxford: Oxford University Press.

Amin, Samir. 1980. *Class and Nation, Historically and in the Current Crisis*, New York and London: Monthly Review Press.

Anderson, B. 1983. *Imagined Communities: Reflections on the Origin and Spread of Nationalism*, London: Verso.

Anith, B.K. 2000. *Village, Caste and Education*, Jaipur: Rawat Publications.

Appadurai, A. 1996. *Modernity at Large: Cultural Dimensions of Globalization*, Minneapolis: University of Minnesota Press.

Augoustinos, M. 1995. *Social Cognition: An Integrated Introduction*, New Delhi: Sage.

Augustine, J.S. (ed.) 1982. *The Indian Family in Transition*, New Delhi: Vikas Publishing House.

Axford, B. 1995. *The Global System: Economics, Politics and Culture*, Oxford: Polity Press.

Axtmann, R. (ed.) 1998. *Globalization and Europe: Theoretical and Empirical Investigation*, London: Pinter.

Back, L. 1996. *New Ethnicities in Urban Culture*, London: UCL Press.

Bamyeh, Mohammed A. 2000. *The Ends of Globalization*, Minneapolis: University of Minnesota Press.

Barber, Benjamin R. 1996. *Jihad vs. McWorld*, New York: Ballantine Books.

Barker, C. 1999. *Television, Globalization and Cultural Identities*, Buckingham: Open University Press.
Baron, S. 1960. *Modern Nationalism and Religion*, London: Routledge.
Baudrillard, J. 1998. *The Consumer Society: Myths and Structures*, London: Sage.
Bauman, Z. 1993. *Post Modern Ethics*, Oxford: Blackwell.
——. 1998. *Globalization: The Human Consequences*, Cambridge: Polity Press.
Beck, U. 1992. *Risk Society: Towards a New Modernity*, London: Sage.
Beyer, P. 1994. *Religion and Globalization*, London: Sage.
Bhabha, H. 1994. *The Location of Culture*, London: Routledge.
Bihari, S. 2013. 'Cultural Capital and School Success', *International Journal of Research in Sociology and Social Anthropology*, 1 (1): 33–38.
Castells, M. 1996. *The Rise of Network Society*, Cambridge: Blackwell.
——. 1997. *The Information Age: Economy Society and Culture*, Oxford: Blackwell.
Chase-Dunn, C. 1998. *Global Formation: Structures of the World Economy*, Lanham: Rowman and Littlefield.
Cheney, D. 1996. *Lifestyles*, London: Routledge.
Dirks, N. 2001. *Castes of Mind: Colonialism and Making of Modern India*, New Jersey, USA: Princeton University Press.
Doshi, S.L. 2003. *Modernity, Postmodernity and Neo-Sociological Theories*, Jaipur: Rawat Publications.
Dubey, S. 1990. *Indian Society*, New Delhi: National Book Trust.
Eschle, C. 2001. *Social Movements and the Globalisation of Democracy*, Boulder: Westview Press.
Featherstone, M. 1990. *Global Culture: Nationalism, Globalization and Modernity*, London: Sage.
Giddens, Anthony. 1990. *The Consequence of Modernity,* Cambridge: Polity Press.
Harvey, D. 1989. *The Condition of Post-Modernity: An Enquiry into the Condition of Cultural Change,* Cambridge: Blackwell.
Ohmae, K. 1990. *The Borderless World: Power and Strategy in the Global Market Place,* London: Collins.

3

The World Beyond State and Market

Locating Elinor Ostrom in India[1]

Pampa Mukherjee

Elinor Ostrom, whose work on the commons earned her the 2009 Nobel Prize in economics passed away in 23 June 2012, thereby creating a void for scholars and policy analysts deliberating on sustainable development. While highlighting the significance of common resources, she acknowledges that it is difficult to exclude people from using them and making it vulnerable to overuse. To be sustainable, the usage has to be regulated and coordinated and for this, government or privatization is not the only solution. Ostrom, thus demonstrated through her empirical work in India, Nepal and Sri Lanka that people have the ability to manage shared resources sustainably and hence there is a need to go beyond the state and market solution to halt its overuse. It may be mentioned here that despite its contribution to the rural economy of India and existence of multiple community based institutions managing these resources, common property resources became a subject of academic enquiry, only in the last few years and particularly with the publication of Ostrom's seminal work Governing the Commons (1990). There is, in fact, a need to undertake an in-depth and systematic understanding of Ostrom's study on CPR to understand what makes collective action successful in diverse settings – in irrigation, forests, fisheries and pastures. It is in this larger context that the present paper locates itself. Taking recourse to Ostrom's work on Common-Pool Resource (CPR), it discusses the existing debates on institutional governance of these resources,[2] analyses the central arguments of Ostrom's work, *Governing the Commons*[3] and situates its significance in the context of India. In the first section, the paper discusses the conventional debates on institutional arrangements in governing these resources and in the next section, analyses the central arguments of Ostrom's work and in doing so also reflects on the relevance of Ostrom's thesis in the context of India.

I

State and Market: Solutions to Institutional Management of Commons

As has been discussed above, that for long there have been debates among scholars regarding the best institutional alternatives to manage 'common' resources. In fact, in *Governing the Commons*, Ostrom's main concern is to understand the problem of how institutions are devised to conserve natural resources. She finds that different explanations have been given by analysts, suggesting different solutions to the problem. It is identified that theoretically there are three influential models that are most frequently used to provide a foundation for recommending either state or market solutions to address decline and better management of CPRs – the tragedy of the commons, prisoners dilemma and the logic of collective actions (Ostrom 1990). The common assumption in all these paradigms is that without external forces people are unlikely to cooperate and therefore, the only solution would be private enclosure or state regulation (Wade 2003; Nicita 2002).

The Tragedy of the Commons

The expression first appeared in Garrett Hardin's article in *Science* (1968) denoting degradation of an environment that is held in common. Hardin's description of the situation was simple and captivating. 'Imagine a village common', said Hardin, 'in which a number of herders graze their cattle. What is to stop the herders from adding more animal to their herds? Each herder would find it very attractive to augment his or her herd even if this meant that the carrying capacity of the grazing area would eventually be exceeded. For each herder as an economically rational decision maker, it would be profitable to graze more animal because the herder would take all the profit from the extra animal but would only bear a fraction of the cost of overgrazing. Thus individual rationality would lead to a collective tragedy. There was no escape', declared Hardin.[4] To avoid this tragedy, Hardin (1968) concluded that the commons could either be privatized or kept as public property, under government regulation, to which rights to entry and extraction could be allocated. Hardin argued that if we do not act in one of these ways, we 'acquiesce in the destruction of the commons' (Hardin 1968: 1245). In other words, Hardin traces the source of resource problems to the common-property nature of the resource regime which gives a resource user an incentive 'to increase his use without limit – in a world that is limited' (Hardin 1968: 1244). Hardin saw the solution to tragedies of the commons as top down, autocratic management by the government (Acheson 2011).[5]

It is to be noted that the tragedy of the commons paradigm was published at a time when environmental problems had recently come to the forefront. There was an urgent need for a theory to explain the environmental ills of the world and for a framework to help make sense of observations. The paradigm applied, Hardin thought not only to common pastures, but also to public lands such as national parks, air and water. The commons' paradigm provided the framework in which most environmentalists and those in related social and especially natural sciences understood resource issues (Berkes and Feeny 1990). Few questioned Hardin's assumption of individual interest unchecked by social tensions and his emphasis on competition (rather than on cooperations) as the overriding relationship that shaped interactions among resource users.

The implication of Hardin' formulation influenced policy makers in many of the developing countries including India. They generally accepted Hardin's arguments uncritically and thought that the conclusions made it obligatory for them to take positive actions to impose policies on the users of resources like forests, pastures and other common-pool resources in their domain. Government officials did not even take efforts to enquire whether the users of these resources had developed rules of their own, over time as it was simply assumed that resource users were trapped and helpless. National governments around the world declared that government ownership was the *only* way to save resources from destruction. Forests in India and many countries in Africa were nationalized during the 1970s and 1980s in an immediate acceptance that this was the *only* way to avoid massive deforestation. As many of these policies appeared to accelerate deforestation and overuse of resources, a major debate about the best property-rights system for sustainable management of natural resources was initiated (Nagendra and Ostrom 2008).

Collective Actions and the Prisoners Dilemma

The Prisoners Dilemma can be seen as a formalization of Hardin's expression into a game model.[6] Ostrom explains that collective action problem can be modelled as the prisoner's dilemma game. It is a simple game model in which collective decision produces outcomes harmful to the group as a whole without intervention by some higher authority. It is conceptualized as a non-cooperative game in which all the players possess complete information. In non-cooperative games, communication among the players is forbidden or impossible or simply irrelevant as long as it is not modelled as a part of the game. If communication is allowed, verbal agreements among players are presumed to be non-binding unless it is so mentioned in the game structure.

The parable of the 'Prisoners Dilemma' is quite well known and need only be summarized.

Ostrom uses the classic example of two prisoners who have been caught by the police and they are then separated so that they cannot talk to each other. Since it is a hypothetical game, one needs to imagine that the two suspects are interrogated separately about a crime they have jointly committed. Both know that if they cooperate with each other and neither confesses, they will get suspended sentence; if one defects and give evidence to the State, he will be paid and released and the other will receive a heavy prison term and if both defect each gets a prison sentence. We find, therefore, that from the point of view of either of them, staying silent while the other confesses would give the worst outcome. What is most important in this single period game is that the choice of best strategy is made regardless of the expected choice of the other player. In other words, here confessing is the 'dominant' strategy.

In the context of the use of common pool resources, the option is either to cooperate with others in a rule of restrained access or not to cooperate at all. Each individual has a clear preference and an order of choice.

(a) Everyone else abides by the rules while the individual enjoys unrestrained access (i.e. free rides or shirks)
(b) Everyone, including himself follows the rule (cooperates)
(c) No one follows the rule
(d) He follows the rule while no one else does (he is suckered)

According to the order of preference, if the third alternative (c) is followed, i.e. unrestrained access to all in the group, stable group outcome is achieved. The second alternative, i.e. mutually agreed upon rules, is more desirable. However, this is not a stable equilibrium because each individual has an incentive to cheat and go for his first alternative, i.e. restrained access by all except him. Even if it then turns out that no one else follows the rule, his cheating, at least, ensures that he avoids his own worst alternative following the rule while no one else does.

In this situation, it is argued that the only solution is either coercion from outside the group to force people to reach and maintain social optimum (second preference) or a change in the rule from outside the group to a private property regime. The Prisoners Dilemma was of great interest to social scientists for it seems to justify the conclusions that rational people cannot achieve rational collective outcomes. It seems to be applicable to all situations in which it is possible for some to refuse to cooperate while others are willing to cooperate (Ostrom 1990).[7] There are two assumptions, which are dominant in a situation modelled as prisoner's dilemma game. One, there is a lack of information among the players in the game. The players choose in ignorance of each other's choice. The second is that each player chooses once before the payoffs are received and so cannot change his mind upon finding out what the other has done. The first assumption has the important implications that the players cannot negotiate among themselves to change the rules of the game, so as to secure more desirable collective outcomes. If at all the changes in rules must come, it should be from outside the group.

However, if the situation is a consistent phenomenon or a recurrent one, then in that case the logic changes. If the players in a 'Prisoners Dilemma' know that the game will be played repeatedly into the future, the chance that they will cooperate today in the hope that the others will then do so are much higher where the game is played only once (Taylor 1976; Axelrod 1981; Hardin 1982). In fact, economists in their recent work on game theory have drawn heavily on the theory of repeated games in arguing that cooperative behaviour grows out of self interest (Seabright 1993). The basic question is that if the game is repeated a number of times, can the two players find a way to cooperate? The idea that repetition can sustain cooperation is based on the premise that individuals who are tempted to defect may be dissuaded from doing so from the fear of losing the benefits of cooperation in the future. This dissuasion becomes effective only in the presence of some incentive structure that is accompanied by assurances at different levels provided by the system in which one is operating.

Although 'Prisoners Dilemma' provides important insights into the breakdown of social institution, it gives no explanation of how or why institutions begin and are maintained. Instead, they imply that within any group no internal incentives exist to initiate or maintain institutional arrangements (Runge 1984).

The Logic of Collective Action

Another view is closely related to 'Prisoners Dilemma' model, expressing difficulty in getting individuals to look after their joint welfare rather than individual welfare, was developed by Mancur Olson (1971) in his work *The Logic of Collective Action* where he says that one of the core problems of collective action is free riding. His basic postulate is that a large group, through voluntary contributions, cannot provide collective goods. His contention is that the reason as to why rational individuals will not act in their common interest is that individuals trade in collective or common goods, which are characterized by non-excludability and consequently they want to enjoy the benefits of collective goods without contributing to the cost. It points out the difficulty of collective action because there is little incentive in voluntary action to further collective good when free rider can, in such situations, obtain the benefits of these goods. Olson also contested optimism of group theory which is based on the idea that individuals would voluntarily act to further the common interests (Bentley 1949; Truman 1958). On the first page of his book he summarized that accepted view: Olson observed that economists and political scientists typically took it for granted that individuals in groups with a common interest, or with a prospective collective benefit to be attained, would act to further that common interest; indeed, it was assumed that groups would 'act', much in the same way as individuals act, to pursue their interests (Olson 1971: 1).

The idea that groups tend to act in support of their group interests is supposed to follow logically from this widely accepted premise of rational, self-interested behaviour. In other words, if the members of some group have a common interest or object, and if they would all be better off if that objective were achieved, it has been thought to follow logically that the individuals in that group would, if they were rational and self-interested, act to achieve that objective (Olson 1971: 1).

The core position of his contention is 'unless there is coercion or some other special device to make individuals act in their common interest, rational self interested individuals will not act to achieve their common or group interests' (Olson 1971: 2). His argument rests on the premise that whenever individuals cannot be excluded from the benefits that other provide, thereby individuals are not motivated to contribute to the joint effort, but to free ride on the efforts of others. If all participants choose to free ride, the collective benefit will not be produced. The temptation to free ride, however, may dominate the decision process, and thus all will end up where no one wanted to be. Alternatively, some may provide while others free ride, leading to less than the optimal level of provision of the collective benefit. In other words individual's contribution to a group objective is either by compulsion or through inducements. Without selective punishments or inducements, individuals will free ride and the public good will not be supplied or it will be in inadequate measures. Existing cases of common interest groups are thus to be explained in terms of selective punishments or inducements.

Influenced by the same model, there are policy analysts who have called for the imposition of private property rights whenever resources are held in common (Demsetz 1967; Welch 1983). The case for privatizing property rights in what have been hitherto common property resources rests on the efficiency logic. It is argued that if an individual or a firm owns the resource, it will be allocated in a more efficient way. The basic problem with privatization lies in the fact that private contractual rights can provide effective incentives for only some of the individual actions concerning production. Other necessary action may be unenforceable as they are unobservable by the affected parties or by the enforcing authorities. And at times they are too complex to be specified in a contractual form. This may lead to breakdown of whatever cooperative mechanism that may have evolved among those who share implicit non contractual rights in relation to common property resources. For example, if a forest is being privatized for timber production it may fail to internalize all the externalities that are involved.[8]

Thus the tragedy of the commons, prisoners dilemma and the logic of collective action are closely related concepts in the models that have defined the problems that individuals face when they want to achieve collective benefits. The central issue in each of these models is the free rider problem. As Ostrom points out that whenever one person cannot be excluded from the benefits that others provide, each person is motivated not to contribute to the joint effort but to free ride on the efforts of other. These models are thus extremely important in explaining how perfectly 'rational' individuals can produce under some circumstances outcomes that are not 'rational' when viewed from the perspective of all those involved. At the same time as the foundation for policy, these models have been used more often, other than metaphorically. The limitations that are taken to be fixed for the purpose of analysis are taken as faith and as being definite in empirical settings and continues to be so unless one takes recourse to external authorities to change them. In the process, certain key questions remain unanswered like how individuals who are jointly using a common property resource might achieve an effective form in governing and managing their own commons.[9] The need is therefore to go beyond the state and market solution to solve the commons dilemma by exploring alternative strategies that are feasible and sustainable in a given socio-economic set-up at the micro level.

II

Beyond State and Market: Ostrom on Governing the Commons

Contesting the justification provided by the three models for state control, Ostrom in her work on irrigation system in Nepal argues that the theoretical presumption that an external central government is necessary to supply and organize forms of collective such as providing irrigation works has been reinforced by the colonial experience. In most parts of Asia and Africa, during the colonial period, governmental organizations were created to develop unirrigated areas. This centralized government power over the supply of irrigation water continues till date. In most cases these are governments that were created when the colonial powers left the scene. However, irrigation developments in much of the developing countries 'have been highly distorted by the process of state concentration on investments and governance and the concomitant demise of local rights and initiatives' (Ostrom and Gardner 1993).[10]

Ostrom and her colleagues in the Workshop in Political Theory and Policy Analysis at Indiana University, Bloomington, USA, demonstrated that community-based institutions need to be recognized in addressing issues concerning degradation of natural resources. She proves empirically that human communities have crafted number of informal institutional arrangements for regulating access to common resources and are thus successful in creating a balance between use and resource renewal. Her work primarily highlighted the significance of collective action for sustainable use of natural resources and for which she finally received the Nobel Prize in Economics in 2009.

In other words, Ostrom demonstrated, both theoretically and empirically, that legal regulation is not the only possible solution to public goods problem and that community-based solutions to common property resource problems have been successful over multiple generations. According to her, these are quasi-voluntary arrangements through which a community of users (fishers, grazers and irrigators) are able to manage resources collectively and control violators, in such a way as to preserve the resource over time. She points out that these institutions can be self-maintaining, in that participants have an incentive to monitor and sanction the rule breaker and shirkers.

Essentially, her research points out that there are multiple property systems through which access to natural resources can be mediated. Apart from the state and private property, Ostrom contends that it is possible to conceive forms of 'social property' or community property through which at least some assets are held in common, and there exists fair and well-defined procedures for providing rights of access to the use and enjoyment of the social property. She demonstrated in depth that there exists number of socially feasible arrangements in which a 'common property resource' such as a fish stock is exploited by a number of independent producers within the context of a stable community. In this context, one finds that there is a combination of private property (nets and boats) and social property (the waterway and fish stock) and Ostrom documents several different sets of social rules that establish the terms of access and use that individuals will have to the common property resource.

The point here is not that social property regimes are superior, but rather that they are possible. And Ostrom's research illustrates a great variety of common-property resource regimes that appear to be efficient and stable. It is, therefore, a matter of public debate which particular rules and institutions ought to govern the use of property. And there may be something in this finding that provides new ways of thinking about economic arrangements in the twenty-first century as well. In this context she states:

> We will all be the poorer if local, self-organized institutions are not a substantial portion of the institutional portfolio of the twenty first century. Many indigenous institutions developed to govern and manage local common-pool resources have proven themselves capable of enabling individuals to make intensive use of these resources over the long run – centuries or even millennia – without destroying the delicate resource base on which individuals and their future offspring depend for their livelihood. (Ostrom 1994: 2)

Ostrom's main purpose in *Governing the Commons* was to identify the conditions or circumstances that result in effective and ineffective rules to manage common-pool resources at the local level. For this, she analysed a number of cases where robust, long-lasting resource-management systems existed like the common-pasture management in Switzerland and Japan, irrigation systems in Spain and the Philippines, along with systems that have failed – fisheries in Turkey, Sri Lanka, and Nova Scotia; and irrigation systems in Sri Lanka. She, however, could not find specific rules that worked in large number of cases. This prompted her to move to a higher level of generality and specify 'the broader institutional regularities among systems that were sustained over a period of time' (Poteete et al. 2010: 99; Acheson 2011). The consequence of such an exercise resulted in drawing up a set of design principles for a stable local common resource management. She argues that successful common-pool resource management systems of different kinds like forests, pastures, irrigation are characterized by eight traits. These includes the need for clear boundaries to exclude outside parties; arrangements to involve all beneficiaries in the decision-making process; effective policing; a scale of graduated sanctions for those who violate community rules; simple mechanisms of conflict resolution; and the need for community self-regulation to be recognized by higher-level authorities

Thus refuting Hardin's (1968) pessimistic 'tragedy of the commons', her publications have highlighted a variety of conditions under which collective action in resource management operates effectively. However, this is not to ignore the fact that certain critical concerns still remained unattended by Ostrom and others on CPR. By not paying enough attention to the nuances of community dynamics, CPR work has often underplayed questions of social difference and the diverse and sometimes conflicting interests of resource users. In addition, the focus on collective action has tended to direct attention away from the fact that, while institutions can enhance cooperation, they can also be beset with conflict, factional divisions and power politics. The design principles, while very useful, can also appear to be a 'blueprint' for collective action, something that risks ignoring uncertainty and unpredictability, both in terms of human action and in the ecological world.

It is in this context, Elinor Ostrom's recent work[11] demands attention where she has attempted to engage more fully with some of these issues, moving away from the early rigid approach to design principles and embracing questions about the learning and resilience needed for successful collective action in today's rapidly changing societies and environments (Mehta 2009). Ostrom argues that human-induced causes of change in natural resources, like forest, takes place at multiple scales. But in reality, most governance mechanisms are designed at a single level – whether international, national, regional or local and consequently they fail to provide effective solutions in an overall manner. Efforts to 'decentralize' governmental arrangements frequently do not recognize the importance of complex, polycentric arrangements and are based on a presumption of a single government at one level taking charge of a policy arena, often ignoring the existence of many vibrant self-governed institutions. 'Polycentric institutions provide a useful framework for governance, enabling aspects of preferred solutions to be used together in efforts to protect the long-term sustainability of diverse forested social-ecological systems. By considering the interaction between actors at different levels of governance, polycentricism contributes to a more nuanced understanding of the variation in diverse governance outcomes in the management of common-pool resources based on the needs and interests of citizens and the complexity of resources and governance systems at local, regional, national, and global levels' (Nagendra and Ostrom 2012: 1).

Significance of Ostrom for India

Ostrom's research is particularly important for India, struggling to manage its commons – forests, village common land, groundwater etc. Even though common land in India are often officially designated 'wasteland', rural households depend on them for water, fuel, fodder, and for large range of non-timber forest products for their survival and livelihoods. As most of these are not directly marketed, they are ignored in conventional calculations of output, though their contributions to livestock production and overall household welfare are high.

Institutionally, India also remains an important site with reference to Ostrom's work, primarily for two reasons. One, there exists in India, multiple informal institutional arrangements at the local level to monitor use and access to these resources. These are community based and are primarily governed by customary laws and rules. Crafted collectively and enjoying the sanctity similar to that of formal legal order within the community, these institutions are successful in conserving and preserving these resources, whether they are man-made or natural. For example, in the arid region of Rajasthan, *bawris* or water tanks that were constructed hundreds of years back are still maintained by local communities. Similarly there are institutions like the sacred groves in many forested areas of Uttarakhand, Himachal Pradesh or the Van Gujjars, the pastoral communities in, Gujarat and Rajasthan, who have evolved their own context specific rules and norms surrounding the resources in question.

Second, and more importantly is the role of the Indian state that controls the use and access of most of the common resources, like land, water and forests and over the years they have witnessed steady decline. Despite the fact, there exists hundreds of community-based institutions, they have fallen apart due to lack of legitimacy from the state and the government. Even though the decade of the seventies, it has introduced the popular slogan of people's participation, empirical evidences show that it did not recognize the ability of people to manage natural resources and more so if the resource commercially lucrative like the village common land, pastures etc. A classic example in this context is the story of Mangar Bani – a sacred grove located in the Aravallis in Faridabad. This is a 600 acre pristine forest areas in the outskirts of Delhi which has become a site for Haryana government's dream project, known as Mangar Bani Development Plan 2030.The state government proposes non-forest activities in the ecologically sensitive Aravalli hills and is one of the last patches of natural forests with its native tree species that has been protected for years by the village residents as sacred grove. According to Ostrom, given such context, the best possible solution to manage Indian forests, water and land is to recognize institutional diversity-polycentric governance. There is a need for state to recognize diversities of context – both cultural and economic which proposing institutional solutions. She advocates that equal measures of community, government and private involvement are necessary to manage the commons in India.

Writings on common property have been instrumental in developing a theoretical justification for decentralization of environmental policies around the world (Agrawal 2001). In India, we find that efforts towards decentralized governance of forests had been initiated way back in the 1980s through Joint Forest Management programme in the Midnapore district of West Bengal and were later replicated in different states. There has been a huge amount of literature on the subject that such initiatives were not successful as it should have been. Ostrom's work provides policy makers an understanding as how best to incorporate different dimensions of livelihood that concerns people in their policy frameworks. She argues that most of government policies fail to have desired outcomes because of the 'simplistic, prescriptive and inflexible design' (Nagendra and Ostrom 2012: 4–5).

The spate of writings on common property resources initiated by Ostrom and followed by large body of empirical literature generated by scholars working on commons in India is of relevance for making of policies on commons in India.[12, 13] It may be noted that following a Supreme Court order, asking states to take measures to ensure that common land and property, like pastoral land and tanks are freed from encroachment and acquisition on January 28, 2011, number of states have started taking initiatives in this regard. For instance in April 2012, Gujarat government circulated a draft policy on pastoral land which prohibits handing over of pastoral or *gauchar* land for industrial and other commercial purposes. This is unusual for a state, which had earmarked the same land for industrial uses. Similar step has been taken by Rajasthan to protect village common land. In fact, in November 2011, it became the first state to draft a policy in this regard and before the draft was finalized, the state has issued orders to demarcate, develop and protect the commons. Along with Andhra Pradesh and Karnataka, fifteen other states are in various stages of making policies and programmes in this direction (Mahapatra 2012).

Ostrom and her colleagues were instrumental in initiating the International Forestry Resources and Institutions Programmes (IFRI) since 1992. The programme today is an important source for providing a huge set of baseline data on social, economic, ecological and institutional aspects of forests. Initially focussed itself on few sites including India and Nepal, the studies from IFRI examine how a governance arrangement affects forests and also people who are dependent on them. Large number of scholars and policy makers associated with IFRI attempt to study the role of both formal and informal institutions in enhancing resource sustainability and livelihoods of people among other things. The extensive body of research initiated by Ostrom is, therefore, significant in demonstrating the benefits of self-rule and local monitoring for sustainable management of community forests and such insights are extremely important for Indian settings where there exist both formal and informal arrangements for managing forests.

Conclusion

Ostrom's theoretical formulation and empirical evidences which goes beyond the dominant debate between the state and market, thus strongly recommends institutional diversity, i.e. looking to individuals to solve problems rather than relying on top down, one-size-fits-all solutions. In fact, her work has shown that creative solutions to problems such as the depletion of common-pool resources exist outside of the sphere of national governments. She demonstrates how common resources like forests, fisheries, oilfields, grazing lands and irrigation systems can be managed successfully by people who use them, rather than by governments and private companies. According to her 'what we have ignored is what citizens can do and the importance of real involvement of the people as opposed to just having somebody in the country's capital make a rule'. To put it differently, in a world where attention too often focuses solely on the state or private sector as the driving force, Ostrom helped us understand institutional diversity, and ways of governing the commons that build on people's capacity for collective action. Her research thus offers valuable lessons on how to craft rules and provide incentives for cooperation at the many levels needed for a sustainable future.

Notes

1. A revised version of this paper was published in *Man and Development*, XXXV (1), CRRID, March 2013.
2. One of the most critical aspects concerning common resources is regarding governance and in identifying appropriate institutions to manage them. There has been much debate on this issue among social scientists, particu-

larly among the mainstream economists. There are those who recommend 'the state' as the best possible alternative when market arrangements fail, while there are others who see privatization as the only solution. It may be mentioned here that till the period of the 1980s, scholars mostly had presumed that the users of such resources could not self-organize to manage them and hence recommended the imposition of government or private ownership, based on the theories of Gordon (1954), Demsetz (1967) and Hardin (1968). These debates tend to assume that boundaries between different forms of resource tenures – public, private and communal are neatly defined. Despite the fact that there is now a huge body of literature that contests such neat demarcation between resource tenures and argue that they are in fact blurred and overlap with each other, still continue to frame most of the academic debates and policy decisions on the issue of governance of natural resources. The environmental debates in India reflect such institutional dilemmas and there is a tendency to repose faith on the state as the sole custodian of natural wealth. However, there has been a partial and incremental shift from this kind of thinking and this is due to two important and interconnected reasons – the failure of centralized policies to bring about targeted development outcomes and emergence of numerous people's movements that questioned the role of the state. These two factors forced successive governments in India since the 1970s, to recognize people as catalyst of change and eventually brought them to the centre stage of several programmes in the name of participatory development

3. Governing the Commons: The Evolution of Institutions for Collective Action, in the opinion of many, is the most important published work in the past few decades on the social-science aspects of resource management. Although it appeared 20 years ago, it is worthwhile reviewing because the book contains concepts, analytic, methods, case studies and conclusions that have strongly influenced social scientists interested in the institutional dimensions of resource management ever since.
4. Hardin was not the first to notice the tragedy of the commons. In fact Aristotle in his famous treatise, *Politics: Book II*, chapter–3 observes, 'What is common to the greatest number has the least care bestowed upon it. Everyone thinks chiefly of his own, hardly at all of the common interest'.
5. It is to be noted that the tragedy of the commons paradigm was published at a time when environmental problems had recently come to the forefront. There was an urgent need for a theory to explain the environmental ills of the world and for a framework to help make sense of observations. The paradigm applied, Hardin thought not only to common pastures but also to public lands such as national parks, air and water. It provided the framework in which most environmentalists and those in related social and especial natural sciences understood resource issues (Berkes and Feeny 1990). Few questioned Hardin's assumption of

individual interest unchecked by social tensions and his emphasis on competition (rather than on co-operations) as the overriding relationship that shaped interactions among resource users.

6. There are number of writings on the game situation – Wagner 1983; Runge 1984, 1992; Sindal 1985; Ostrom 1990.
7. Campbell in *Paradoxes of Rationality and Cooperation* explains this issue – 'Quite simply, these paradoxes cast in doubt over understanding of rationality and, in the case of Prisoner's Dilemma, suggest that it is impossible for rational creatures to cooperate. Thus, they bear and threaten the foundations of social sciences. It is the scope of these consequences that explain why these paradoxes have drawn so much attention and why they command a central place in philosophical discussions' (Campbell and Sowden 1985: 3).
8. Those who live nearer to the forests may be in a position to easily monitor the rate of degradation.
9. Most of the writings on common property resource have uncritically taken to the original models and seem to take it for guaranteed the existence of a remorseless tragedy. They presume that CPR appropriators are trapped in a Hobbesian State of Nature and cannot by themselves come out of the situation. The logical sequence according to some analysts was to recommend an external agency, i.e. 'the government' to take over the common. In other words 'if ruin is to be avoided in a crowded world; people must be responsive to a coercive force outside their individual psyches, a "Leviathan", to use Hobbes's term' (Hardin 1968).
10. Similar argument is also put forward by Carruthers and Stoner (1981) who argue that public control is required vis-à-vis common property resources if economic efficiency is to result from their development. Further, where large economies and technical knowledge is concerned, it is argued that this external force should be large central governments. They are viewed as necessary agents of change in rural areas where powerful individuals control major resources by discouraging collective action. It is seen that policy prescriptions of this nature, i.e. control and monitoring by a centralized external agency of natural resources such as forests, pastures, fisheries have been implemented quite extensively particularly in developing countries like India.
11. Elinor Ostrom. 2010. 'Beyond Markets and States: Polycentric Governance of Complex Economic Systems', *American Economic Review*, 100 (3), June: 641–672.
12. Some of the important works are by Wade (1986; 1990; 1995), Agrawal (2001) and Ghate, Nagendra and Ostrom (2008).
13. Dilip Athavale. 2009. 'Nobel Winner Ostrom's Work Relevant for India', *The Times of India*, 15 October.

References

Acheson, J. 2011. 'Ostrom for Anthropologists', *International Journal of the Commons*, 5 (2), August: 319–339.

Agrawal, Arun. 2001. 'Common Property, Forest Management and the Indian Himalaya', *Contributions to Indian Sociology*, 35 (2), June: 181–212.

——. 2002. 'Resources and Institutional Sustainability', in E. Ostrom et al. (eds), *Drama of the Commons*, Washington, D.C.: National Academy Press.

Axelrod, E. 1981. 'The Evolution of Cooperation among Egoists', *American Political Science Review*, 75, June: 306–318.

Bentley. 1949. *The Process of Government*. Evanston, Ill, Principia Press.

Berkes, Fikret and David Feeny. 1990. 'Paradigm Lost: Changing Views on the Use of Common Property Resources', *Alternatives*, 17 (2): 48–55.

Campbell, R. and L. Sowden. 1985. *Paradoxes of Rationality and Cooperation*, Vancouver: University of British Columbia Press.

Carruthers, I. and R. Stoner. 1981. 'Economic Aspects and Policy Issues in Ground Water Development', World Bank Staff Working Paper No. 496, Washington, D.C.

Demsetz, H. 1967. 'Toward a Theory of Property Rights', *American Economic Review*, 57: 347–359.

Gordon, H.S. 1954. 'The Economic Theory of a Common Property Resource: The Fishery', *Journal of Political Economy*, 62: 124–142.

Hardin, Garrett. 1968. 'The Tragedy of Commons', *Science*, 162 (13), December: 1243–1248.

Hardin, Garrett. 1978. 'Political Requirements for Preserving Our Common Heritage' in H.P. Brokaw (ed.) *Wildlife and America,* Council on Environmental Quality, Washington DC.

Jodha, M.V. and S.A. Pasha. 1991. 'Developing Uncultivated Lands: Issues from Karnataka's Experience in Social Forestry', *Indian Journal of Agricultural Economics*, 46 (4), October–December: 543–554.

Jodha, N.S. 1986. 'Common Property Resources and Rural Poor in Dry Regions of India', *Economic and Political Weekly*, 21 (27): 1169–1181.

——. 1990. 'Rural Common Property Resources: Contributions and Crisis', *Economic and Political Weekly*, 25 (26), 30 June: A65–A78.

——. 1995. 'Common Property Resources and the Environmental Context Role of Biophysical Versus Social Stress', *Economic and Political Weekly*, 30 (51), 23 December: 3278–3283.

Mahapatra, Richard. 2012. 'Uncommon Swings for Commons', *Down to Earth*, 15 June.

Mehta, Lyla. 2009. 'Elinor Ostrom's Nobel Prize for Economic Sciences', http://www.ids.ac.uk/go/news/elinor-ostrom-s-nobel-prize-for-econom ic-sciences, last accessed on 14 March 2013.

Nagendra, Harini and Elinor Ostrom. 2008. 'Governing the Commons in the New Millennium: A Diversity of Institutions for Natural Resource Management', http://www.eoearth.org/article/governing_the_commons_in_the_ new_millennium:_a_diversity_of_institutions_for_natural_resource_mana gement, last accessed on 9 March 2013.

Nicita A. 2002. 'Economic Institutions and Common Pool Resources: The Role of Exit Costs in Enforcing Full Cooperation' in M. Frenzini and A. Nicita (eds), *Economic Institutions and Environmental Policy*, Hampshire, England: Ashgate Publishing Limited.

——. 2012. 'Polycentric Governance of Multifunctional Forested Landscapes', *International Journal of the Commons*, 6 (2): 104–133.

Olson, M. 1971. *The Logic of Collective Action*, Massachusetts: Harvard University Press.

Ostrom, E. and Roy Gardner. 1993. 'Coping with Asymmetries in the Commons: Self-Governing Irrigation Systems Can Work', *Journal of Economic Perspectives*, 7 (4), Fall: 93–112.

Ostrom, Elinor. 1990. *Governing the Commons: The Evolution of Institutions for Collective Action*, Cambridge: Cambridge University Press.

——. 1994. 'Neither Market, Nor State: Governance of Common Pool Resources in the Twenty First Century', Lecture presented at IFPRI, Washington, D.C.

——. 2005. *Understanding Institutional Diversity*, Princeton, NJ: Princeton University Press.

——. 2010. 'Beyond Market and States: Polycentric Governance of Complex Economic Systems', *American Economic Review*, American Economic Association, Vol. 100(3), 641–672, June.

Ostrom, Elinor (ed.) 2002. *The Drama of the Commons*, Washington, D.C.: National Academic Press.

Poteete, A.R. et al. 2010. *Working Together: Collective Action, the Commons, and Multiple Methods*, Princeton, NJ: Princeton University Press.

Quiggin, J. 1995. 'Common Property in Agricultural Production', *Journal of Economic Behavior and Organization*, 26: 179–200.

Runge, C. Ford. 1984. 'Institutions and the Free Rider: The Assurance Problem in Collective Action', *The Journal of Politics*, 46 (1): 154–181.

Schmid, U. 2001. 'Legal Pluralism as a Source of Conflict in Multi-Ethnic Societies: The Case of Ghana', *Journal of Legal Pluralism*, 46: 1–48.

Sindal, D. 1985. 'Coordination Versus Prisoners' Dilemma: Implications for International Cooperation and Regimes', *American Political Science Review*, 79 (4): 923–942.

Seabright, P. 1993. 'Managing Local Commons: Thwarted Issues in Incentive Design', *The Journal of Economic Perspectives*, Vol. 7 No. 4 (Autumn, 1993), 113–134.

Taylor, M. 1976. *Anarchy and Cooperation*, London: John Wiley & Sons.

Truman 1958. *The Governmental Process*, New York: Knopf.

Wade, R. 2003. *Governing the Market: Economic Theory and the Role of Government in East Asian Industrialisation*, New Jersey: Princeton University Press.

Wagner, R. 1983. 'The Theory of Games and the Problem of International Competition', *American Political Science Review*, 77: 330–346.

Welch, W.P. 1983. 'The Political Feasibility of Full Ownership Property Rights: The Cases of Pollution and Fisheries', *Policy Sciences*, 16 (2): 165–180.

4

Socio-economic Dimensions of Globalization in India

Challenges and Prospects

Rashi Krishna Sinha

Globalization connotes to the holistic approach in which one finds confluence of political, social, economic and cultural linkages at different levels – individual, organizational, national and international. It has an intrinsic tendency for universal application of economic, institutional, legal, political and cultural practices. It initiates when there is an integration of social policies at the global level to combat poverty and hunger. It implies intermingling of consumer's tastes and industries addressing their concerns by narrowing dissimilarities across nations and the globe. Globalization facilitates dissemination of science, technology, knowledge, values and information to one and all internationally. It happens when the standardization and regulatory institutions increasingly reach across borders. Greater interdependence facilitates negotiating strategies for designing trade and investment. C. Rangrajan puts it, 'Broadly speaking, the term 'globalization' means integration of economies and societies through cross country flows of information, ideas, technologies, goods, services, capital, finance and people. The essence of globalization is connectivity. Cross border integration can have several dimensions – cultural, social, political and economic. In fact, some people fear cultural and social integration even more than economic integration ...' (*The Economic Times* 2005). Globalization means a brave new world with no barriers for few people, for others, it spells doom and destruction. However, there are drawbacks as well as advantages to development via globalization. There are major concerns regarding globalization. It may be with respect to iniquitous distribution of income, opportunities and progress among nations and within the nation. We frequently use phrases like 'two India", 'Digital divide' 'Bharat and India' which are an expression of the iniquitous distribution of opportunities to move forward. There is inhibition that globalization culminates into loss of national sovereignty and the probable problems that may erupt in the independent decision making of the domestic policies which have wider socio-economic ramifications. It is also acknowledged that there are fears associated with globalization with respect to security and volatility as when the countries are interrelated a small spark can lead to large conflagration.

Global exchanges are not a new phenomenon for India. India has a long historical background of trade and cultural relations across the globe. But since the term has become ubiquitous in the contemporary world and so in India for everything our economic, social, cultural and political spheres draw more critical review with it in mind. It is usually referred with reference to the policy of openness initiated in July, 1991.

The human face of neo-liberal prescript for growth, the structural adjustment programme, has already revealed its severe limitations to one and all in India. Economic reforms were initiated with much fanfare in India. They were propagated as the harbingers of progress and prosperity. More than two decades have passed but yet one cannot claim that what was propagated has actualized. It has been established through a number of studies that since the dawn of the policy of laissez-faire the chasm between the haves and the haves-nots has increased manyfolds. The 'elephant' might be awake, economic prosperity in terms of per capita income, growth rate of Gross Domestic Product (GDP) etc. envisaged as an index of good economic health of our economy and its performance during the recent global recession may be cited as an example of our sound economic fundamental base, but all this along with around 275 million people living below poverty line. Few experts advocate that these figures do not reflect reality and the actual figure may be much more. Notwithstanding the respectable rates of growth in Gross National Product (GNP) and Gross Domestic Product (GDP) experienced in India the nation has witnessed persistent increase in absolute poverty. According to the Report of the National Commission for Enterprises in the Unorganized Sector (NCEUS 2009), headed by Arjun Sengupta, about 77 per cent of the population in India continues to live on less than ₹ 20 per capita per day on an average; obviously, considering various poverty norms, this much of population suffers from poverty, hunger and malnutrition. So, it may be safely assumed that substantial fraction of Indian population suffers from hunger and malnutrition to varying degrees. We are erecting jungles of concrete without providing basic amenities to the sea of humanity.

The pertinent question of well-being of the masses is the want of basic amenities: freedom from hunger, access to food and nutrition, health care, hygiene, sanitation, safe drinking water, what to say about electricity and good connectivity with adjoining towns, still haunts us in the era of globalization. In a healthy body resides a healthy mind is equally true for a nation. The contemporary socio-economic scenario reminds us of the state of one step forward and two steps backward. Initiating measures for the upliftment of the vulnerable groups without appropriate precautions are efforts in a void. If the voiceless and resourceless remain where they were prior to the initiatives then one may well imagine what and how the things are moving. If this situation lingers then one may not be incorrect to call the process of integrating economies through the mechanism of globalization a hopeless, directionless and inhuman way to growth and development. Most important is the fact that we were among the 23 nation in 1947 who signed the General Agreement on Tariffs and Trade (GATT), actively participated in the different rounds of negotiation which culminate into World Trade Organisation (WTO) but the effort to bring the benefits of integrating at the global level are yet to see the light of the day for the masses. The phenomenon of inclusiveness in the approach to growth and development seems to see the light of the day. The fruits of integration are yet to trickle down to the ground zero level, i.e. the masses. The so called level playing field is conspicuous by its absence. It has raised certain very basis curiosities about the very dynamics of inclusive growth where the inclusion is of India and not of Bharat. The mask of human face fails to hide the bitter reality of increasing disparities along with increasing prosperity that is selective.

'Our national mission must continue to be what it was ... to eliminate the curse of poverty ... . There is no humiliation more abusive than hunger ... . We must lift those at the bottom so that poverty is erased from the dictionary of modern India' (Acceptance Speech, President of India 2012), is a very apt description of the state of affairs at the ground level and the priority that should be on the top. To begin with, one must examine how far we have been able to fill the belly of the millions. This article makes an effort to analyse the situation as it prevails in Bharat not India. This article primarily focuses on very few socio-economic concerns in the contemporary India. The analysis concentrates on the post-reform era on the issues of food and nutritional security, women and women empowerment. It tries to analyse the trend on the very basic aspect of socio-economic dimensions, i.e. hunger, state of food and nutritional security, well-being, anaemia, body mass index and empowerment of women-on certain specific issues, in India briefly during the post-reform period.

I

State of Hunger

One may wonder about the state of food and nutritional security for the millions of Indians. But before we get into the conceptual framework of food and nutritional security it would be better to look at the concept of hunger in brief. Hunger has been conceptualized at two different frames.

One is what we usually feel after specific interval of time if we do not get something to eat. That is usually referred as self-reported hunger. The other is about the contents of what we actually eat. The composition of what we consume and if it becomes a part of our routine, knowingly or unknowingly or by compulsion of economic factors, to consume less than what is the standard requirement of the body it is said to be hidden hunger. It is important to make clear that one can fill up the stomach with non-nutritious food, which do not provide the required calories or micro-nutrients-vitamins, iron, iodine, zinc, and calcium that are required in tiny amounts, and another situation could be when the essential calories, proteins, fats and micronutrients are not absorbed in the body due to ill-health and poor hygiene. In all such cases, hunger is not articulated. This second kind of hunger may be termed chronic or endemic hunger, as it is not felt, recognized or voiced by children or adults.

To gauge the state of hunger on the basis of equally weighted indicators – under nourishment, child underweight and child mortality, Global Hunger Index (GHI) is developed by International Food Policy Research Institute (IFPRI). The global averages reveal dramatic differences among regions and countries. Compared with the 1990 score, the 2012 GHI score was 16 per cent lower in Sub-Sahara Africa, 26 per cent lower in South Asia, and 35 per cent lower in the Near East and North Africa. Progress in South Asia and Latin America and the Caribbean was particularly remarkable, with the GHI scores decreasing by 46 per cent and 44 per cent respectively (although the score was already low in the later region). In Eastern Europe and the Commonwealth of Independent States, the 2012 GHI was 46 per cent lower than the 1996 score (Global Hunger Index 2012).

It is significant to take note of the fact that notwithstanding the high growth rate achieved by India during 1996 to 2001, India's score on GHI fell and returned back to 1996 level in 2012. This stagnation in GHI scores occurred during a period when India's gross national income (GNI) per capita almost doubled, rising from about 1,460 to 2,850 constant 2005 international dollars between 1995–97 and 2008–10 (World Bank 2012). Generally, higher incomes are associated with less hunger. This pattern is shown by the black line, which was predicted from a regression of the GHI on GNI per capita for 117 countries with available data.

It is evident from Figure 1 that India's data points fall consistently above the predicted line. This result means that given India's per capita income, it has higher GHI scores than would be expected. Between 1990 and 1996, India's trend line moved in parallel with the predicted line, indicating that its GHI score was falling commensurate with economic growth. After 1996, however, the disparity between economic development and progress in the fight against hunger widened, and India moved further away from the predicted line (World Bank: 2012).

Figure 1

Global Hunger Index and Income in India and Elsewhere

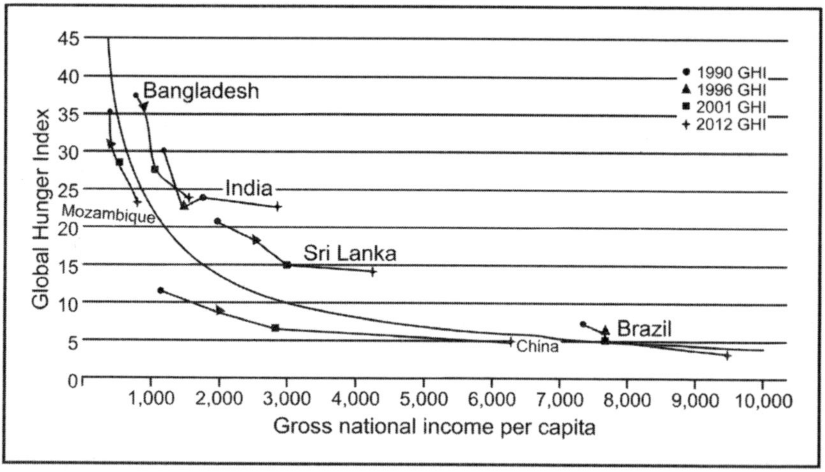

Source: Based on data on per capita GNI from World Bank (2012).

Note: Data on Gross National Income (GNI) per capita are based on purchasing power parity and expressed in constant 2005 international dollars. The pack trend line was predicted from a regression of 1990, 1996, 2001 and 2012 country level GHI scores on GNI per capita for all countries with available data. Data points for 1990 GHI, 1996 GHI and 2012 GHI correspond with GNI per capita data for 1989–91, 1995–97, 2000–02, and 2008–2010 respectively.

Courtesy: Global Hunger Index-2012, IFPRI, Box 2.1, page 12.

According to the latest data on child undernutrition, from 2005–10, India ranked second to last on child underweight out of 129 countries – below Ethiopia, Niger, Nepal, and Bangladesh. Only Timor-Leste had a higher rate of underweight children. By comparison, only 23 per cent of children are underweight in Sub-Saharan Africa (although India has a lower proportion of undernourished in the population than Sub-Saharan Africa.) (World Bank: 2012). The pertinent question is why we are facing contrast-abundance in foodgrain production and hunger existing simultaneously. The management of surplus production, its delivery mechanism and execution of policies adopted, so as to ensure foodgrain availability to the deprived masses needs serious review and revamp. The issue is not the diagnosis of the problem but the right strategy to overcome it. The grey areas are evident to one and all, but the puzzle lies in nailing it at the right spot.

Food Security

The idea of 'food and nutritional security', as we know it in the contemporary scenario, to consider freedom from want with respect to food and agriculture was articulated in 1943. The basic concept of 'freedom from want' meant a secure, adequate and suitable supply of food for every man, woman and child, where 'secure' referred to the accessibility of the food, 'adequate' referred to the quantitative sufficiency of the food supply and 'suitable' referred to the nutrient content of the food supply (CFS 2012).

Amartya Sen's vision on food security gave it a new dimension as it emphasized on 'access' to food through what he called 'entitlements'; a combination of what one can produce and exchange in the market along with the state or other socially provided supplies. Subsequently, he and other scholars emphasized on adequate nourishment and adequate healthiness, which should be of concern for food security. Sen highlighted that food related problems are influenced not only by food production and agricultural activities, but also by the structure and processes governing entire economies and societies. There has been a substantial shift in comprehending food security and in 1995 World Food Summit (WFS) where it was declared that 'Food security at the individual, household, national, regional and global levels exists when all people, at all times, have physical and economic access to sufficient, safe and nutritious food to meet their dietary needs and food preferences for an active and healthy life' (FAO 1996: 3). Within this global strategy, one can distinguish between strategies of growth-mediated security and strategies of support-led security (Dreze and Sen 1989; Von Braun et al. 1992). The current terminology in use, as adopted from the 1996 World Food Summit, emphasizes the multidimensionality of food security: food security exists when all people at all times have physical and economic access to sufficient, safe and nutritious food to meet their dietary needs and food preferences for an active and healthy life (World Food Summit 1996). This definition has widely established the four pillars of food security: availability, accessibility, utilization and stability.

Nutritional Security

Nutrition security emerged with the recognition of the necessity to include nutritional aspects into food security. Unlike food that is mostly defined as any substance that people eat and drink to maintain life and growth, nutrition adds the aspects of health services, healthy environment and caring practices. More precisely, ... 'a person is considered nutrition secure when she or he has a nutritionally adequate diet and the food consumed is biologically utilized such that adequate performance is maintained in growth, resisting or recovering from disease, pregnancy, lactation and physical work' (Frankenberger et al. 1997: 1). Nutrition (*The Lancet* 2008) was regarded as one of the most important parts in development priorities but underemphasized by both donor and developing countries. This notion is now widely shared and triggered a broad framework for collective action among key stakeholders. Based on the initiative of these stakeholders through the Road Map for Scaling-Up Nutrition (2010), nutrition security is deemed to be achieved when secure access to an appropriately nutritious diet is coupled with a sanitary environment, adequate health services and care, to ensure a healthy and active life for all household members. Recently, FAO has defined nutrition security as a condition when all people at all times consume food of sufficient quantity and quality in terms of variety, diversity, nutrient content and safety to meet their dietary needs and food preferences for an active and healthy life, coupled with a sanitary environment, adequate health and care (CFS 2012). Weingärtner (2010) further developed a definition of food and nutrition security as a condition under which adequate food (quantity, quality, safety, socio-cultural acceptability) is available and accessible for and satisfactorily utilized by all individuals at all times to live a healthy and happy life. As the term of food and nutrition security has combined both security concepts in a more integrated way as a single goal of policy, this term is more widely used.

II

State of Global Food Production in Brief

Security, whether of food or nutrition, is associated with the production of foodgrains. Global production of cereals has almost trebled in the past 50 years, outpacing the twofold rise in world population. Yet over a billion people in the world remain hungry. Cereals form the staple diet of poor people and are also their main imported food item. In 2010, cereals made up 40 per cent of the food imports of least developed countries. Increasing consumption of vegetables and meat is indicative of growing incomes, and these items typically account for half of the food imports by developed countries. Wheat, maize, and rice account for the majority of trade in cereals; maize and other coarse grains are not only consumed by humans but are also used as animal feed in the production of meat and for the manufacture of biofuels. Most cereal production is for domestic consumption, with just 10 per cent of world production traded globally: over the past decade, only one-fifth of all wheat produced globally was traded, while rice trade accounted for 6 per cent of global rice production (Kshirsagar and Baffes 2011). In value terms, approximately two-thirds of world food exports go to developed countries, and just under one-third to middle income ones, with the poorest countries being insignificant in world food trade: the share of the least developed countries in world food trade is just 1 per cent. However, food trade forms a higher share of the *total* trade basket of developing countries compared with developed countries.

The progress to halve the proportion of people suffering from hunger, according to the Millennium Development Goal (MDG) and achieve the basic right to food by 2015 remains slow (MDG 2010). On the other hand hunger may have increased as a consequence of global food and financial crises. There are also large regional differences in the distribution of hungry people in the world. According to Food and Agriculture Organisation the highest number of 642 million hungry people live in Asia and the Pacific.

State of Food and Nutritional Security in the Post-Reform Period in India

Food Security: Empirical evidence from the post-liberalization phase, after 1991 in India suggests that the situation of food security deteriorated in this phase: 1990–91 to 2003–04. The availability of foodgrains declined since the growth rate of production which was 2.85 per cent per annum during 1980–81 to 1990–91 decelerated significant to only 1.16 per cent per annum during the post-liberalization period 1990/91 to 2003/04. Important here is to point out that this growth rate of foodgrains production was lower than the population growth rate of 1.95 per cent per annum during this period, thereby leading to the reduction in the per capita availability of foodgrains from 492 kg per capita during Triennium Ending 1992 to 449 kg per capita during the TE 2004. Stability of production of foodgrains has shown fluctuations but increase in the buffer stocks is where one finds solace. It is another matter that notwithstanding comfortable situation in buffer stock people died out of hunger! It is an acknowledged fact that since growth has not been uniform, so in spite of increase in the per capita incomes, economic access to food for the millions of people has posed a grave challenge before the planners. It is quite natural to point out that during 1980s the growth rate of area in rice was marginal at 0.41 per cent but growth in production and yield was above 3 per cent. From 2000–01 to 2009–10 the situation changed with growth in area turning negative and production and yield standing at 1.59 per cent and 1.61 per cent respectively. In wheat too, during the 1980s the growth in area was marginal at 0.46 per cent but the production and yield was above 3 per cent. During 2000–01 to 2009–10 the growth in area in wheat was 1.21 per cent and production and yield was 1.89 per cent and 0.68 per cent respectively. This indicates towards the fact that in case of these two crops the yield levels reached plateau and there is an urgent call for adequate steps before things becomes worst from bad. It is also important to take note that coarse cereals which are high in nutritional value constitute only 15 per cent of the total foodgrains production (2009–10). Even though area under coarse cereals has shown a significant decline over the years, its yield has shown a significant increase; there was no breakthrough in these crops. The growth rate in area of total coarse cereals, in both the periods, 1980–81 to 1989–90 and 2000–01 to 2009–10, was negative reflecting either shift to other crops or relatively dry area remaining fallow. Maize was an exception to this. In case of pulses, which are high in protein content, there was a negative growth in area during 1980s and production and yield for the period was 1.52 per cent and 1.61 per cent respectively. During 2000–01 to 2009–10, whereas area and production grew at 1.17 per cent and 2.61 per cent respectively the growth in yield has remained at about same level thereby reflecting increase in the level of production to be driven by area and a need for technological breakthrough (Economic Survey 2010–11). If one looks at the aggregate of cereals and pulses taken together for the per capita net availability per day (grams) for 1991 and 2009 (provisional) one finds a decline from 510.1(grams) and 444.0(grams) respectively (Economic Survey 2010–11).

The MSSRF report (2009) states that despite high rates of economic growth, levels of food insecurity remain elevated across most states in both periods.[1] The rankings of the states have been worked out for two time periods – 1998 to 2000 and 2004 to 2006. The reduction of food and nutrition insecurity remains a daunting challenge in rural India and requires policies that improve availability, access and absorption. The government needs to play a key role. Per capita rather than per household should be the basis of ration allocation through Public Distribution System (PDS), is what many experts believe.

Countrywide surveys revealed that when it comes to qualitative aspect of quantitative food our diet is found to be highly deficient in vitamins and minerals (hidden hunger) than protein due to low intake of income elastic foods like vegetables, fruits, pulses and foods of animal origin. Nutritional millets are disappearing. The decline in calorie consumption of the top quartile could be the result of a more sedentary lifestyle or of increasing diversity in food intake, but the decline for the bottom quartile since 1987, as shown in Table 1, cannot be interpreted as a sign of prosperity.

Several inferences can be drawn from Table 1. First, at any given point in time the calorie intake of the poorest quartile continues to be 30 to 50 per cent less than the calorie intake of the top quartile of the population, despite the poor needing more calories to compensate for harder manual work. Second, the calorie consumption of the bottom 50 per cent of the population has been consistently decreasing since 1987, which is a matter of concern. And last not the least, whereas the top quartile derived only 58 per cent of its calories from cereals in 2004–05, the bottom quartile still depended on cereals for 78 per cent of its calorie consumption.

Table 1

Total cereal calorie consumption by decile and quartile of per capita expenditure, rural India, 1983 to 2004–05 (in kcal) mean per capita consumption of calories, proteins and fats

	Bottom Decile	Bottom Quartile	Second Quartile	Third Quartile	Top Quartile
Total Calories					
1983	1,359	1,580	2,007	2,328	3,044
1987–88	1,488	1,683	2,056	2,334	2,863
1993–94	1,490	1,659	2,000	2,251	2,702
1999–2000	1,496	1,658	1,978	2,250	2,707
2004–05	1,485	1,624	1,900	2,143	2,521
Cereal Calories					
1983	1,150	1,309	1,589	1,738	1,974
1987–88	1,221	1,359	1,598	1,715	1,894
1993–94	1,203	1,316	1,504	1,591	1,690
1999–2000	1,197	1,289	1,591	1,509	1,566
2004–05	1,189	1,259	1,690	1,430	1,471

Source: Deaton and Dreze (2008).

Courtesy: N.C. Saxena (nd). 'Hunger, Undernutrition and Food Security in India', CPRC-IIPA Working Paper 44, 13.

Table 2

Trends in Cereal Consumption across Expenditure Groups (K per month)

Rural	*Percentile*				
	Lowest 5%	*5%–10%*	*40%–50%*	*90%–95%*	*95%–100%*
1993–94	9.68	11.29	13.33	14.98	15.78
1999–2000	9.78	11.15	12.89	13.73	14.19
2004–05	9.88	10.87	12.16	12.77	13.5
Urban	*Percentile*				
	Lowest 5%	*5%–10%*	*40%–50%*	*90%–95%*	*95%–100%*
1993–94	8.91	10.11	10.99	10.19	10.29
1999–2000	8.99	10.15	10.8	9.94	9.72
2004–05	9.25	10.04	10.25	9.5	9.1

Source: NSSO 2007, 61st Round Report.

The NSSO data on consumer expenditure on food indicates a declining trend in the annual per capita consumption of cereals, for all classes of people, else than the lowest 5 per cent group in the rural and urban category as shown in Table 2. The rate of decline in the rural areas in all the expenditure groups, else than the lowest 5 per cent, was higher than the urban areas. The table clearly shows that, as India has moved to greater prosperity in the past 20 years, the cereal consumption of the rich has gone down, but there has been no increase for the poor. At any given point in time the cereal intake of the bottom 10 per cent in rural India continues to be at least 20 per cent less than the cereal intake of the top decile of the population, despite better access of the latter group to fruits, vegetables and meat products. This group's sedentary lifestyle should also be taken into account when assessing the difference between the two groups. For the upper segment of the population the decline may be attributed to a diversification in food consumption, easy access to supply of other high value agricultural commodities, changed tastes and preferences, and to consumption of more expensive non-foodgrain products (Mittal 2008). Higher economic growth and per capita incomes thus contribute to reduction in per capita demand for cereals for the rich. However, for those who are around the poverty line, this has to be understood as a distress phenomenon, as with only a marginal increase in their incomes over time they are forced to cut down on their food consumption to meet other pressing demands that were not considered important in the past.

A survey by Mander (2008) of 474 destitute people in eight villages found that intense food shortages often demand the most unreasonable choices, such as between food and medicines, or between eating to save a life and relieving unbearable pain. Most hungry people reported that their most hazardous fall into pauperization was because they, or a loved one, had fallen gravely ill. Many old people simply try to wait out an attack of illness, and if that does not work they consult a local untrained practitioner, who demands his fees in advance, never guaranteeing cure. They do this by cutting back their food intake even further.

Between 1972–73 and 2004–05 the share of food in total consumer expenditure fell from 73 per cent to 55 per cent in rural areas and from 64 per cent to 42 per cent in urban areas (NSSO Report 2007–08: 508).

As stated earlier, consumption patterns are changing toward non-cereal food. Thus, diet diversification needs to be achieved, particularly in states with high rates of undernutrition. A recent National Sample Survey (NSS 2011)[2] on consumer expenditure showed that the share of cereals in total consumption declined further in 2009–10, when it was only 15.6 per cent in rural areas and 9.1 per cent in urban areas (Table 3). Despite the decline in the total share of food expenditure over time, the decline in the share of non-cereal foods has been lower than that of cereals.

The recent food inflation in non-cereal food is therefore a serious concern. It should be noted, however, that the Indian poor still get more than 55 per cent of their daily calories from cereals, and thus, there is still

Table 3

Changes in Consumption Pattern: All of India, 1987–88 to 2009–10

Food Items	*Rural (Percentage share of consumption)*					*Urban (percentage share of consumption)*				
	1987–88	*1993–94*	*1999–2000*	*2004–05*	*2009–10*	*1987–88*	*1993–94*	*1999–2000*	*2004–05*	*2009–10*
Cereals	26.3	24.2	22.2	18	15.6	15	14	12.4	10.1	9.1
Grains	0.2	0.2	0.1	0.1	0.2	0.2	0.2	0.1	0.1	0.1
Cereals substitutes	0.1	0.1	0.1	0.1	0.1	0.1	0.1	0	0	0
Pulses and Products	4	3.8	3.8	3.1	3.7	3.4	3	2.8	2.1	2.7
Milk and Products	8.6	9.5	8.8	8.5	8.6	9.5	9.8	8.7	7.9	7.8
Edible oil	5	4.4	3.7	4.6	3.7	5.3	4.4	3.1	3.5	2.6
Egg, Fish and Meat	3.3	3.3	3.3	3.3	3.5	3.6	3.4	3.1	2.7	2.7
Vegetables	5.2	6	6.2	6.1	6.2	5.3	5.5	5.1	4.5	4.3
Fruits and Nuts	1.6	1.7	1.7	1.9	1.6	2.5	2.7	2.4	2.2	2.1
Sugar	2.9	3.1	2.4	2.4	2.4	2.4	2.4	1.6	1.5	1.5
Salt and Spices	2.9	22.7	3	2.5	2.4	2.3	2	2.2	1.7	1.5
Beverages and the like	3.9	4.2	4.2	4.5	5.6	6.8	7.2	6.4	6.2	6.3
Food total	64	63.2	59.4	55	53.6	56.4	54.7	48.1	42.5	40.7
Non-food total	36	36.8	40.6	45	46.4	43.6	45.3	51.9	57.5	59.3
Total expenditure	100	100	100	100	100	100	100	100	100	100

Source: NSS (2011).

a need for diet diversification to improve nutrition among the poor. Achieving this would require diversification of agriculture, particularly among small and marginal farmers and those in resource-poor areas.

The proportion of income that goes to consumption expenditure and of the total consumption expenditure, the proportion spent on the purchase of food is significant from the point of view to analyse the well-being of the masses in India. The NSSO report (Economic Survey 2010–11) of the unit level data revealed that of the monthly consumption expenditure, based on the uniform period, bottom quintile of rural India devotes approximately 67 per cent of their aggregate household expenditure on food. One may not find it much perplexing to comprehend the state of people under Below Poverty Line (BPL) and those just above in case of food inflation spiral. The reduced purchasing power on one hand makes it difficult for the poor to somehow fill the belly, what to talk about nutritional aspect and on the other hand, it increases the absolute numbers within the BPL. Purchasing power is necessary to access food and nutrition security. Thus, poor tend to be undernourished than the rich. However, malnutrition is not confined only to the people below the poverty line (BPL). Many people above the so called poverty line are also undernourished. Hidden hunger is also seen in the so-called well-to-do and 'safety net' is needed for them as well. FAO estimates that the rising prices have plunged an additional 75 million people below the hunger threshold.

One can comprehend the plight of these people by having a glance over Table 4 which depicts income generated by farmer households during 2002–2003 (NSS Report No. 497: 2003). On average a farmer household on all India basis, earns ₹ 2,115 per month of which (₹ 969 per month, i.e. approximately 46% of total income) comes from cultivation, followed by wages (₹ 819 per month, i.e. 39%) non-farm business (236 per month, i.e. 11%) and animals (₹ 91 per month, i.e. 4%). The above monthly income estimates exclude income from rent, interest, dividend etc. The State wise disaggregated data shows considerable variations in per family income as well as the components of the income.

The empirical facts from the aforestated Table, point out that of 18 states 11 have the cultivation and the remaining 7 have wages as the main source of income generation, significantly enough one must not forget to pinpoint that one among the two sources-cultivation and wages, not only remained the primary source of income generation but also had a substantial percentage composition in the total income generated through various sources. On all-India basis these two sources of income-wages and cultivation, comprises of about 84.54 per cent of the total income generated from different sources o income.

The State-wise break up of monthly incomes is depicted in the table below:

Table 4

Break-up of Average Monthly Income (excl. rent, interest, dividend etc.) per farmer household by source in each of the major states during the agricultural year 2002–03

States	*Average monthly income (₹) per family household from*				
	Wages	*Cultivation*	*Farming of animals*	*Non-farm business*	*Total*
Andhra Pradesh	643 (–39.35)	743 (–45.47)	93 (–5.69)	155 (–9.49)	1,634 (–100)
Assam	973 (–30.78)	1792 (–56.69)	141 (–4.46)	255 (–8.07)	3,161 (–100)
Bihar	497 (–27.46)	846 (–46.74)	265 (–14.64)	202 (–11.16)	1,810 (–100)
Chhattisgarh	709 (–43.82)	811 (–50.12)	–3 (–0.19)	101 (–6.24)	1,618 (–100)
Gujarat	925 (–34.46)	1,164 (–43.37)	455 (–16.95)	140 (–5.22)	2,684 (–100)
Haryana	1,268 (–44)	1,494 (–51.84)	–236 (–8.19)	356 (–12.35)	2,882 (–100)
Jammu and Kashmir	2,060 (–37.54)	2,426 (–44.21)	382 (–6.96)	620 (–11.29)	5,488 (–100)
Jharkhand	924 (–44.66)	852 (–41.18)	86 (–4.16)	207 (–10)	2,069 (–100)
Karnataka	1,051 (–40.18)	1,266 (–48.39)	131 (–5.01)	168 (–6.42)	2,616 (–100)
Kerala	2013 (–50.27)	1,120 (–27.97)	154 (–3.85)	717 (–17.91)	4,004 (–100)
Madhya Pradesh	560 (–39.16)	996 (–69.65)	–227 (–15.87)	101 (–7.06)	1,430 (–100)
Maharashtra	799 (–32.44)	1,263 (–51.28)	144 (–5.85)	257 (–10.43)	2,463 (–100)
Odisha	573 (–53.95)	336 (–31.64)	16 (–1.51)	137 (–12.9)	1,062 (–100)
Punjab	1,462 (–29.48)	2,822 (–56.9)	236 (–4.76)	440 (–8.87)	4,960 (–100)
Rajasthan	931 (–62.15)	359 (–23.97)	5 (–0.33)	203 (–13.55)	1,498 (–100)
Tamil Nadu	1,105 (–53.33)	659 (–31.81)	110 (–5.31)	198 (–9.55)	2,072 (–100)
Uttar Pradesh	559 (–34.23)	836 (–51.19)	53 (–3.25)	185 (–11.33)	1,633 (–100)
West Bengal	887 (–42.66)	737 (–35.45)	77 (–3.7)	378 (–18.19)	2,079 (–100)
All India	819 (–38.72)	969 (–45.82)	91 (–4.3)	236 (–11.16)	2,115 (–100)

Source: Calculated from Serving Farmers and Saving Farming, Fifth and Final Report, 4 October 2006, National Commission on Farmers, 78.

Note: Figures in parentheses are the percentages with respect to the total.

Since nearly 83.6 per cent of the operational holdings belongs to marginal, marginal or small category, according to Agriculture Census 2010–11, improving the small farm productivity and profitability as a single development strategy could make the greatest contribution to improve food and nutritional security at the household level. Besides the general steps for raising the productivity and incomes of the farmers the specific steps which focus on benefitting small and marginal farmers could be, increased investment in human resources, adequate and timely availability of quality inputs including credit, a fine tuned insurance system to cover production risks, an effective and efficient extension system, arrangements for custom hiring of agricultural machinery and development of low cost machinery for use by these farmers. It is also of paramount importance that the unfinished agenda of land reforms; distribution of ceiling surplus land, attention to common property and wasteland resources and consolidation of holdings etc. is vigourously and meticulously pursued. However, these would require a strong political will and determination.

Table 5

Food Inflation for Major Groups and Sub-groups, 2007–08 to First Quarter of 2011–12 (year-on-year changes)

	2007–08	*2008–09*	*2009–10*	*2010–11*	*2011–12 (Average of April–July)*
All Commodities	4.7	8.0	3.8	9.6	9.5
Primary (non-food) items	8.3	11.1	12.7	17.7	12.9
Food items	7.0	9.1	15.3	15.6	8.9
Foodgrains	6.9	11.0	14.5	4.8	2.3
Cereals	9.5	11.9	12.6	5.3	5.0
Rice	11.02	14.8	12.3	5.9	2.6
Wheat	7.3	9.9	12.8	3.0	0.4
Maize	6.3	6.8	10.2	10.1	1.9
Pulses	–2.3	7.5	22.4	3.2	–8.18
Urad	–16.5	0.0	43.0	19.0	–9.23
Arhar	16.4	14.4	48.8	–4.5	–16.3
Mung	–11.9	6.1	55.4	19.9	–21.4
Fruits and Vegetables	11.5	8.2	9.7	16.4	16.0
Milk	5.1	7.6	18.8	20.1	8.0
Eggs, Meat and Fish	3.2	7.7	20.8	25.5	9.1

Source: Office of the Economic Advisor to Government of India, Ministry of Commerce and Industry, various years.

In recent years, the increase in food prices has been a significant negative feature of India's economic environment. This trend has had a tremendous impact on quality of life, as people struggle to maintain the nutritional standards they had previously achieved, possibly giving up some other forms of consumption to keep themselves well-fed. As shown in Table 5, food prices were very high in the past two years, before declining moderately in the first quarter of fiscal year 2011–12. The inflation for food items rose from around 7 per cent in 2007–08 to 9 per cent in 2008–09, 15 per cent in 2009–10, and 16 per cent in 2010–11, before declining to 9 per cent in the first quarter of 2011–12.

Food consumption patterns have diversified in recent years. The predominance of cereals in the typical household diet has given way to greater balance, with a consequent increase in the demand for proteins – pulses, milk, meat, fish, and eggs – and for vegetables and fruit. Thus, it is no surprise that these items have been the primary causes of food inflation in the recent past. Table 5 shows that inflation for three major pulses (*urad, arhar,* and *moong*) was nearly 50 per cent, although they showed negative inflation in recent months. Similarly, inflation for milk, eggs, meat, fish, fruits, and vegetables was very high. One can well image the impact on the common man's budget and their well-being.

III

Women and their Empowerment in India

Women and their Well-being

It is a widely held belief that the responsibility of child bearing and rearing comes naturally to women from which they cannot run away. Their families responsibilities in particular and social obligations are posed as reasons for not letting them enter into economic activities. Education or no education hardly tampered this attitude. Reason and scientific temper still eludes the mindset of masses in general. The attitude towards girl child in States like Punjab, Rajasthan, UP, MP, Bihar, Haryana, Delhi and Chandigarh strengthens this idea. Females have been denied right to live through female infanticide, cultural neglect and blatant use of violence, a study reported (Dagar 2002). According to her, modern technology has been used to ensure that the girl child does not get birth. No doubt the disturbing sex ratio can be viewed as an index of failure of education to develop scientific temper so that male and female child can be seen and viewed from the same yardstick.

The report (NNMB TR No. 21) states that the children in the age group of 1–3 to 18 years of age suffered from a huge deficiency in the recommended dietary intakes and the actual intake. Actual intake lags behind the recommended intake. As for the male child it improves while for the girl child gap widens. For girls, adolescence is a period of growth with an increased nutrient intake. 'Girl Child in India: The Situational Analysis' (India Country Paper 1999), points out that a large number of adolescents are under nourished and the problem is among girls (45%) than boys (20%), primarily due to deep rooted gender discrimination. The most visible manifestation of nutritional deficiency is high prevalence of anaemia and stunting among adolescent girls. Studies suggest that as many as 55 per cent of adolescent girls may suffer from anaemia. The only states in which less than half of children are anaemic are Goa (38%), Manipur (41%), Mizoram (44%), and Kerala (45%). The prevalence of anaemia among children 6–35 months has increased from 74 per cent in NFHS-II to 79 per cent in NFHS-III. This increase is largely due to a sharp increase in anaemia among young children in rural areas.

Table 6

Certain Indicators from NFHS-II and NFHS-III

	*NFHS-3 (2005–06)**	*NFHS-2 (1998–99)*
Ever-married women age 15–49 who are anaemic (%)	56.2	51.8
Pregnant women age 15–49 who are anaemic (%)	57.9	49.7
Currently married women who usually participate in household decisions (%)	52.5	N A
Ever-married women who have ever experienced spousal violence (%)	37.2	N A
Women whose Body Mass Index is below normal (%)	33	36.2

* Provisional Data

Source: Key indicators from NHFS-3 (2005–06), Ministry of Health and Family Welfare, Government of India.

NFHS-III collected information on the height and weight of women age 15–49 and men age 15–54. The height and weight measurements provide an estimate of the body mass index (BMI), a measure of nutritional status.[3] The proportion of ever-married women who are thin (33%) has decreased slightly from 36 per cent in NFHS-II. Women who are undernourished themselves are also much more likely than other women to have children who are undernourished. The proportion of women who are undernourished is highest in Bihar (45%), Chhattisgarh (43%), Madhya Pradesh (42%), and Orissa (41%). It is lowest in Sikkim (11%) and Mizoram (14%). This is higher than the males for the respective periods and respective groups. It may not be inappropriate to state that with the slightest provocation the casualty is the girl child in case of nutrition. As for more than one reason, if the choice is to be made between a girl child and a boy, the weightage is inclined in favour of the male child. The sacrificial goat is the female. Hindrance in education and discrimination in caring of the girl child does not stop here but the separatist attitude prolongs in the matters of giving space and freedom for taking personal decisions-choice of lifestyle, vocation etc. But one must not forget that education, awareness and empowerment is the only way to meet this challenge.

Women Empowerment

Mahatma Gandhi once said, 'Man the law-giver will have to pay a dreadful penalty for the degradation he has imposed upon the so-called weaker sex'. Empowerment is for empowering. It may connote different meanings to different people but the point of agreement lies on the broader perspectives. Precisely, it implies right and freedom to make choice in the matters related to personal life, expression of self without hitch, prerogative in decision making at all levels – family, society and nation, at large. It also implies right and ability to control wealth and assets, to be an integral part of the process that involves change – social, economic and political. There can be many ways to look at it. A number of studies have pointed out that economic strength is significant. Sen (1993) looks at empowerment as 'altering relations of power ... which constrain women's options and autonomy and adversely affect health and well-being'. For Batliwala (1994) empowerment is 'how much influence people have over external actions that matter to their welfare'. Kabeer (2001) sees it as 'the expansion in people's ability to make strategic life choices in a context where this ability was previously denied to them'. To Bennett (2002) empowerment is 'the enhancement of assets and capabilities of diverse individuals and groups to engage, influence and hold accountable the institutions which affect them'.

Hence, control over resources – physical, human, intellectual, financial (Kabeer 1994) and ideology – values, beliefs, and attitudes (Batliwala 1994), is one of the most important dimensions of most views. The second dimension shared by most definitions is women's agency, i.e. that women themselves have a right to make choices and should be involved in determining which choices make the most sense for them and their families (Kabeer 1999). Third, since empowerment implies a move from being without power to having power, most definitions also imply that it is a process that involves change over time. Finally, empowerment is also understood as outcome, such as improvement in education, health, economic and political participation. Very often, however, women's empowerment and gender equality are used interchangeably. While they are related, they are not the same. Gender equality is about women's status, relative to men while women's empowerment is about women's ability – in an absolute sense – to exercise control, power, and choice over practical and strategic decisions (Grown 2008).

In a patriarchal society the indicators, Table 7, are very relevant. If one recalls the state of freedom of movement for men on all these parameters, as stated in the respective Tables and then does a comparative one finds what is left and how far we have to move forward. The figures collected through the survey, NFHS-III, do pinpoint towards urban-rural divide.

Table 7

Women's Freedom of Movement

(in percentage)

Place	*Alone*	*With someone else*	*Not at all*	*Total*
	Urban			
To the Market	66.2	26.8	7.0	100
To the Health Facility	60.3	36.2	3.5	100
To the place outside village/community	45.5	48.0	6.6	100
	Rural			
To the Market	44.3	40.4	15.3	100
To the Health Facility	41.5	53.0	5.5	100
To the place outside village/community	34.0	56.6	9.4	100
	Total			
To the Market	51.4	35.9	12.6	100
To the Health Facility	47.7	47.5	4.8	100
To the place outside village/community	37.7	53.7	8.5	100

Source: NFHS-III, Table 14.12, p. 470.

It would be better to have a glance over the state of autonomy in decision making on few selected issues which may be termed as personal decisions an individual makes in his day-to-day life even without giving a thought of the freedom one enjoys, especially men. The say of currently married women in her family to take decision on her own is not satisfactory. Table 8 puts forth specifically four issues, and one finds that the respondent's say in taking decisions on her own is not satisfactory.

Earning of a woman, her command over her earnings and the channels in which it has to be spent, her earnings as a proportion to her husband's earnings and that to the family are significant aspects of the issues which bring to the surface the true face of the society and the nation. Her control over other resources in the family, freedom of movement and freedom in the decision making of the family, access to and use of microcredit programmes are significant indicators reflecting the ground reality. It is significant to point out that even in the matter of her own health the percentage of women making decision by herself is very low from any standards, what to talk about other issues.

It is quite interesting to note that the percentage of women who have money and can decide how to use is highest in Uttar Pradesh. At 59.9 per cent it is the highest across the states (refer Table 9), while in a state like Kerala the percentage stands at 20.7. More significantly in Kerala 82.6 per cent of women know of a microcredit programme but only 8.0 per cent of women take loan from microcredit programme.

Table 8

Per cent Distribution of Currently Married Women by Person who usually makes Decisions about Four kinds of Issues, India, 2005–06

	Women's Participation in Decision Making						
Decision	*Mainly Respondent*	*Respondent and Husband Jointly*	*Mainly Husband*	*Someone Else*	*Other*	*Missing*	*Total*
		Urban					
Own Health Care	29.6	39.1	26.5	3.5	1.1	0.1	100
Major Household Purchases	10.4	51.5	26.8	8.7	2.5	0.1	100
Purchase of Daily Household Needs	39.9	28.9	19.8	8.8	2.5	0.1	100
Visit to her Family and Relatives	12.2	57.3	22	6.6	1.8	0.1	100
		Rural					
Own Health Care	26.0	33.4	31.7	7.6	1.3	0.1	100
Major Household Purchases	7.6	41.2	34.6	13.5	2.9	0.1	100
Purchase of Daily Household Needs	29.1	27.1	26.9	13.9	2.9	0.1	100
Visit to her Family and Relatives	10.0	46.4	28.9	12.1	2.4	0.1	100
		Total					
Own Health Care	27.1	35.1	30.1	6.3	1.3	0.1	100
Major Household Purchases	8.5	44.4	32.2	12	2.8	0.1	100
Purchase of Daily Household Needs	32.4	27.7	24.7	12.3	2.8	0.1	100
Visit to her Family and Relatives	10.7	49.8	26.8	10.4	2.2	0.1	100

Source: NFHS-III, Vol. 1. I Chapter 14, Table 14.6, p. 461.

Table 9

Women's Access to Money and Credit by States

	Percentage who have money that they can decide how to use	Percentage who have a bank or saving account that they can themselves use	Percentage who know of microcredit programme	Percentage who have taken loan from a microcredit programme	Percentage of women allowed to go to three specified places alone
India	44.6	15.0	38.6	4.0	33.4
North					
Delhi	43.2	30.3	29.0	1.0	36.6
Haryana	35.5	12.4	36.8	0.6	40.7
Himachal Pradesh	28.8	22.2	20.2	2.2	64.0
Jammu and Kashmir	55.4	21.9	13.1	0.2	51.0
Punjab	26.5	14.6	40.8	0.9	39.0
Rajasthan	32.7	7.5	12.3	0.6	31.6
Uttaranchal	36.1	20.1	13.7	1.6	42.8
Central					
Chhattisgarh	34.3	8.1	29.1	1.3	17.8
Madhya Pradesh	36.8	8.9	30.6	1.0	25.7
Uttar Pradesh	59.9	13.2	14.1	0.4	23.4
East					
Bihar	58.6	8.2	27.0	1.0	25.2
Jharkhand	60.2	14.4	25.5	1.9	36.6
Odisha	36.2	9.8	69.4	6.9	18.7
West Bengal	37.4	14.1	41.4	2.7	32.3
Northeast					
Arunachal Pradesh	38.7	19.0	31.4	2.5	40.2
Assam	26.9	11.7	41.7	1.4	35.3
Manipur	32.0	8.0	65.1	4.2	53.5
Meghalaya	35.3	16.9	22.8	1.5	28.1
Nagaland	36.5	7.4	23.8	1.3	25.4
Sikkim	36.9	20.9	18.3	0.8	30.9
Tripura	26.9	18.7	38.7	2.3	36.8
West					
Goa	56.7	42.3	56.6	2.5	56.8
Gujarat	57.8	19.9	45.[illegible]	1.7	47.3
Maharashtra	40.8	20.3	35.5	2.1	40.2
South					
Andhra Pradesh	48.6	18.0	59.6	16.3	37.3
Karnataka	60.3	22.1	55.1	9.2	30.6
Kerala	20.7	27.0	82.6	8.0	34.7
Tamil Nadu	25.4	15.8	79.0	13.4	54.2

Source: NFHS III, Table 14.14, p. 473.

In the contemporary socio-economic scenario where much of the decisions at the family level are dictated by economic consideration, there is a need to build up confidence among women; that they are no less to any one in performing in any sphere of social, political and economic activities. Repeated psychological bombardment of prejudiced social mindset has somewhere affected their thought process and the confidence level. The process of globalization has facilitated a positive channel in making us think constructively about the issues concerning women.

Conclusion

The most obvious inference that one may draw from the aforestated analysis is about how to nail the inertia among people and the society at large so that the benefits of any policy or programme focused to the vulnerable section and women trickles down to them, still confronts us all. The influence and impact of the literacy and education has been constrained in inculcating scientific temper, to developing confidence among the vulnerable section of the society and women. They face a situation where everything is for them, with them, without them. The fruits do not reach the desired destinations and the whole hue and cry is for them – the vulnerable masses in general and women. Governments come and go but these people remain where they were. It is not that the issues, problems and challenges are unknown to anyone but yet remedies are conspicuous by their absence. The core issue is how we are preparing ourself for the positives along with the challenges that come with Globalization at different levels. Globalization is done and it cannot be undone. The need of the hour is to dwell upon how to enable the masses reap its benefits. The societal attitude, political manoeuvring and economic hiccups shall be taken to advantage for the larger section in India. The need of the hour is to concentrate on the execution part of any strategy that is oriented towards these wonderful people for whom governments are formed and dethroned.

Notes

1. The report is divided into two parts. The first part has seven indicators: 3 are *input* indicators: percentage of population consuming less than 1890 Kcal per day; percentage of households with access to safe drinking water; and percentage of households with access to a toilet inside the premises. The remaining 4 are *outcome* indicators: percentage of women with anaemia; percentage of women with chronic energy deficiency; percentage of children under three years of age who are underweight for age; and percentage of children under three years of age who are anaemic. The second part looks into food delivery systems like PDS, ICDS and MDMS.

2. URP refers to Uniform Reference Period. MRP refers to Mixed Reference Period. NSS usually has URP of 30 days for food and non-food items. In 1999–00 onwards, MRP (30 days for some items and 365 days for some items) was introduced. URP estimates shown except for 1999–2000, for which only MRP estimates are available.
3. The Body Mass Index (BMI) is defined as weight in kilograms divided by height in metres squared (kg/m^2). A cut-off point of 18.5 is used to define thinness or acute undernutrition and a BMI of 25 or above indicates overweight or obesity. More than one-third (36%) of women aged 15–49 in India have a BMI below 18.5, indicating chronic nutritional deficiency, including 16 per cent who are moderately to severely thin.

References

Batliwala, Srilatha. 1994. 'The Meaning of Women's Empowerment: New Concepts from Action', in G. Sen, A. Germain and L.C. Chen (eds), *Population Policies Reconsidered: Health, Empowerment and Rights*, Cambridge, MA: Harvard University Press, 127–138.

Bennett, Lynn. 2002. 'Using Empowerment and Social Inclusion for Pro-poor Growth: A Theory of Social Change', Working Draft of Background Paper for the Social Development Strategy Paper, Washington, D.C.: World Bank.

Braun, Joachim Von et al. 1992. 'Improving Food Security of the Poor: Concept, Policy and Programs,' IFPRI Washington DC.

Caldwell, Wayne, Anneliza Collett, Therese Ludlow, Ian Sinclair and Jenny Whitehead. 2011. 'Planning and Food Security within Commonwealth: Discussion Paper', Commonwealth Association of Planners.

CFS. 2012. http://www.fao.org./docrep/meeting/026/md776e.pdf

Chattapadhayay, Gopa and Seevalingum Ramasawmy. 2012. *Integrating Food Security Information in National Statistical Systems: Experiences, Achievements, Challenges*, Rome, Italy: Food and Agriculture Organization of the United Nations.

Dagar, Rainuka. 2002. 'Identifying and Controlling Female Foeticide and Infanticide in Punjab', sponsored by DWCD, MHRD, GOI.

Deaton, Angus and Jean Dreze. 2008, Nutrition in India: Facts and Interpretations. Electronic copy available at : http://ssm.com/abstract=1135253, accessed on 14 February 2015.

Desai, Manisha. 2010. 'Hope in Hard Times: Women's Empowerment and Human Development', United Nations Development Programme, Human Development Research Paper 210/2014.

Dev, S. Mahendra and Alakh N. Sharma. 2010. 'Food Security in India: Performance, Challenges and Policies', Oxfam India Working Papers Series (OIWPS) VII, September.

Dreze, J. and A. Sen. 1989. *Hunger and Public Action*, Oxford: Clarendon Press.

Duflo, Esther. 2012. 'Women Empowerment and Economic Development', *Journal of Economic Literature,* 50 (4): 1051–1079, http://dx.doi.org/10.1257/jel.50.4.1051.

Economic Survey 2010–11. Statistical Appendix-A22, GOI.

Family Welfare Statistics in India 2011. Statistics Division, Ministry of Health and Family Welfare, GOI.

Frankenberger, T.R. et al. 1997. A Definition of Nutrition Security. CARE. Mimeo.

Gill, Gerard J., John Farrington, Edward Anderson, Cecilia Luttrell, Tim Conway, N.C. Saxena and Rachel Slater. 2003. 'Food Security and the Millennium Development Goal on Hunger in Asia', Working Paper 231, Overseas Development Institute, London, December.

Global Hunger Index 2012. IFPRI.

Grown, Caren. 2008. 'Indicators and Indexes of Gender Equality: What Do They Measure and What Do They Miss?" pp. 93–146 in *Equality for Women: Where Do We Stand on Millennium Development Goal 3?* edited by Mayra Buvinic, Andrew R. Morrison, A. Waafas Ofosi Arnaah and Mirja Sjoblon Washington DC: The World Bank.

Kabeer, Naila. 1994. *Reversed Realities: Gender Hierarchies in Development Thought*, London: Verso.

Kabeer, Naila. 1999. 'The Conditions and Consequences of Choice: Reflections on the Measurement of Women's Empowerment', UNRISD Discussion Paper No. 108.

——. 2001. 'Reflections on the Measurement of Empowerment', in A. Sisask (ed.), *Discussing Women's Empowerment: Theory and Practice,* SIDA Studies No. 3, Stockholm: Novum Grafiska AB, 17–57.

Kshirsagar, V. and J. Baffes. 2011. *The Nature and Structure of Global Food Markets*, Washington, D.C.: World Bank.

Mander, Harsh. 2008. 'Living with Hunger: Deprivation Among the Aged, Single Women and People with Disability', *EPW*. 26 April 2008.

Millennium Development Goals Report 2010. United Nations, New York.

Mittal, Surabhi. 2008. Demand Supply Trends and Projection of Food in India, Working Paper No. 209, March, ICRIER. New Delhi.

M.S. Swaminathan Research Foundation (MSSRF) Report 2009, supported by the United Nations World Food Programme.

Nandakumar, T., Kavery Ganguly, Pravesh Sharma and Ashok Gulati. 2010. 'Food and Nutrition Security Status in India: Opportunities for Investment Partnerships', ADB Sustainable Development Working Paper Series No. 16.

NCEUS 2009. The Challenge of Employment in India: An Informal Economy Perspective, Volume 1 – Main Report.

National Family Health Survey – II, 1998–99.

National Family Health Survey – III, 2005–06. Volume I and II, September 2007.

National Nutritional Monitoring Bureau's Technical Report No. 21.

National Statistical Organisation. 2012. *Women and Men in India*, 14th Issue, Central Statistics Office, Ministry of Statistics and Programme Implementation, GOI.

NSS Report No. 497: Income Expenditure and Production Assets of Farmer Households 2003.

NSSO 2007–08. Household Consumer Expenditure in India, 2007–08. MOSPI.

NSSO 2011–12. Key Indicators of Household Consumer Expenditure in India, 2011–12. MOSPI, GoI.

Pangaribowo, Evita Hanie, Nicolas Gerber and Maximo Torero. 'Food and Nutrition Security Indicators: A Review', Joachim von Braun, Manfred Denich, Solvay Gerke, Anna-Katharina Hornidge and Conrad Schetter (eds), ZEF Working Paper Series, ISSN 1864–6638 Working Paper No. 108, Department of Political and Cultural Change Center for Development Research, University of Bonn.

Road Map for Scaling-Up Nutrition (SUN) First Edition. 2010

Saxena, N.C. nd. 'Hunger, Undernutrition and Food Security in India', CPRC-IIPA Working Paper 44.

Sen, Amartya. 1999. *Development as Freedom*, Oxford: Oxford University Press.

Sen, Geeta. 1993. 'Women's Empowerment and Human Rights: The Challenge to Policy', paper presented at the Population Summit of the World's Scientific Academies, *Smallholders Farmers in India: Food Security and Agricultural Policy*: RAP Publication 2002/03, Food and Agriculture Organization of the United Nations, Regional Office for Asia and the Pacific Bangkok, Thailand.

The Economic Times, 15 January 2005.

The Lancet. 2008. 'Material and Child Undernutrition', Special Five Paper Series, 371 (9608–9612), January–February: 340–357.

United Nations System Standing Committee on Nutrition. 2010. *Sixth Report on the World Nutrition Situation: Progress in Nutrition*, Geneva, Switzerland.

World Bank 2012. World Bank Development Indicators Database. Washington, DC.

World Food Summit. 1996. Rome Declaration on World Food Security and World Food Summit Plan of Action.

Weingartner, Lioba. 2010. The Concept of Food and Nutrition Security, Background Paper No. 1.

5

From the Silent Spring to the Globalization of the Environmental Movement

Habibul Haque Khondker

> The world has enough to meet everyone's need but not everyone's greed.
>
> —Gandhi

> The environment is humanity's first right.
>
> —Ken Saro-Wiwa

Environmental degradation is one of the small numbers of key issues that have been truly global as it has affected the lives of billions of people around the world. The recent (in the middle of year 2013) spate of unusual climatic events, i.e. floods in Alberta, Canada, Northern India, and Europe (Germany, facing the brunt) is reminder of the changes in global climate system for the worse. The non-believers of climate change, notwithstanding the changes in the climate and its far-reaching consequences are not unreal. The gradual degradation of the global environmental system is an integral part of the centuries of industrialization rooted in a world-view of progress; therefore, an appropriate answer to this huge challenge lies in a reconceptualization of development. Various names of the new mode of development have been proposed. Whether we call it sustainable development, green development, eco-friendly development, inclusive development or development as freedom and/or dignity, the point is, we need to re-evaluate our development paradigms in light of the eco-consciousness or what Frances Lappe calls, eco-mind. The only redeeming thing is that in the last half century we have also witnessed a rise in the awareness of global environmental condition and a growing and incipient environmental movement. The movement, in order to be successful, must be holistic and global at the same time.

The two important concepts used in this chapter that need to be clearly defined are: globalization and environmental movement. We take the publication of Rachel Carson's *Silent Spring* in 1962 as a point of departure and try to recount the making of the environmental movement as a global movement in the past half a century. First, we try to provide with a clearer understanding of globalization as a concept and a phenomenon.

I

Globalization and Environmental Movement

Since the introduction of the concept in sociology in the late 1970s and early 1980s, by Roland Robertson, globalization has become a contentious concept, especially in the wake of the collapse of the Soviet socialism. As the globalization literature has expanded with the diversities of interpretations reaching a new height of complexity, it makes sense to turn to a simple minded, yet useful, taxonomy provided by Leslie Sklair.

There are three modes of globalization: (1) Generic globalization; (2) Capitalist globalization, and (3) Alternative globalization, according to Sklair (2009: 526). The generic globalization incorporates emancipatory potential as it includes electronic revolution creating opportunities for anti-corporatist social movements from below (Sklair 2009: 532). To view globalization as capitalism rests on a world-scale under the cover of a neo-liberal economic ideology and nothing else presents a highly reductionist view which ignores alternative forms of globalization with critical and emancipatory spaces. When we juxtapose environmental movement with globalization, we emphasize both the generic globalization as well as alternative globalization.

In the present chapter, as we explore the globalization of environmental movement, we look at globalization as a generic social process, what historians would call a secular movement of history. This view is consistent with the definitions provided by Roland Robertson and others who shared a similar neo-rationalist position. A number of writers on globalization in the 1980s – the present author included – were influenced by the ideas of Roland Robertson and Immanuel Wallerstein, although the former took a culturalist yet pragmatic position and the latter took a position that combined Marxian logic with Braudelian historical reach. Both positions were pragmatic and critical of the dominant and competing ideological positions during the Cold War. In a deeper sense both of them developed a critical yet a realist view of the world as the markers of globalization were clearly visible in the post-World War II world, and more specifically in the 1960s marked by assorted social movements that continued into the 1970s against various aspects of global capitalism and the American hegemony. Globalization, we argue is a complex and contingent historical process and should not be reduced to western economic domination (Turner and Khondker 2010). As we try to look at the complexity of the world and try to understand the predicament human society is faced with, we are compelled to question the taken-for-granted world and its institutions. Social change is mediated by the rising consciousness of the people originating with key ideas. The idea of questioning the status quo with regard to environment was one of the game-changers in recent history.

It is common to see a general mood of despondence and a tone of pessimism in the discussion of environmental or lopsided economic growth and quality of life. Here we need to follow what Antonio Gramsci recommended: 'Pessimism of the Intellect, Optimism of the Mind' and sociologist Fred Block once quipped, 'We also need Optimism of the Intellect'. A crying need of the hour is to develop and spread the optimism of the intellect by critically engaging with the dominant ideologies of the time.

It may be old-fashioned but Platonic idea of 'knowledge is virtue' or more germane to the present discussion, Socratic tradition that 'knowledge is power' an idea echoed by Sir Francis Bacon is more valid today than in the past as we live in a much-touted 'knowledge society'. But knowledge society based on a consensus view of knowledge is yet to be constructed since there are great debates on epistemologists as what is knowledge. Knowledge itself becomes a contested terrain, a battle-ground. We continuously debate, what happened, how did it happen? Even as we begin to emphasize the idea of evidence-based knowledge, the question as to what constitutes evidence is far from settled. Is 'global warming' a fact? Is climate change for real? And what should we – as citizens – do? What should be the role of the experts? Since understanding the climate or the ecosystems depends on a good grasp of science and expert knowledge, what should be the role of scientists and experts or knowledge-brokers in shaping the opinion of the general public.

We may start with the general Marxist epistemological position that the ruling (dominant) ideas are the ideas of the ruling class, but there is no reason to be contented with this status quo. Having stated the obvious, the ground reality, if you wish – we need to move beyond and develop an argument following another illuminating idea of Marx that, it is the social existence that determines social consciousness. If so, then the new circumstances featured by a perilous ecological system and a crisis-ridden economic system must per force produce new consciousness and new knowledge.

Social existence determines (social) knowledge. This broad notion was shared by not only Karl Marx and Karl Mannheim, sociologists of knowledge such as Robert Merton and the contemporary practitioners would add to the view that although there is a material or social basis of knowledge, knowledge does not emerge in a linear fashion. They also emphasize the autonomy of science and the scientific knowledge to avoid the pitfall of economic determinism. The role of experts remains a vexed issue in a democratic society.

Social existence is more than the economic structure (base/super structure model) it is the ensemble of ideas, challenges to previously held assumptions as laid by Thomas Kuhn, socio-economic-political and ideological conditions, and the conditions of social movements. All these moments reinforce one another. Of the various strands of Marxist reinterpretations, the Frankfurt School has been the most productive and nuanced, especially in matters of problems of modernity and the problematics of knowledge. The take away point of the Frankfurt School is to expose knowledge as control to champion the emancipatory potential of knowledge through a sustained critical engagement. The distinction made by Jürgen Habermas (1971) between knowledge and interest is as relevant today as it always has been. One can read in it a reformulation and refinement of the earlier distinction between knowledge and ideology. The central issue, however, is not just interest but the interest of a dominant section of the society who would side with interest-laden knowledge rather than unbiased knowledge for selfish reasons.

The reason for revisiting the subjects of knowledge-power linkage and the role of knowledge is because despite strong scientific evidence there are skeptics with vested interests who continue to deny facts such as global warming. I follow the definition of global warming as 'the gradual heating of Earth's surface, oceans and atmosphere. The rise in average temperatures worldwide has been documented by scientists since the late 1800s. The Earth's average temperature has risen by 1.4 degrees Fahrenheit (0.8 degrees Celsius) over the past century, the Environmental Protection Agency (EPA) reports. Though the existence of global warming was once considered controversial, it is now acknowledged as real by an overwhelming majority of researchers throughout the international scientific community, according to the National Aeronautics and Space Administration (NASA).' (http:// www.livescience.com/37003–global-warming.html#sthash.x7W1av4o.dpuf)

The contentious nature of knowledge related to environmental issues is illustrated in the fact that as Rachel Carson's *Silent Spring* (1962) was celebrated as a true precursor to the environmental awareness, and subsequently, movement; the Cato Institute, a Washington-based conservative Think Tank came forth with a critique of Carson's contentions. In 2012 many celebrated the half a century of Rachel Carson *Silent Spring* but also the death of Barry Commoner who devoted a life-long research and activism for environmental protection. The ecological movement began in the US in the 1970s following the declaration of the Earth Day in April 1970. That was a time of protests, social turbulence, a time of social movements and a time of hope. A variety of social movements emerged in the USA, the most prosperous economy of the world at that time. It was a moment of self-reflection and criticism. The Civil Rights movement, the anti-Vietnam war movement, the feminist movement timed with the celebration of the Earth Day in April, 1970.

Rachel Carson's *Silent Spring* published in 1962 was an eye-opener. The book turned controversial from the very beginning yet the claims could not be readily dismissed for the author was not only a writer but also a professional having a long career at the department of agriculture. However, some scientists pointed fingers at some of the claims and the book became controversial. President John F. Kennedy asked his science advisors to look into the claims made in the book and being convinced of the merits, passed laws protecting the environment. A rethinking of environment and ecology, a renewed understanding of what the world is turning out to be, grabbed the attention of a new generation. That was the generation of experimentation, innovation of ideas and questioning the everyday taken-for-granted assumptions.

One, however, should not be misled to believe that there was no concern with environment before the 1960s or 1970s. In fact, piece-meal legislations and writings on environment can be traced back far into history. Laws have been passed as early as the fourteenth century England, to prohibit water pollution. In India, Manu's Laws in post-Vedic period included guidelines of environmental protection covering air pollution, water pollution, and disposal of waste, quality of food, cleanliness, purity and ethical principles of environment (Padhy et al. 2006).

We examine in the following how the ecological crisis became global in terms of its understanding and how movements arose worldwide to protect the ecosystem. In fact, environment provides us a point of entry to the subject of globalization. Some analysts examine the functioning of the worldwide economic transactions, i.e. international trade, transfer of capital, financial transactions as evidence of economic globalization, others point to the flows of ideas, values and so on as markers of globalization, yet there are others who would look at technology, especially communication technology and its spread from satellite television, to long distance telephony and internet, 'world wide web' as concrete technologies of globalization.

Yet, the physical environment that we live in can hardly be bounded in a geographical boundary. Not only weather patterns, rivers, oceans and mountain ranges overlook national borders and seamless the recent threats to ecosystems too are unguided by environmental pollution.

The environmental movement as it unfolded, it brought government, scientists, civil society on the same platform although at an uneven speed and degree of enthusiasm. It led to serious questioning of taking for granted assumptions about economic growth. Sustainability brought to clear focus debates about growth that were limited to experts.

There was a time of optimism and progress in the 1960s, when there was greater hope for a braver new world, and for progressive international ideas. Colonies blessed with natural resources were becoming nations. The locals of cooperation and sharing seemed to be seriously pursued. Paradoxically, the 1970s slid slowly into moods of reaction and isolation while at the same time a series of UN conferences offered hope for greater cooperation on major issues. The 1972 UN Conference on the Human Environment brought the industrialized and developing nations together to delineate the 'rights' of the human family to a healthy and productive environment. A string of such meetings followed: on the rights of people to adequate food, to sound housing, to safe water, access to means of choosing the size of their families.

According to the Brundtland Commission Report: 'The present decade (the 1980s) has been marked by a retreat from social concerns. Scientists bring to our attention urgent but complex problems bearing on our very survival: a warming globe, threats to the Earth's ozone layer, deserts consuming agricultural land. We respond by demanding more details, and by assigning the problems to institutions ill-equipped to cope with them. Environmental degradation, first seen as mainly a problem of the rich nations and a side effect of industrial wealth, has become a survival issue for developing nations. It is part of the downward spiral of linked ecological and economic decline in which many of the poorest nations are trapped. Despite official hope expressed on all sides, no trends identifiable today, no programmes or policies, offer any real hope of narrowing the growing gap between rich and poor nations. And as part of our 'development', we have amassed weapons, arsenals capable of diverting the paths that evolution has followed for millions of years and of creating a planet our ancestors would not recognize' (Brundtland Commission 1987). The Report continued: 'We live in an era in the history of nations when there is greater need than ever for coordinated political action and responsibility, (Brundtland Commission 1987).

Citizens were brought at the centre of these debates. The idea of active citizenship took a new meaning. This was also consistent with the idea of participatory or deliberative democracy. But the movement has only begun for changing consciousness and the ways of thinking is a long-drawn out process. 'To this end, we appeal to citizens' groups, to non-governmental organizations, to educational institutions, and to the scientific community. They have all played indispensable roles in the creation of public awareness and political change in the past. They will play a crucial part in putting the world onto sustainable development paths, in laying the groundwork for Our Common Future' (Brundtland Commission 1987).

The 'modern' notion of sustainability has its origin in the international debate that began in 1972 in Stockholm and was consolidated twenty years later in Rio de Janeiro (Guimaraes 2004). The Brundtland report of 1987 was a manifesto for sustainable development.

The report stated: 'The satisfaction of human needs and aspirations is the major objective of development. The essential needs of vast numbers of people in developing countries for food, clothing, shelter, jobs – are not being met, and beyond their basic needs these people have legitimate aspirations for an improved quality of life. A world in which poverty and inequity are endemic will always be prone to ecological and other crises. Sustainable development requires meeting the basic needs of all and extending to all, the opportunity to satisfy their aspirations for a better life' (Brundtland Commission 1987).

In 1972, under the stewardship of Maurice Strong and under the sponsorship of the United Nations, a conference on environment was held in Stockholm. The 1972, UN Conference on environment recognized the importance and need for environmental management and paved the way for the formation of a specialized UN Agency, United National Environmental Protection Agency (Johnson 2013). The publication that followed the Stockholm conference was titled *Only One World* capturing the interdependence between ecology and economy, environment and society. In fact, this can be taken as a manifesto of ecological movement. Ecopolitics is thus a short-hand word for ecological politics of sustainable development. Maurice Strong in a recent book recounted that, 'The organizational capacity of the new program was kept weak. A group of countries which supported its establishment, including Britain, the US, Germany, Italy, Belgium, the Netherlands and France, had agreed secretly to ensure that it would not have the support required. The group was concerned that any new environmental regulations would have an impact on trade. They also wanted to ensure that UNEP did not have a large budget as it would then be restricted on what it could do' (Dodds et al. 2012: 16).

It would be important to recount the genealogy as well as the intellectual context of the environmental movement. Both in Europe and the United States the growth of such organizations as bird watching societies were the precursors to the rise of environmental consciousness. The growing concern with nature and nature conservation in the West in the late nineteenth century and early twentieth century foresaw the rise of environmental movement. From nature conservation to the idea of sustainable development has been a meandering course of social and economic developments spanning the last two hundred years or so.

In the late eighteenth century there were reactions to industrialization in Europe which may have played a role to conservation of a nostalgic past and may account for the rise of romanticism which was in essence an intellectual and artistic movement. More to the point, it was a reaction against scientific rationalization of nature. English romantics, especially poets such as William Wordsworth (1770–1850), Robert Byron (1788 – 1824), P.B. Shelley (1792–1822,) and John Keats (1795 – 1821), displayed a growing concern with nature in their creative works. The name of John Ruskin (1819 – 1900) and his work *Unto the Last* (1860) deserves special mention because the Indian nationalist leader and a champion of conservation of nature, M.K. Gandhi was so impressed that he translated this book into Gujarati in 1908. Ruskin as well as other naturalists left an indelible impression on Gandhi's thoughts.

In Germany, known more for its philosophical than industrial might in the late eighteenth and early nineteenth century provided the intellectual milieu for the development of Marxist ideas. Marx was a poet and romantic in his youthful writings, a true legatee of German idealism, which he later criticized. The German idealism and romanticism was rooted in the works of Herder (1744–1803), and Goethe (1749 – 1832), two of the prominent exponents. Following this intellectual heritage, in late 19th and early 20th century 'Life reform' (Lebensreform) social movement emerged in Germany. The Wandervogel movement in 1901 was an archetypical representation. Spiritual alienation of the masses from the land and the nature (nature objectified, commodified) was the theme of the movement with a focus on its rejoining.

The 19th Century American writers such as Ralph Waldo Emerson (1803 – 1882), Herman Melville, *Moby Dick* (1851), Henry David Thoreau, (1817 – 1862) *Walden, or Life in the Woods* (1854) Thoreau's writings influenced Tolstoy, Gandhi, Martin Luther King among others. The nineteenth century United States also saw the emergence of associations dedicated to nature appreciation and conservation. In 1883 American Ornithologists' Union was set up. The Audubon Society was established in 1886. In 1892 Sierra Club was established. Rachel Carson was a long-time member of the Audubon Society. Carson's concern arose from and a true heir to the earlier nature conservation.

The environmental movement in the United States began in the context of the 1960s and 1970s. These two decades were truly a season of social movements. What is interesting in the period of the 1960s and 1970s was that several social movements surfaced at the same time. The interlocking and overlapping nature of social movements created a ferment of social change. The policy makers could hardly overlook the rumblings on the ground. New periodicals and journals emerged and a public discussion took the centre stage. Environmental movements as Civil Rights movement, Feminist Movement and Anti-Vietnam war movements brought intellectuals, academics, students and various civil society organizations into a sustained conversation that helped to create a favourable public opinion in favour of the issues at hand. Environmental Movements came in the wake of such other lifestyle movements as Alternative lifestyle, alternative medicine, religion, Back to Nature, Hippie movements and so on. This should not be construed as a movement of and for a certain progressive, non-conformist segments in society.

Environmental movement was predicated upon not only the rise of a new educated, affluent class with interest in leisure and happiness, but there were other objective factors. It was not simply a pastime of the new leisure class. It was predicated upon the rise of a new knowledge society.

II

Knowledge – Social Movement – Institutional Changes

New scientific knowledge about environment played an important role in setting off environmental movements. In this context it would be important to remember the contributions of writers such as Barry Commoner, a biologist who championed the cause of environment and environmental protection through his research and promotion of public interest. Barry Commoner (1917 – 2012) was 'a founder of modern ecology and one of its most provocative thinkers and mobilizers in making environmentalism a people's political cause, ... scientist-activists who recognized the toxic consequences of America's post-World War II technology boom, and one of the first to stir the national debate over the public's right to comprehend the risks and make decisions about them', according to the *New York Times* (1 October 2012).

His pioneering research on the environmental impact of nuclear test explosions in the United States of America was an important factor for changes in the public policies in the US. In the late 1950s, Commoner was conducting research on baby tooth to examine the impact of radioactivity due to atmospheric test of nuclear weapons. His research raised a public awareness and played an important role in accomplishing a Nuclear Test Ban Treaty in 1963.

Barry Commoner in *The Closing Circle* (1971), argued that a sustainable economy, which may include corporations, government, and consumers, needs to be aware of the 'four laws of ecology':

1. Everything is connected to everything else.
2. Everything must go somewhere.
3. Nature knows best.
4. There is no such thing as a free lunch.

These four laws that he proposed became the catch phrase of ecological activists and supporters alike. *The Closing Circle* helped introduce the idea of sustainability, a notion that grew in popularity over time and became a mainstream idea. In his time, however, the idea of sustainability was controversial. 'Commoner pointed out, there is only one ecosphere for all living things. What affects one, affects all. He also noted that in nature, there is no waste. We can't throw things away. Therefore, we need to design and manufacture products that do not upset the delicate balance between humans and nature. We need to utilize alternative forms of energy, such as wind, solar and geothermal power. And we need to change our consumption habits accordingly – to use fewer products with plastics (which are based on oil), aerosol cans (which harm the atmosphere) and industrial-grown food (which is produced with harmful chemicals).' (http://www.thenation.com/ article/170251/remembering-barry-commoner#ixzz2aJ5yma5G)

Barry Commoner's, *The Poverty of Power* (1976) was a remarkable achievement linking energy crisis with the environmental and the economic crises. As a scientist with a deep understanding of Marxist political economy, Commoner presented a holistic, albeit reformed Marxist framework. For him, even prior to economic base lay ecological base on which economic base and subsequently political and cultural layers stood. Barry Commoner could link global poverty not to so-called 'population explosion' a major concern for many of his contemporaries, but to imperialism and capitalist exploitation. Rejecting the Neo-Malthusian explanation in vogue in the 1970s, Commoner argued that poverty is the cause of so-called overpopulation or population explosion. Demographic historians would be more inclined to side with Barry Commoner on that point.

Interestingly in the 1970s, a paradigm shift was taking place in the thinking of economic development. Ideas like economic growth without development were getting tractions. Writers such as Dudley Seers (1970) famed for his essay 'The Meaning of Development' began to influence a new generation of economists who challenged the conventional paradigm of economic growth accepting the Kuznets curve of initial inequality. The idea of holistic development began to emerge.

Following the schema of 'Knowledge – Social Movement – Institutional Changes' the new scientific ideas which were at the basis of environmental movements began to spread in other parts of the world. The newly emerging countries embraced the idea of industrial development equals to economic development. Outside of the so-called Western developed countries, Soviet Union embarked on heavy industrialization since the 1930s. The creed of industrialization at any cost had huge environmental costs in the Socialist countries. Later in the 1970s, when the so-called Tiger economies of East and South-East Asia (South Korea, Taiwan, Hong Kong, and Singapore) started to emerge, environmental concerns were not at the centre-stage. It took some time, for environmental awareness to seep in. The initial cost of environmental degradation in some of these countries was huge. Later in the 1980s and 1990s, when the socialist China in the post-Mao era embarked on a catch-up industrialization and became the 'factory of the world', initially environmental fallouts were not issues of central concern for the champions of economic development. Initially, the pressure and critique came from outside China and with the development of local and homegrown concerns, the Chinese leadership paid heed to environmental concerns and now China has come to terms with environmental issues but only after significant damage has been made. No matter, how careful the leaders are in their strategies of environmental protection, if the present consumption patterns remain unchanged, China may not avoid the curse of ecological meltdown. A feeble homegrown environmental movement may not match the rising tide of consumers of private cars and other industrial amenities. The Chinese thirst for energy has taken China to the world over from Africa to Latin America with environmental consequences far beyond China's shores.

The rise of the so-called BRICS (Brazil, Russia, India, China and South Africa) and the emergence of Indonesia as a rising economic power are celebrated in terms of shifting economic gravity from the west to the rest of the world, heralding a new pluralistic world economic order; the environmental consequences are not always adequately addressed or understood. At the present moment, China with its 1.3 billion people leave a relatively small ecological footprint on a per capita basis, which is neither cause for celebration nor consolation. It will not take long before China matches the American standard of high per-capita footprint if the growing consumption patterns in China remain unchanged. The rise of India as a regional economic power sends mixed signal with regard to ecological fate of this large and populated nation.

In *Ecofeminism* (1993) Maria Mies and Vandana Shiva presented a forceful argument by combining the two emergent paradigms in social sciences: feminism and environmentalism. In India, even before the rush of neo-liberal economic globalization, Indian civil society and its conscious intellectuals took up the cause of ecology. India has a deep tradition of valorizing nature. The harmonious relationship between the nature and society is entrenched in Indian traditions, religions and world-views. Indian mythology and even cosmology is based on a relationship of respect between human society and nature. The idea of caring for nature in India is a homegrown idea. The new combined and recombined paradigms steeped in (western) scientific knowledge found a sympathetic audience in India. Edward Schumacher's celebrated contribution, *Small is Beautiful* (1973) was inspired in part by Gandhian philosophy and Buddhist traditions.

As scientists, policy-makers and citizens become more aware of the key issues of social well-being, a redefinition of conventional economic growth is bound to happen. Especially, with a growing understanding of the importance of preserving the ecological balance amidst reckless economic growth, a redefinition of economic growth as sustainable and inclusive economic development is in progress. More and more economists, social scientists and the public would be able to persuade the politicians and leaders to see the world in a new light.

We do not see things as they are, we see things as we are – Reality is always filtered. Human knowledge is embedded in the epistemological and ideological presuppositions of a given social and temporal context. It is not always easy to untangle the ideological underpinnings. Even at the 50 years celebration of Rachel Carson's *Silent Spring*, the Washington-based conservative think-tank, Cato Institute published a critique of Carson's book using selective scientific evidence. No one is oblivious to the contentious nature of science and scientific facts, but to deny the trends of global warming and believing that nature will fix its own problem without human interventions is to go against the grain of scientific spirit. Those who are in denial of the coming intensification of ecological crises are steeped in the ideology of neo-liberal economism with blind trust in market-led solutions to all the problems of human existence.

A number of economists are challenging some of the basic assumptions of economics. Many economists are now inclined to measure progress by establishing a new set of criteria that track well-being and not just economic performance. Economic growth is a means and not an end in itself (Sen 1984: 1999). Development is a means to achieve well-being and happiness. Happiness is seen as a goal, and the new idea of flourishing rooted in Aristotelian philosophy is capturing the imagination of many. The discussion of economic growth is beginning to address not only quantity which is measureable, but also non-measurable quality of economic growth. It is time to focus the discussion less on what is growth and more on why growth and to what cost? Surely, development at any cost is no longer the only dominant paradigm; contesting views are enforcing environmental impact and viability of the development projects. The nurturing of eco-mind is a necessary condition for the development of an inclusive global society. 'We need to change the way we think to create the world we want' (Lappe 2012). Whether we privilege the ideas of Frances Moore Lappe's *EcoMind* (2012) or traverse the path of environmental movement with Arundhati Roy's *The Cost of Living* (1999), remaining neutral to the issues of environment is no longer an option. In mapping the interrelated issues of knowledge creation, social awareness, institutional changes compel us to examine the changing definitions and understandings of 'development'.

The Crown Prince Haakon Magnus of Norway delivered a talk at Zayed University in which the crown prince dealt with the link between development and dignity. Development is not just the goal – it is a means to enhance and retain our dignity as human beings. The crown prince's point drew on the philosophical question of the meaning of life. A life worth living is an examined life.

He began his talk by saying let us think about two things:

1. What are you for?
2. What do you want to do next year to help enhance someone's dignity?

Development can be defined as absence of war, rule of law, democratic governance etc. However, development can also be seen as a wave. The key question is how to remain stable amidst this wave. The world we live in today is highly unequal. The richest 20 per cent of the population have 74 per cent of the wealth. The rest of the 80 per cent have 26 per cent of the wealth. Of them, the bottom 20 per cent (the poorest) has only 2 per cent.

Top 20 % of the rich	74% of wealth	They must share
60% in the middle	24% of wealth	They need better distribution
20% who are the poorest	2% of wealth	They need help

The crown prince's daughter was born in 2004, at that time Norway's Human Development indicators were comparable to those of Singapore.

	Year of Birth	*Standard of living in Norway comparable to …*
Daughter	2004	Singapore
The crown prince	1973	Malaysia
His mother	1937	Peru
Grandmother	1898	Uganda
Great grandmother	1868	Mozambique
Great grandmother	1831	Burundi

The Crown Prince said: 'just because my great grandmother lived in conditions comparable to Mozambique does not mean he is anyhow less than my great grandmother.' Why should we care about people we do not know? The answer: because, we want to enhance our dignity. We enhance the dignity of others so that we can increase our own dignity. We can look back and see what have we achieved. How am I going to navigate my life? Meeting people eye-to-eye helps us challenge our morality; that's how our ethics is formed. True globalization is global emotional interconnectedness.

The United Nations reported that results of achievement of the Millennium Development Goals are mixed. While important gains have been made in a number of goals, achievements in environmental sectors have lagged behind. 'The target of reducing extreme poverty by half has been reached five years ahead of the 2015 deadline, as has the target of halving the proportion of people who lack dependable access to improved sources of drinking water. Conditions for more than 200 million people living in slums have been ameliorated – double the 2020 target. Primary school enrollment of girls equaled that of boys, and we have seen accelerating progress in reducing child and maternal mortality … biodiversity loss continues apace, and greenhouse gas emissions continue to pose a major threat to people and ecosystems' (United Nations 2012: 3).

Conclusion

In the last, we can turn to the philosophy of Peter Singer who has been championing the idea of global ethics which is premised on the notion of obligation and duty to the larger humanity not just citizens of a particular nation. He correctly pointed out that people are more inclined to be moved by the immediate and spectacular and not by the distant and indirect, although the indirect and less visible may be more harmful; for example, the effects of environmental pollution. The idea of true globalization is dependent on the reinstating the idea of a common humanity. Following the call of Ken Saro-Wiwa, a play writer and a defender of environment who was executed by the Military regime and echoing the Stockholm Declaration of 1972, we can reaffirm that clean environment is humanity's fundamental right and global institutions must bear the responsibility of ensuring that right.

References

Brundtland Commission. 1987. *Report of the World Commission on Environment and Development: Our Common Future*, http://www.un-documents.net/wced-ocf.htm

Carson, R. 1962. *Silent Spring*, New York: Fawcett Crest.

Cato Institute. 2012. *Silent Spring at 50: The False Crises of Rachael Carson*, Washington, D.C.: Cato Institute.

Commoner, B. 1971. *The Closing Circle*, New York: Bantam.

——. 1976. *The Poverty of Power*, New York: Knopf.

——. 1990. *Making Peace with the Planet*, New York: Pantheon Books.

Dodds, Felix, Michael Strauss and Maurice F. Strong. 2012. *Only One Earth: The Long Road via Rio to Sustainable Development*, London and New York: Routledge.

Gore, Albert. 2006. *An Inconvenient Truth*, Emmaus, PA: Rodale Press.

Guimaraes, R.P. 2004. 'The Political and Institutional Dilemmas of Sustainable Development', in M. Geodon (ed.), *Handbook of Developmental Policy Studies*, Boca Raton, FL: CRC Press.

Habermas, Jurgen. 1971. *Knowledge and Human Interests*, Boston: Beacon Press.

Johnson, Stanley. 2013. *UNEP: The First 40 Years*, Nairobi: UNEP.

Khondker, H. 2001. 'Politics of Environment', *Special Issue: Asian Journal of Social Sciences*, 29 (1).

Lappe, F.M. 2012. *EcoMind*, New York: Nation Books.

Mies, M. and V. Shiva. 1993. *Ecofeminism*, London: Zed Books.

Padhy, S., S. Dash and R. Mohapatra. 2006. 'Environmental Laws of Manu: A Concise Review', *Journal of Human Ecology*, 19 (1): 1–12.

Roy, Arundhati. 1999. *The Cost of Living*, London: Flamingo.

Schumacher, E.F. 1973. *Small is Beautiful: Economics as if People Mattered*, New York: Harper and Row.

Seers, Dudley. 1970. 'The Meaning of Development', *Ekistics*, 30 (180): 353–355.

Sen, Amartya. 1984. 'Development: Which Way Now?', in Amartya Sen (ed.), *Resources, Development and Values*, Cambridge, Mass.: Harvard University Press.

——. 1999. *Development as Freedom*, New York: Anchor Books.

Singer, P. 2002. *One World: The Ethics of Globalization*, New Haven, CT: Yale University Press.

Sklair, L. 2009. 'The Emancipatory Potential of Generic Globalization', *Globalizations*, 6 (4): 525–539.

Speech by the Crown Prince Haakon Magnus of Norway at Zayed University, Abu Dhabi, UAE on 13 April 2010 (Author's notes).

Turner, Bryan S. and Habibul H. Khondker. 2010. *Globalization: East and West*, London: Sage.

United Nations. 2012. *The Millennium Development Goals Report.*

Part II

Ecological Dimensions

6

A Review of the Climate Change Debate, its Causes and Necessary Action in the Context of Developed and Developing Countries

Chamira Gamage

Climate change is a subject that has attracted the attention of governments, private sector corporations, non-governmental organizations, the scientific community and general population in the last few decades and has exponentially risen in prominence over the past few years. The term 'Climate Change' commonly refers to the effects on climate resulting from natural causes and human practices and the effects of global warming, through increases in the concentration of greenhouse gases within the Earth's atmosphere. These increases in greenhouse gases (GHGs) result from human activities such as the burning of coal, oil, and natural gas, as well as deforestation and various agricultural and industrial practices. The major contributing gases include carbon dioxide (CO_2), methane, nitrous oxide, chlorofluorocarbons, and ozone in the lower part of the atmosphere (NSW Government: Primary Industries Science and Research 2010).

According to the Intergovernmental Panel on Climate Change (IPCC), climate change is defined as:

> a change in the state of the climate that can be identified (e.g. using statistical tests) by changes in the mean and/or the variability of its properties, and that persists for an extended period, typically decades or longer. It refers to any change in climate over time, whether due to natural variability or as a result of human activity (Pachauri and Reisenger 2007).

The IPCC further states

> This usage differs from that in the United Nations Framework Convention on Climate Change (UNFCCC), where climate change refers to a change of climate that is attributed directly or indirectly to human activity that alters the composition of the global atmosphere and that is in addition to natural climate variability observed over comparable time periods. (Pachauri and Reisenger 2007)

There is robust debate as to what causes climate change and what remedies are necessary to address these causes. As the global community continues to develop a greater appreciation of these, place a higher priority on the subject and implement further legislation to combat climate change, an understanding of the contributing factors becomes necessary. This is particularly relevant for governments constructing effective strategies to manage carbon emissions, and their contributors. In recent years, we have witnessed the growing global consciousness of the problem and necessary steps through the advent of such legislation as the carbon tax (Australia), the essentially unanimous global signing of the Kyoto Protocol in 2005 and its subsequent ratifications, the UNFCC, the Rio+20 summit and the GLOBE. All these initiatives were established with the intention to lower the emission of greenhouse gases and work towards a more sustainable future.

According to the IPCC (2001), climate change has unique characteristics and it interacts with several processes with broader implication:

> It is global, long-term (up to several centuries), and involves complex interactions between climatic, environmental, economic, political, institutional, social and technological processes. This may have significant international and intergenerational implications in the context of broader societal goals such as equity and sustainable development. (IPCC 2001)

Notably, two areas of climate change have become focus areas for global debate: overpopulation and overconsumption. They are both considered as contributing factors to atmospheric pollution and climate change. Overpopulation is a concept based on the assumption that human population figures exceed the carrying capacity of the Earth. Overconsumption refers to a situation in which resource use by the world population has outpaced the sustainable capacity of the ecosystem. Overconsumption and over-population, whilst being closely linked to one another, have contributed in varying degrees to climate change as causal factors singularly and jointly. Some academic and scientific discussions focus on whether it is overconsumption or overpopulation that contributes more to climate change (Green and Minchin 2010).

In the following pages, I review respective arguments, facts and figures that have been presented by various authors and agencies on climate change as well as necessary actions and its causes. The paper looks at the global debate on these and why it is important to discuss this topic. The role of developed and developing countries in taking adaptation or mitigation measures is also scrutinized with a view to ascertaining their respective positions in the debate, and more importantly, what action is to be taken by each group of countries?

I

Climate Change and Its Causes: Global Debate

The debate over climate change and its causes is undoubtedly a controversial global issue with repercussions for both the developed and developing world. In this section, a brief review of this debate is undertaken, detailing relevant facts, statistics and various arguments/counter arguments advanced by proponents of sustainable climate measures as well as climate sceptics.

Notable global organizations such as the United Nations body, Intergovernmental Panel on Climate Change (IPCC) have reflected and championed the views of many scientific organizations and societies by asserting that 'An increasing body of observations gives a collective picture of a warming world and other changes in the climate system ... . There is new and stronger evidence that most of the warming observed over the last 50 years is attributable to human activities' (Pachauri and Reisenger 2007). The observable physical causes of climate change are described as warming, melting land-based glaciers, reduced snowpack and Arctic Sea ice, shifts in rainfall, and responses in ecosystems worldwide (Desonie 2008). It is stated that 'Warming of the climate system is unequivocal, as is now evident from observations of increases in global average air and ocean temperatures, widespread melting of snow and ice and rising global average sea level' (Pachauri and Reisenger 2007).

The Encyclopaedia of Earth (2007) states that, 'Over the past three centuries, the concentration of carbon dioxide has been increasing in the Earth's atmosphere because of human influences. Human activities like the combustion of fossil fuels, conversion of natural prairie to farmland, and deforestation have caused the release of carbon dioxide into the atmosphere' (EoE 2007). The following graph (Figure 1) taken from the Encyclopaedia illustrates the exponential growth of carbon dioxide levels in the atmosphere between 1744 and 2005.

According to the following graph (Figure 2) on the projected carbon emissions between 1990–2030, the highest emitter is China and the next highest emitter is the USA. The Third group of countries with the highest emissions is recorded as OECD – Europe.

Thus, there is evidence that climate change is affected by human activities/actions, which contribute directly to the warming of the planet. The

Figure 1

Atmospheric Concentration of Carbon Dioxide (1744–2005)

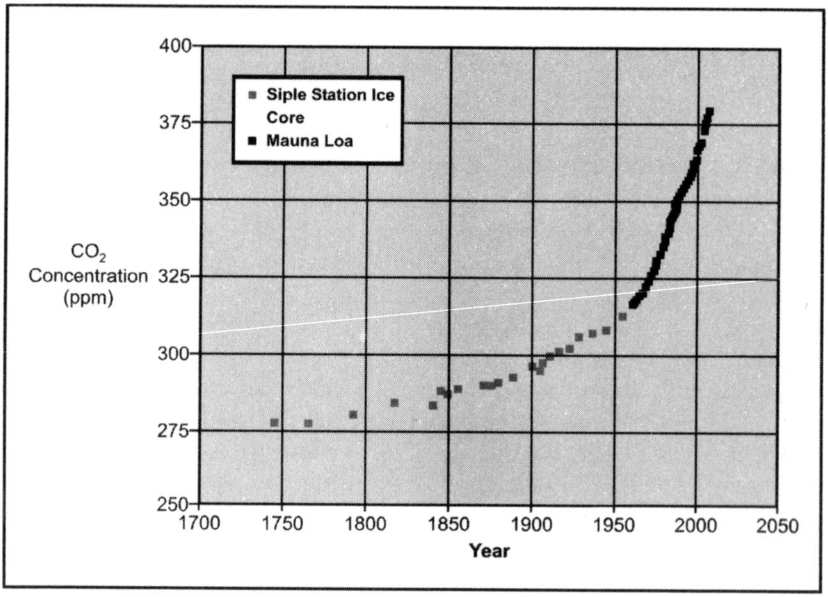

Source: The Encyclopaedia of Earth (2007).

main contributing factors include the burning of fossil fuels and biomass, deforestation, urbanization, and air pollution (Desonie 2008: 33). The fact that human activity contributes to greenhouses gases is further corroborated by various reports and authoritative sources in addition to the IPCC e.g. *Climate Change in Australia, Technical Report* (2007), *The Encyclopaedia of Earth* (2007) and Stern (2006). According to Stern (2006), the main sources of greenhouse gases, in order of global importance, are: electricity generation, land use changes (particularly deforestation), agriculture and transport (Stern 2006: 1). He describes global emissions of greenhouse gases by their sources based on the World Resource Institute 2000 estimate. Thus, agriculture – mostly from soils and livestock, contributes 14 per cent, land use changes – primarily deforestation 18 per cent, and energy-consuming fossil fuels 61 per cent. These are important indicators of the causes of climate change.

Sceptics

Irrespective of the scientific evidence and various technical reports produced by reputable organizations over the last few decades, there are climate change sceptics in policy areas, academia, media, politics and government who deny the existence of climate change or its causes.

Figure 2

World Carbon Dioxide Emissions by Country, 1990–2030

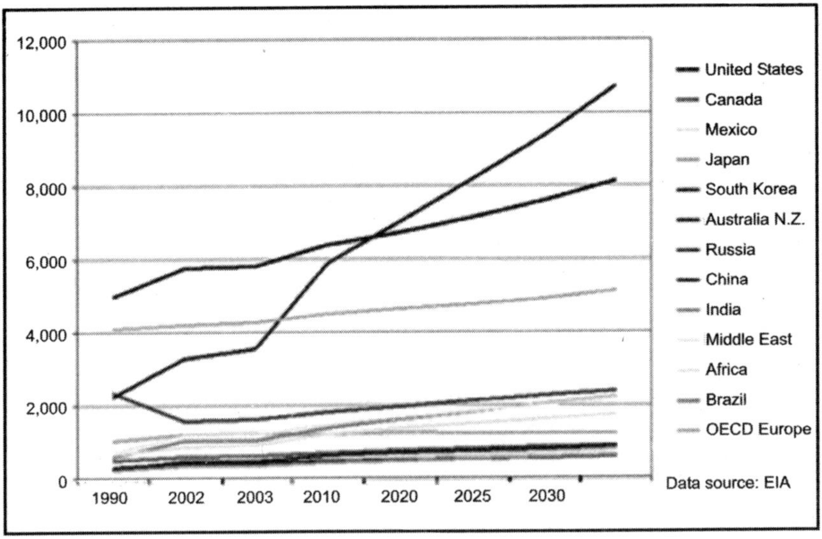

Source: Human Population Growth and Climate Change (2013).

Lockwood (2010) classifies these sceptics who are outside the scientific realm as 'economically, politically and ideologically motivated'. Some climate change sceptics doubt the level of human influence on the changing of climate and also debate whether the earth is even warming in light of historical trends such as previously recorded fluctuations in weather temperature.

Some of the arguments advanced by climate sceptics can be summarized as follows:

- Global average temperatures have not risen since 1998.
- Climate researchers are engaged in a conspiracy; global warming is a hoax.
- Climate models are defective and therefore cannot provide reliable projections of future climate trends.
- The greenhouse gas signature (tropospheric hot spot) is missing; The Medieval Warm Period was just as warm as, or warmer than, today
- Carbon dioxide levels increase after temperatures rise in the ice core records.
- Earth's climate is driven only by the sun.
- Water vapour is the most prevalent greenhouse gas.
- CO_2 in the atmosphere is already absorbing all of the infrared radiation heat it can.
- Climate sensitivity is overestimated in current climate models.

- Increasing carbon dioxide will stimulate plant growth and improve agricultural yield.
- Human society and natural systems have adapted to past climate change. (DB Climate Change Advisory Group 2010: 4–5).

These claims have been rejected by the scientific community whose 'clear conclusion is that the primary claims of the sceptics do not undermine the assertion that human made climate change is already happening and is a serious long-term threat' (DB Climate Change Advisory Group 2010: 3). Giddens (2009) observes that, 'Many scientists believe (the sceptics') writings are irresponsible, since they convey to the public that there is extensive space for doubt about the origins, and probably consequences, of warming when in fact there is little' (Giddens 2009: 24).

In response to the claim that Global average temperatures have not risen since 1998, the DB Climate Change Advisory Group argue that in fact, 'the decade of 2000 to 2009 is the warmest since measurements have been made. Multiple factors affect global average temperatures, including the long-term warming trend from GHGs' (2010: 4).

In response to the sceptics argument that human society and natural systems have adapted to climate change in the past, the DB Climate Change Advisory Group state that 'past climate changes have often been accompanied by migration, war, and disease. The growing human population will inevitably make environmental change more disruptive in the future, even in the face of increased technological prowess' (2010: 5).

These examples show that the arguments of climate sceptics have been responded to sufficiently by scientific advisors.

II

Why is it Important to Discuss Climate Change and What are its Causes?

The consequences of climate change are universal. If not addressed by governments, corporations and human activity now, it can be more costly to address it later. Climate change, unaddressed, can have a significant damaging impact on developing countries where most lower income populations live. Developing countries face a higher degree of vulnerability and lack adequate technical knowhow, infrastructure and other resources to deal with climate change, which can become a serious problem. Furthermore, the extremity of weather fluctuations and patterns can lead to heavy heat waves, storms and floods, drought and the changing of rainfall patterns. As a result, these environmental effects can impact on the well-being and safety of people, leaving the elderly, disabled and the very young vulnerable to health risks.

According to information provided by the European Commission – Climate Action Website concerning Europe's climate change implications:

'warming of 1.5°C – 2.5°C degrees Celsius beyond today's levels would put as many as 20–30 per cent of plant and animal species at increased risk of extinction' (The European Commission 2012). Furthermore, the Commission highlights the impact on property and infrastructure and the imposed costs on the economy and society. Sectors such as agriculture, forestry, energy and tourism, which rely heavily on particular temperatures and precipitation levels, will specifically be affected.

In consideration of overpopulation and overconsumption, one must be aware that both of these factors directly affect the stability of the planet and environment and their ability to exist sustainably. Recommendations have been made by many leading organizations and societies regarding the potential dangers overpopulation and overconsumption can cause. The world needs to both stabilize population and cut consumption:

> The number of people living on the planet has never been higher, their levels of consumption are unprecedented and vast changes are taking place in the environment. We can choose to rebalance the use of resources to a more egalitarian pattern of consumption ... or we can choose to do nothing and to drift into a downward spiral of economic and environmental ills leading to a more unequal and inhospitable future. (The Royal Society Science Policy Centre Report 2012: 11)

Thus, avoiding future impacts of climate change is paramount for human survival and it is important to understand the nature of causes and effects adequately based on scientific research. Impacts of climate change can be:

> climate variability and extreme events, land degradation and loss of biological diversity ... . They result in loss of human lives; as threats to livelihoods; in increases in the prevalence and severity of diseases; as constraints on and shocks to economic development; in increases in the magnitude and frequency of floods, droughts and other disasters; as recurring or persistent famine; as human displacement; and as disruptions to social and political systems. (United Nations Framework Convention on Climate Change 2011)

Preventative or mitigation steps need to be taken in order to minimize the impact of climate change with improved understandings.

III

Role of Overpopulation and Overconsumption in Climate Change

In this section, the contribution of population growth and consumption to climate change is explored on the basis of available writings, statistics and other information.

Products and services are essential for human survival – even though the level and quality of consumption can vary from country to country or region to region. People – whether they are infants, children, youths, middle aged, or elderly – all consume the products of the environment and those produced or manufactured by human beings. Human sustenance is dependent on the availability of needs and wants. Essential needs include: income, food, medicine, clothing, housing, water and education. Basic consumption needs can be distinguished from secondary needs, e.g. going to the theatre, eating out, overseas travel. Even if we are talking about the basic needs of a given population, their manufacture and distribution involve the use of energy – which in turn contributes to greenhouse gases such as carbon dioxide or methane.

To most of us, given the limited resources available on the planet, having an exponentially growing human population is not sustainable. Citing *Living Planet Report* (2006), Guillebaud (2008) notes that, 'The world's population now exceeds 6.7 million, and humankind's consumption of fossil fuels, fresh water, crops, fish, and forests exceeds supply. The annual increase in population of about 79 million means that every week an extra 1.5 million people need food and somewhere to live' (Guillebaud 2008: 337). Globally the current population sits around 7 billion and it is set to increase to 9.1 billion by 2050, putting greater strains on our natural resources (Cohen 2010: 160).

A growing population requires more land for agriculture and the inputs used in modern agriculture is not that friendly for the environment:

A growing population has meant more and more mouths to feed. Because the land area available for agriculture is limited (and in fact, is actually shrinking as a result of ecological degradation!), high-yielding varieties of crop are being grown to increase the agricultural output from a given area of land. However, such high-yielding varieties of crops require large quantities of fertilizers; and more fertilizer means more emissions of nitrous oxide, both from the field into which it is put and the fertilizer industry that makes it. Pollution also results from the run-off of fertilizer into water bodies' (EduGreen 2007).

According to Cohen, there has been a fifteenfold increase in emissions by human beings between 1900–2000:

> The human population of Earth in 1900 was 1.6 billion. The population in 2000 was 6.1 billion. The population increased by less than a factor of four. A fifteenfold increase in carbon emissions accompanied a less than fourfold increase in numbers of people. Population growth alone, with constant rates of emissions per person, could not account for the increase in the carbon emissions to the atmosphere. (Cohen 2010: 178)

On the basis of these assertions, one can see how population growth is a major contributory factor to climate change. What are embedded in these views are also the effects of technologies used, strains on natural environment and the impacts of development and consumerist lifestyle needs.

However, population numbers alone cannot contribute to the emission of greenhouse gases and climate change unless they are combined with modern consumption patterns and the production and marketing of goods and services that are needed to meet such consumption. When they do, they add to the greenhouse gas emissions significantly polluting the atmosphere.

Some authors believe that the importance of population in driving climate change is regularly exaggerated (Green and Minchin 2010: 45). It is rather the amount of resource use in industrialized countries that matter:

> Even when these countries have very low (or even negative) rates of population growth, their overall consumption and waste production tends to grow. This situation supports the vehement insistence of low-consumption countries with higher levels of population growth that the rich nations must address their own consumption before insisting on population curbs. (Planet Thoughts 2008)

Arguing that consumption is the major cause in driving climate change as opposed to population growth, Satherthwaite states that the greatest human contributor to climate change is the 'numbers of consumers on the planet and their consumption levels' (Satherthwaite 2010: 37). Very poor households with no access to modern energy contribute very little or nothing to climate change, i.e. 1 in 5 people. His argument is that their use of fossil fuels are limited and they rely on traditional fuels such as fuel wood, charcoal and agricultural wastes including dung. Lower middle-income nations[1] contribute more to the CO_2 emissions compared to high-income nations (see Figure 3). In high-income nations, the population growth is low, whereas in the lower middle-income nations, the population growth is high. The graph shows that low-income nations with high population growth contribute less CO_2 emissions. This shows that 'population growth can only be a significant contributor to greenhouse gas (GHG) emissions if the people making up the population growth enjoy levels of consumption which cause significant levels of emissions per person as in highly industrialized, developed countries. If most growth in the world's population is in low-income households who never get out of poverty, then there is and will be little connection between population growth and GHG emissions growth' (Satherthwaite 2010).[2]

Thus, population growth can be a significant contributor to GHG emissions if the people making up the population growth enjoy levels of consumption, which cause significant levels of emissions per person. If most growth in the world's population is in low-income households who never get out of poverty, then there will be little connection between population growth and GHG emission's growth.

Figure 3

Contribution to the growth in world population and CO_2 emissions by groups of nations classified according to their average per capita income levels, 1980–2005

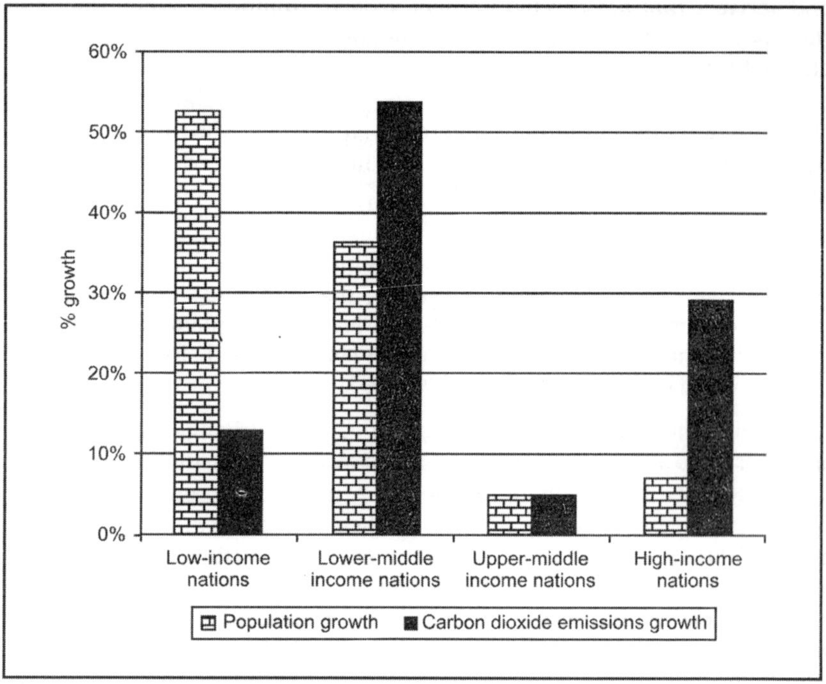

Source: CAIT Indicator. Satherthwaithe (2010).

IV

Climate Change and its Causes: Developed and Developing Countries

A main division exists between economically developed and developing countries in dealing with climate change at an international level. A key issue dominating discussions is which countries should take more responsibility for atmospheric pollution and climate change, and initiate necessary action. In this section I examine this division in some detail.

On the basis of the review of literature, it could be argued that countries or sections of populations within given countries that follow a high consumption-oriented pattern of life (therefore, high energy use), contribute to greenhouse gases and global warming more than those countries or population segments that depend on primitive technologies and the satisfaction of basic needs only. Countries that utilize primitive technologies in food production or consumption cannot be put in the same basket as those using ultramodern technologies for meeting human needs both primary and secondary. For example, digging a well or an irrigation canal in an African or Asian village by people using small tools consumes less energy as compared to the use of heavy machinery in countries like Australia, USA, Germany or France. Thus, it could be assumed that the countries of the developed West, i.e. Europe, North America, and Australia contribute to greenhouse gases and global warming more so than economically developing countries of Asia, Africa and Latin America with the exception of China and to a lesser extent India, Russia and Japan. This is a very powerful argument that the policy makers have to take into consideration.

The director of the Princeton Environmental Institute, Stephen Pacala, calculated that the wealthiest 7 per cent of the people in the world have been responsible for generating around 50 per cent of global carbon dioxide emissions. In comparison, the poorest 50 per cent of people have been responsible for just 7 per cent of emissions. These findings confirm that people living in countries like Australia and the US have a disproportionately large role in determining whether global emissions continue to rise or fall (Pacala 2007).

Economically developed, industrialized countries generally use heavy machinery in agriculture, manufacturing, transport and distribution, and consumption. While these technologies are being employed increasingly by developing countries also. It is said that in China they open a coal-fired gas station every month. Being the largest emitter of greenhouses gases, the figures relating to the transport sector in the US tell the story about its high emission levels. 'Transportation accounts for a full third of CO_2 emissions in the United States, and that share is growing as others shrink in comparison, rising from 31 per cent in 1990 to 33 per cent today. It is hard to envision a 'solution' to the global warming crisis that does not involve slowing the growth of transportation CO_2 emissions in the United States' (Ewing et al. 2007: 11). This expansive development has caused CO_2 emissions from cars to rise even as it has reduced the amount of forest land available to absorb CO_2' (Ewing et al. 2007: 13). In the US, industry was the largest contributor to CO_2 emissions followed by transportation and residential sector in 1997. CO_2 is the by-product of burning fuels such as coal, oil, and natural gas. Following graph (Figure 4) illustrates the contributions of each sector in emissions:

As we noted already, when the population grows the economy has to grow to provide for human needs such as food, housing, health, employment and other services. Given the fact that major industrial countries are now exporting their products and services to other countries in a globalized economy, even if their own populations are not growing at a rapid rate, they keep producing industrial and agricultural products and transporting the same worldwide. This adds to greenhouse emissions –

Figure 4

Emission Sector

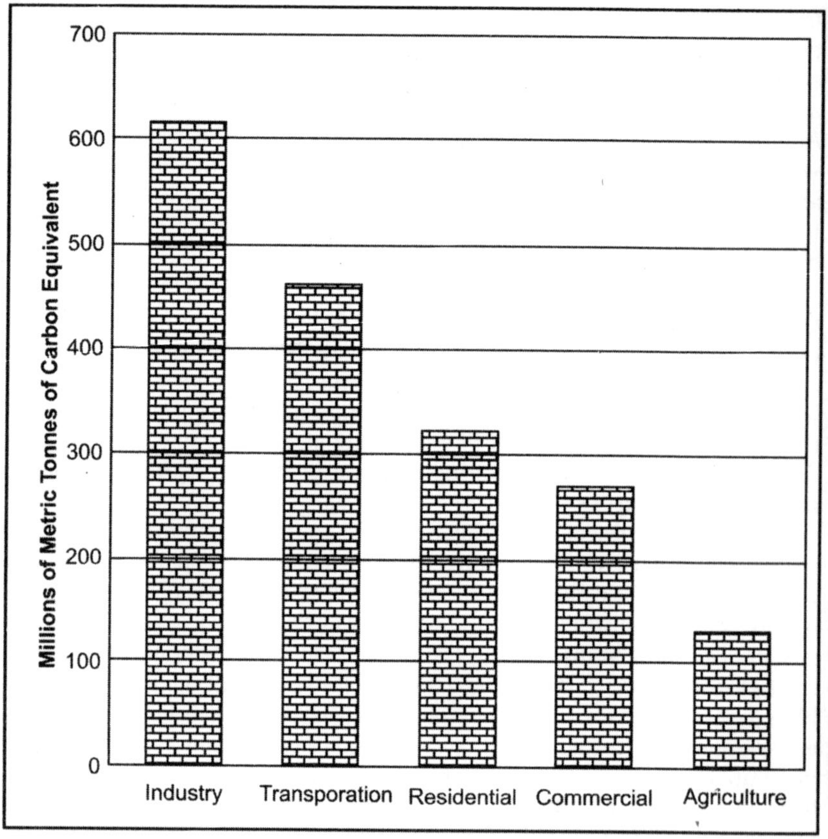

Source: Science Museum of the National Academy of Sciences (2010).

the consumers are spread around the world – because those who can afford to buy such products and services are spread throughout the world due to the rise of middle classes. Therefore, the figures for a given country as given here have to be interpreted from a global perspective. Nonetheless, high emissions by the USA, a country without high level of population growth, support the argument that it is overconsumption that contributes to climate change more than overpopulation. But when we add China to the picture, complexity of the issue increases. Being a highly populated country, it is also amongst the two largest emitters in the world. Between 1990–2004, USA emitted 1616 million tonnes of carbon, while China emitted 1021 tons. Europe's contribution was 955 tonnes, India's 301, and Japan's 338 tonnes.[3] As of 2011 however, China's CO_2

emissions were estimated as nearly double that of the US. However, emissions per capita were still significantly higher in the US (European Commission EDGAR 2013).

To clarify this better, we can look at information for developed West and developing countries in terms of greenhouse gas emissions. According to Green Blog,

> The West is responsible for about 80 per cent of the world's CO_2 increase. An average person living in Great Britain will in only 11 days emit as much CO^2 as an average person in Bangladesh will during a whole year. And just a single power plant in West Yorkshire in Great Britain will produce more CO_2 every year than all the 139 million people combined living in Uganda, Kenya, Tanzania, Malawi, Zambia and Mozambique. (Green Blog 2009)

Another complicating factor we need to consider is immigration. EU countries, North America and Australia keep inviting large numbers of immigrants annually. For example, by December 2007, 'the estimated foreign born population of the USA was 38.1 million' (Alfred 2010: 220). Alfred points out that the immigrant population of today come from both rural and urban backgrounds, underdeveloped regions of the Western hemisphere, and industrially developed areas of East Asia (Alfred 2010: 221). The main source countries for Australia are UK, China, India and New Zealand (Australian Bureau of Statistics 2011). Thus, the immigrant populations are absorbed into high consumption, high-energy use ways of life over time in the host countries. Such movements of people and displacement as immigrants or refugees can have an impact on climate change as many of them give up their traditional ways of low consumption/low energy use lifestyles once they settle in the industrially developed countries.

Moreover, with the increasing levels of globalization and expanding capitalist economy, more and more people hitherto unaffected by such an economy are being attracted to the forces of consumption and waste, both in the developed and developing countries. The logic of capitalist economics dictate that to make more and more profits, producers and manufacturers have to find cheaper sources of labour, raw materials, infrastructure, energy sources on one hand and markets where they can sell the produce at a high price competitively. This means that more and more people are being attracted to a consumerist lifestyle as a result of aggressive marketing and expanding company operations. The best example is fast food chains spreading around the world. Thus, a consumerist lifestyle that is associated with high-energy use can be the most significant detrimental factor for climate change in both types of countries – even though the consumerist lifestyle, waste and high emissions of greenhouse gases are concentrated more in a given set of countries as we discussed earlier.

V

Proposed Solutions to Climate Change and Actions Needed

Notwithstanding which types of countries are the main polluters in absolute terms or in per capita terms, countries need to focus and engage in collaborations toward potential solutions and actions. Like all policy debates, the climate change debate is ultimately based on arguments over action and resources necessary to implement them. How shall we respond to the risks posed by climate change? Who shall bear the responsibility? How much and when? These are other questions relating to solutions and actions.

A key debate focusing on climate change solution policies is based on the question as to whether adaptation or mitigation should be the desired plan of action.

> Climate change mitigation will both be affected by, and have impacts on, broader socio-economic policies and trends, such as those relating to development, sustainability and equity. Climate mitigation policies may promote sustainable development when they are consistent with such broader societal objectives. (IPCC 2001)

Some authors and country officials argue that the main polluting countries such as the USA, China, India, Australia and the EU should take steps to control greenhouse gas emissions by introducing emission trading schemes, carbon taxes or alternative energy solutions first. Some developing countries have their own arguments and even demands claiming that their colonial rulers are responsible for the state of affairs in their countries, leading to high emissions. 'Unfortunately, the central problem remains a deep chasm between the developing world and the developed world on assigning responsibility for climate change and sharing responsibility for action' (Mastel et al. 2009: 53).

There is no doubt, the developed world has contributed the most to current level of global emissions; however, as China has now become the world leader in greenhouse gas emissions, with other developing nations not far behind, this can soon change.

Industrially developed countries want developing countries also to introduce mitigation measures; however, as became evident in recent climate change conferences and conventions it is the argument of developing countries that they cannot take these steps on their own. Developed countries have to help them to do it financially and technically.

Two solutions proposed by The Royal Society centred upon contraceptive availability to stabilize population growth and also cuts in consumption to reduce inequality among the developed and developing world.

Satherthwaite suggests that the 'Responsibility for greenhouse gas emissions should be allocated to individuals and households, not nations. It should be based on the greenhouse gas implication of their consumption' (Satherthwaite 2010). To reduce greenhouse gases, we need to 'phase out the technologies dominating and damaging nature and people – those based on coal, oil, gas and uranium – and phase in the technologies that work with nature – those based on sunshine, wind, biomass, wave, tide and earth's heat' (Green and Minchin 2010: 63). Green and Minchin (2010) look at overconsumption and prescribe ways of consuming less. The authors state that we are not being effectively educated by government campaigns regarding overconsumption. Various methods to cut emissions proposed by Green and Minchin include recycling, exchanging clothes, double checking wardrobes, buying green goods and eating less meat. If we change the way we consume, other solutions will be more effective (2010: 51). The Working Group III Special Report on Renewable Energy Sources and Climate Change Mitigation provides a detailed assessment of six renewable energy sources to the mitigation of climate change. These include: bioenergy, direct solar energy, geothermal energy, hydropower, ocean energy and wind energy (SRREN 2010).

According to Green and Minchin, wind power 'is the fastest growing electricity generating technology in the world' (2010: 65). It is the least expensive new renewable source of electricity and has the lowest measureable environmental impacts.

Solar energy has benefits in heating, cooling and can be spoken of in various forms. The major forms detailed in the IPCC include – solar thermal, photovoltaic (PV), concentrating solar power (CSP) and solar fuels production methods (Edenhofer 2011). Solar thermal electric power is expanding in countries like USA, Spain and Portugal. Dry inland areas are more suited for this technology, which has better storing capacity (Green and Minchin 2010: 69). It can be generated 24 hours a day on demand. The book discusses photovoltaic power and its advantage, geothermal power and heat (low in environmental impacts), marine power, and hydroelectric power. However, 'The potential for additional hydroelectric power from large dams is constrained by the environmental and social impacts of flooding valleys' (2010: 73). Natural gas, coal seam methane, and bio-energy are other technologies discussed. They emit less than half CO_2 emissions compared to conventional coal-fired power stations. Uses of bio-energy to generate electricity and heat are also discussed along with some technical details. These discussions are very informative and helpful for climate action groups.

Diesendorf (2009) focuses on change activists and how the readers can be empowered to make change. He outlines transport and agriculture, sustainable energy scenarios and the need to treat climate action like a wartime emergency. Furthermore, he describes policies for effective greenhouse mitigation, which have received growing consensus within the climate action movement, e.g. setting strong greenhouse targets, setting targets for energy efficiency, renewable energy and renewable heat and bans on new conventional coal-fired power stations. Green and Minchin further describe cost-effective methods for energy efficiency and conservation through examples such as better design for buildings, including insulation, windows located to catch sunshine, water efficient taps and shower heads, and efficient lighting and appliances. Energy conservation involves behavioural change such as buying fewer consumer products and turning off lights, which requires a cultural change (2010: 64).

According to Diesendorf, within the climate action movement 'there is growing consensus on policies and strategies that would achieve an effective greenhouse mitigation' (2009: 90). These include – science-based national greenhouse gas emissions targets, a ban on all new conventional coal-fired power stations, a carbon price that increases over time, termination of subsidies to the producers and users of fossil fuels, levelling of population growth, renewable energy targets, research and innovation, essential infrastructure – railways and transmission lines, regulations and standards for efficient energy use. From these suggestions, it is obvious that a single policy measure cannot solve the climate crisis. Instead, 'we need a wide range of policies that span carbon pricing, regulations and standards, education and training, research and innovation, and institutional change' (2009: 125).

Some of these policies could be implemented quickly and achieve almost immediate results. For example, banning new conventional coal-fired power stations. This is urgently required due to the significant amount of CO_2 emitted by coal-fired power stations per year (6–12 million tonnes) (Green and Minchin 2010: 96). As 'every new base-load power station undermines energy efficiency programs,' the author claims that 'phasing out coal is the key to drawing Earth's climate away from tipping point, to a safe state' (2010: 96). Atmospheric carbon levels would decline after the next few decades if a ban was to be imposed now.

The author asks the climate action movement to be wary of developers and supporters of new coal-fired power stations when they say these will be ready to capture carbon. This claim is not an alternative to the ban. The analysis is clear and instructions are direct. Suggested solutions serve the national and global interest rather than a vested interest. The arguments carry credibility.

Black believes that 'the world likely will see a fundamental shift in how it establishes its priorities for addressing the climate crisis in the next decade, focusing more on adaptation – learning to live with the effects of climate change' (2010: 359). According to her, 'Climate change adaptation – the strategies through which populations respond to climate change, in order to both minimize potential harms and maximize potential benefits – differs from mitigation in that the 'mitigation' refers to efforts to reduce greenhouse has emissions and minimize the effects of climate change whereas adaptation refers to the measures required to address those effects' (2010: 361). Furthermore, 'Unlike mitigation, adaptation effort largely involve local-decision making, making it difficult to ensure that those responsible for creating the problem also play a role in solving it' (2010: 360). Adaptation was a key issue discussed at the 2009 Copenhagen Climate Change Conference.

In the specific case of India, Rosencranz et al. (2010) note that a healthy balance between mitigation and adaptation is crucial in achieving sustainable development benefits against the impact of climate change. This being so, as it is said to improve the economic condition of the poor, who are most vulnerable to climate change.

Conclusion

This paper examined the causes of climate change, the nature of global discussions in the context of developed and developing countries and proposed actions either to mitigate the effects of greenhouse gas emissions or to change consumption and energy use patterns to a low carbon culture. It was argued that the world's richest countries are using a disproportionate amount of the world's energy and they are also the ones who emit the largest amount of greenhouse gases into the world's atmosphere while having high consumption levels. Developing countries are fast moving to adopt high consumption and high-energy use lifestyles, making the distinction between developing and developed countries redundant to some extent. Overconsumption and the need to produce and transport goods and services to meet consumer needs, was considered a major contributor to climate change in comparison to population growth in raw numbers. While population growth and overconsumption contribute to climate change, it was observed that overconsumption has had a greater effect and is an area that must be modified and addressed to sufficiently reduce the risk of irreversible damage to the earth and environment.

The chapter outlined a range of possible climate change actions. These can be adapted by specific countries depending on the level of their needs, wants and constraints, as well as the level of population growth, consumption and GHG emission levels. Whether a country is in the developed or developing world, they all have to make a concerted effort to address climate change and its causes in a timely manner in order to prevent even greater disasters happening due to severe climate change.

Ultimately, countries have to achieve a balance between adaptation and mitigation. Localities will primarily bear the planning and implementation costs of adaptation, but unlike mitigation, the benefits will also be local. In saying this, there exist many market barriers in achieving effective adaptation, in areas such as financial constraints on both personal and governmental levels (Black 2010: 362).

Diesendorf (2009: 5) emphasises that the role of Climate Movement Organizations is also vital for effective climate action. The aim is to help lobby governments to change climate change policies and create better awareness among the population. Usually, the governments and big business groups who hold the power resist policy changes climate change organizations formed in communities, villages and towns can lobby governments to change climate change policies.

Notes

1. For details on income levels by country, see the World Bank Data website, http://data.worldbank.org/country.
2. Satherthwaite (2010) examined data on CO_2 emissions for 185 nations for 1980 and 2005, and identified five sets of 37 nations, 27 of these with the highest CO_2 emissions per person in 2005 were high-income nations, which included North America and much of Europe. They had very low population growth rates, i.e. less than 1 per cent during the period under review.
3. See Appendix 1.

References

Alexander, David. 2008. 'The Ecology of Overpopulation and Overconsumption', http://www.planetthoughts.org/?pg=pt/whole&qid=2431, accessed on 15 April 2013.

Alfred, Mary. 2010. 'Transnational Migration, Social Capital and Lifelong Learning in the USA', *International Journal of Lifelong Education*, 29 (2): 219–235.

Australian Bureau of Statistics. 2011. 'Migration, Australia 2009–10', retrieved on 20 April 2013, from http://www.abs.gov.au/ausstats/abs@.nsf/products/52f24d6a97bc0a67ca2578b0001197b8?opendocument

Black, Elizabeth. 2010. 'Climate Change Adaptation: Local Solutions for a Global Problem', *Georgetown International Law Review*, 22: 359.

Climate Change in Australia, Technical Report 2007. *Encyclopaedia of Earth*, retrieved on 20 April 2013, from http://www.csiro.au/resources/climate-change-technical-report-2007.html

Cohen, Joel. 2010. 'Proceedings of the American Philosophical Society', *Population and Climate Change*, 154 (2): 158–182.

Desonie, Dana. 2008. *Climate: Causes and Effects of Climate Change*, New York: Chelsea House.

Deutsche Bank Climate Change Advisory Group. 2010. 'Climate Change: Addressing the Major Skeptic Arguments', retrieved on 15 April 2013, from http://www.uea.ac.uk/mac/comm/media/press/crustatements/otherreports/deutsche+bank+cru+report

Diesendorf, Mark. 2009. *Climate Action: A Campaign Manual for Greenhouse Solutions*, Sydney: University of New South Wales Press.

Edenhofer, Ottmar, Ramón Pichs-Madruga, Youba Sokona, Kristin Seyboth, Susanne Kadner, Timm Zwickel, Patrick Eickemeier, Gerrit Hansen, Steffen Schlömer, Christoph von Stechow and Patrick Matschoss. *IPCC 2011: Special Report on Renewable Energy Sources and Climate Change Mitigation*, prepared by Working Group III of the Intergovernmental Panel on Climate Change, New York, USA: Cambridge University Press, 1075.

EduGreen Website 2007. Retrieved on 12 April 2013, from http://edugreen.teri.res.in/explore/climate/causes.htm

European Commission: Emission Database for Global Atmospheric Research (EDGAR) Website. 2013. Retrieved on 16 April 2013, from http://edgar.jrc.ec.europa.eu/overview.php?v=co2ts1990-2011.

Ewing, Reid, Keith Bartholomew, Steve Winkelman, Jerry Walters and Chen Don. 2007. *Growing Cooler: The Evidence on Urban Development and Climate Change*, Washington, D.C.: Urban Land Institute.

Giddens, Anthony. 2009. *The Politics of Climate Change*, Cambridge: Polity Press.

Green Blog. 2009. Retrieved on 13 April 2013, from http://www.green-blog.org/2009/07/14/overpopulation-is-not-the-problem-overconsumption-by-the-rich-few-is/

Green, Donna and Liz Minchin. 2010. *Screw Light Bulbs*, Crawley, WA: UWA Publishing.

Guillebaud, John. 2008. 'Editorial: Population Growth and Climate Change', *British Medical Journal*, 337: 247–248.

Human Population Growth and Climate Change 2013. Retrieved on 26 March 2013, from http://www.biologicaldiversity.org/campaigns/ overpopulation/climate/

IPCC. 2001. *Climate Change 2001: Working Group III Assessment Report: Mitigation of Climate Change*.

Living Planet Report 2006. Worldwide Fund for Nature, Global Footprint Network and Zoological Society of London, Gland, Switzerland.

Lockwood, Alex. 2010. 'Seeding Doubt: How Sceptics have Used New Media to Delay Action on Climate Change', *Geopolitics, History, and International Relations*, 2 (2): 136–164.

Mastel, Greg, Stephen Kho and Bernd Janzen. 2009. 'The Developing World and Climate Change: The Disappointing Truth', *The International Economy*, 23 (2): 53.

NSW Government: Primary Industries Science and Research 2010. Retrieved on 14 April 2013, from http://www.dpi.nsw.gov.au/research/topics/climate-change/causes

Pacala, Stephen. 2007. *Equitable Solutions to Greenhouse Warming: On the Distribution of Wealth, Emissions and Responsibility Within and Between Nations*, Princeton: Princeton University Press.

Pachauri, Rajendra and Andy Reisenger (eds) 2007. *Climate Change 2007: Synthesis Report*, Contribution of Working Groups I, II and III to the Fourth Assessment Report of the Intergovernmental Panel on Climate Change, Geneva, Switzerland.

Planet Thoughts. 2008. 'My Thoughts on the Great Planet Debate', The Planetary Society, Pasadena, CA, USA

Rosencranz, Armin, Dilpreet Singh and Jahnavi Pai. 2010. 'Climate Change Adaptation, Policies, and Measures in India', *Georgetown International Environmental Law Review*, 22: 575.

Satherthwaite, David. 2010. 'How Consumption Not Population Growth Drives Climate Change', *Pacific Ecologist*, 19: 37.

Science Museum of the National Academy of Sciences 2010. Retrieved on 28 April 2013, from http://www.koshland-science-museum.org/exhibitgcc/causes03.jsp

SRREN (Special Report on Renewable Energy Sources and Climate Change, Mitigation). 2010. Inter Governmental Panel on Social Change (IPCC), Geneva: UNEP.

Stern, Nicholas. 2006. 'What is the Economics of Climate Change?', *World Economics*, 7 (2): 1–10.

——. 2006a. 'The Stern Review on the Economic Effects of Climate Change', *Population and Development Review*, 32 (4): 793–798.

The European Commission – Climate Action Website 2012. Retrieved on 24 March 2013, from http://ec.europa.eu/clima/policies/brief/consequences/index_en.htm

The Royal Society Science Policy Centre Report 2012. *People and the Planet*, retrieved on 21 April 2013, from http://royalsociety.org/ uploadedfiles/ royal_society_content/policy/projects/people-planet/2012–04–25 peopleplanet.pdf

United Nations Framework Convention on Climate Change. 2011. Least Developed Countries: Reducing Vulnerability to Climate Change, Climate Variability and Extremes, Land Degradation and Loss of Biodiversity, Environmental and Developmental Challenges and Opportunities.

Appendix 1

Carbon Dioxide: The Lengthening Shadow of Coal and Oil

Coal and oil are carbon-based fuels that are the major contributors to the 7.2 billion tonnes of carbon that were released to the atmosphere as a result of industrial activities worldwide in 2004. China is already the world's second largest emitter of this climate-altering carbon at 1.0 billion tonnes annually or 14 per cent of the world total, while India fourth. (See Table.) Measured in per capita terms, China's carbon emissions are still only one seventh the U.S. level, while India's are one eighteenth as high.

Compared with Europe, Japan, and the United States, China and India have contributed far less to the heightened carbon dixoxide concentrations now in the atmosphere. But their emissions have increased by 67 per cent and 88 per cent respectively since 1990, and their shares are projected to grow steadily in the decades to come, making it clear that no serious solution to the world's climate problem is possible without their active participation.

Neither China nor India is covered by the greenhouse gas emissions restrictions included in the Kyoto Protocol. But with the international community now working on the next round of limits, pressure is mounting on these two nations – as well as on the United States – to limit their emissions. Some of that pressure is internal, since both countries are vulnerable to rising sea levels and the more violent weather fluctuations that are likely to accompany climate change.

Carbon Emissions in China, India, Europe, Japan and the United States, 2004 and Increase, 1990–2004

Country or Region	*Carbon Emissions*	*Carbon Emissions per Person*	*Carbon Emissions per Unit of GDP, PPP*	*Increase in Carbon Emissions, 1990–2004*
	(million tonnes)	*(tonnes)*	*(tonnes per million dollars)*	*(per cent)*
China	1.021	0.8	158	+ 67
India	301	0.3	99	+ 88
Europe	955	2.5	94	+ 6
Japan	338	2.7	95	+ 23
United States	1.616	5.5	147	+ 19

Source: See endnote 22.

Source: Worldwatch Institute. *State of the World 2006: The Challenges of Global Sustainability 23rd Edition.* London: The Cromwell Press.

7

India's Climate Policy and Barriers to Low-Carbon Inclusive Growth

Venkatesh Dutta

India's economy is one of the fastest-growing economies in the world, eleventh largest in the world by nominal GDP and the third largest by purchasing power parity (IMF 2012). However, it faces the dual challenge of substantially raising quality of life of its people while also restricting carbon emissions. Even though India's per capita emission and emission intensity of economy is very low compared to other countries, future emissions are predicted to rise sharply considering the amount of infrastructure required to address widespread poverty, energy shortages and regional inequality (WRI 2012). Several researchers have suggested that India should sustain an 8 to 10 per cent economic growth if development and poverty eradication goals are to be met (Parikh 2012; Parikh and Parikh 2011). Higher economic growth will entail higher energy demand calling for higher supply side augmentation, given the current demand-supply gap. Many researchers and development practitioners argue that if India is to continue on its existing course of carbon-intensive development, its absolute emissions will rise substantially as most of the energy demands are met from coal, the emissions are bound to grow in future.

The Government of India has already initiated some ambitious policy measures such as Integrated Energy Policy and National Action Plan on Climate Change (NAPCC) to catalyse large-scale GHG emission reduction. The 12th Five-Year Plan which targets 'low-carbon inclusive growth' as one of the important priority mentions that 'India's per capita emission level will never exceed the average per capita carbon emission level of developing countries'. However, one of the key questions for policymakers is: can India deliver on its promise of inclusive growth and simultaneously reduce future emissions by moving to a development model based on low carbon technology?

This paper attempts to examine India's climate policy and the imperative bottlenecks in transition to low-carbon inclusive growth strategies through an expert survey and review of related literature. It emerges from the expert survey, that current policy architecture is yet to generate a coherent long-term strategy and implementation plan to put India on a low-carbon path of development. Substantial scope still remains to exploit co-benefit potential of mitigation and inclusive growth. Since current investment in infrastructure and technology will decide the future emissions due to higher locked-in emission pathways, the priority of investment is very important. With targeted allocation emphasizing on more social sector outlays such as transport, housing and power (and prioritized investment within these sectors), the growth can be more inclusive, at the same time carbon-friendly. Although the low-carbon growth strategy as envisaged under 12th Plan would be able to catalyse a reasonable share of emission reductions both in the near-and longer-term, there is huge gap between the scale of current energy market reforms, international finance mechanisms and opportunities for global partnership, technology transfer and deployment – and the scale of transformative change needed to shift India's emission pathways towards cleaner options.

I

Data Sources and Methodology

The methodology adopted for the research in this paper includes: (a) a literature review; (b) review of relevant secondary data; (c) structured interviews with experts with a series of pair-wise comparison matrices; and (d) analysis of the gathered information through analytic hierarchy process (AHP). The review of literature draws on available national policy papers and goals for climate change, growth, and sector development outlined in the interim report on low-carbon inclusive growth, National Action Plan on Climate Change (NAPCC 2008), Integrated Energy Policy (2006); and the 11th Five-Year Plan (2007–12).

The rest of the paper proceeds as follows: the next section presents a brief review of literature articulating the research questions, followed by India's position on energy and emission intensity and briefly outlines several initiatives currently being undertaken for reducing emissions. It is followed by results from the expert survey. Low-carbon growth strategies in power, transport and housing sectors are discussed. Finally concluding remarks are given.

II

Review of Literature

The strategy for low-carbon growth is placed according to India's domestic development interests as well as foreign policy strategy in the context of contributing to the international negotiations (Michaelowaa and Michaelowaa 2012). According to a World Bank (2009) report, among the biggest challenges India faces in implementing low-carbon interventions are (a) establishing a cohesive institutional framework and supporting policies and regulations for effective implementation across many sectors, (b) financing the upfront costs of low-carbon interventions, and (c) creating partnerships for implementation. While going through several reports and policy documents, we find that a number of policies are in place to realize low-carbon growth in India. However, it is observed that some of them need to be fine-tuned for effective implementation. It appears from the review, the ongoing discussions in the diplomatic, policy and development sectors for low-carbon growth in India are not equally matched by the possibilities offered on business and technology. It also seems at the moment, a clear articulation of what India needs on technological front and how it might achieve those technology goals, is missing. It is also not clear how high potential technologies will be leveraged to provide technical solutions for developing and sustaining a low-carbon economy. Some experts felt that huge amount of work is to be done in achieving solar and nuclear power and clean coal technologies particularly in terms of costs, timelines, resource constraints, social and environmental implications. The current policy architecture does not encompass the full scope of the challenges and are indeed steps and not the destination. None of these measures has been exploited to its full potential, and it is believed that many could lead to further reductions in emission's growth in the future. There are also several important areas where inclusive growth and climate benefits can coincide, but policy framework has ignored to give adequate attention.

Expert elicitation is a structured approach to gathering expert opinion. Several methods for expert elicitation are used in research and decision making such as Delphi approach, AHP, choice modelling that elicit expert opinion and tries to obtain consensus in a number of iterative steps. Subjective judgments by climate experts have been used in policy analysis by several researchers (McDaniels et al. 2012; Biesbroek et al. 2011; Mathew et al. 2011; Markantonis and Bithas 2010; Doriaa et al. 2009; Morgan and Keith 1995). The complexity, ambiguity and subjectivity that envelop low-carbon growth strategies in India necessitate stakeholder engagement in both policy making and programme development. Expert's opinions have not been elicited for assessing possible barriers towards implementing low-carbon growth policies and programmes in India. This paper fills this gap and attempts to examine India's climate policy and the bottlenecks in transition to low-carbon inclusive growth strategies through an expert survey and review of policy documents.

III

India's Energy and Emission Challenges

India has one of the lowest levels of energy consumption by world standards, with huge shortages. Currently, the installed generation capacity is about 202,979 MW with a per capita consumption of about 750 kilowatt hour (kwh) per year (Ministry of Power 2012). The world average per capita energy consumption is about 2752 kwh/yr and USA stands at 13,000 kwh/yr (IEA 2009). According to a report by TERI (2008), 50 per cent of India's rural population lacked access to electricity. Even in 2030, with all the planned power augmentation achieved as stated by Integrated Energy Policy, per capita gross generation in India will be 1/7th of the current per capita power generation in US. As power generation is unevenly distributed in the country with requirements of transmission to great distances through regional grids, the *'Mission 2012: Power for All'* has not delivered its earlier promise of universal electricity access till 2012. Energy efficiency and reduction of network losses are among the important priorities for reducing emissions in a low-carbon transition. In the power sector, the network losses exceeded 32 per cent in 2010 including non-technical losses, against the world average of less than 15 per cent. An enabling policy environment was initially set up by enacting the Energy Conservation Act 2001 which resulted in many energy efficiency projects, star rating and energy conservation codes, Perform-Achieve-Trade mechanism, accreditation of energy service companies (ESCOs), and other energy efficiency initiatives.

The government is also mandating the retirement of inefficient coal-fired power plants and supporting the research and development of IGCC and supercritical technologies. Therefore, to raise the quality of life of people and to allow them to benefit from economic development, it is crucial that both energy consumption and electricity consumption level are substantially enhanced.

The per capita energy emission in India is in the mid-range for the least developed countries. The world average emission is around 5.8 tCO_2 per capita, and India stands at 1.7 tCO_2 per capita (WRICAIT 2011). At present, the largest contribution of global GHG emissions in percentage term is from China: 19.5 per cent, followed by USA: 19.2 per cent, India: 5.3 per cent, Russia: 5.1 per cent, Japan: 3.6 per cent and Germany: 2.6 per cent (IEA 2012).

Between 1994 and 2007, India's GHG emissions rose by approximately 56 per cent with the energy sector contributing over half of the emissions; however, emission intensity declined by 30 per cent (MoEF 2012). Following the global trend, the largest bulk of emission in India comes from the energy sector. According to the 2007 data (Planning Commission 2011), among the energy sector in India, electricity generation

Figure 1

India GHG emission by sectors in 1994, 2000 and 2007 (based on MoEF 2004 and MoEF 2012)

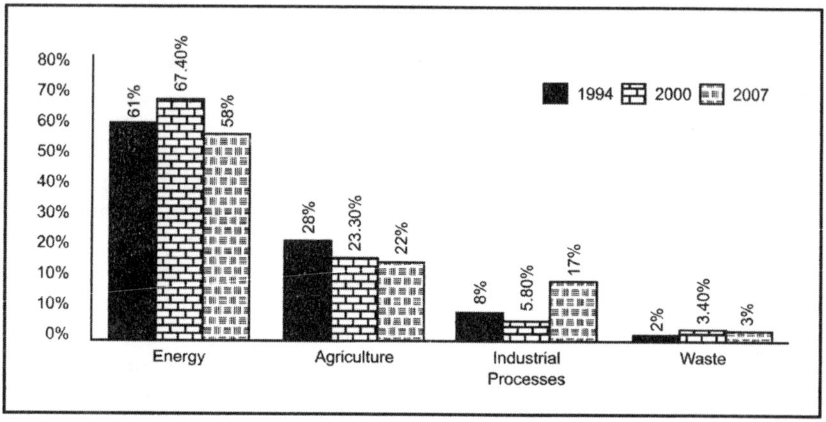

contributed 65 per cent of the GHG emissions, transportation 14 per cent and residential and commercial 13 per cent.

A comparison of emission by sectors reveals that overall share of emission from the energy and agriculture sector has declined slightly, whereas there is an increasing trend from industrial processes and waste (Figure 1). In 2004, energy sector had accounted for about 61 per cent of total CO_2 eq emissions, while, in 2007, energy sector accounted for about 58 per cent of total CO_2 eq emissions of which almost 65.4 per cent came from the electricity generation (MoEF 2010). India's per capita CO_2 eq emissions including LULUCF were 1.5 tonnes/capita in 2007 (MoEF 2010). According to IEA (2008), the emission intensity of India's economy was 0.34 kg CO_2 per US$ GDP (at PPP/2000 price) in 2006, which was much below the world average of 0.49 kg CO_2.

Over the past five years, India has undertaken a systematic revision of its own energy and climate policies, producing a number of action plans and new initiatives. The National Action Plan on Climate Change (NAPCC) outlines existing and future policies and programmes addressing climate mitigation and adaptation through eight core 'national missions' running through 2017 (NAPCC 2008). These missions target solar energy deployment, energy efficiency, sustainable habitat, water resource management, conservation of the Himalayan ecosystem, afforestation, sustainable agriculture and strategic knowledge building for climate change.

Renewable Energy (RE) capacity in India has grown 6 times in last 10 years and is ranked 7th worldwide in the amount of installed RE capacity. According to the Ministry of New and Renewable Energy (MNRE), the share of RE capacity is 12.25 per cent (excluding large hydro) of the total installed capacity of about 202 GW in the country, up from 2 per cent at the start of the 10th plan period (2002–2007). The MNRE has recently introduced a Generation Based Incentive (GBI) scheme to help more independent power producers (IPPs) enter the RE sector. Under the Electricity Act 2003 and the National Tariff Policy 2006, the central and the state electricity regulatory commissions must purchase a certain percentage of grid-based power from renewable sources. India attracted $4 billion in private investments for clean energy, ranked 10th among the G-20 countries. The energy policy intends to support more of renewable energy sources, and scaling up efficiency enhancing efforts with continuous up gradation of technology as well as finance.[1]

The Jawaharlal Nehru National Solar Mission (JNNSM), launched in January, 2010, has set the ambitious target of deploying 20,000 MW of grid connected solar power by 2022. The mission also intends to create an enabling policy framework aiming to reduce the cost of solar power generation through (i) long-term policy; (ii) large scale deployment goals; (iii) aggressive R&D; and (iv) domestic production of critical raw materials, components and products, as a result to achieve grid tariff parity by 2022. Currently sufficient demand is generated through JNNSM and states' policies, and tremendous growth opportunity exists in off-grid segment. The cost of solar equipment has fallen by a third since 2010 and conventional power at relatively higher cost would imply that solar will soon be competitive without subsidies. It is estimated that by 2016, Indian solar will match the price of conventional electricity (*The Economist* April 2012).[2]

Wind energy is the fastest growing renewable energy sector in the country. With more than 13 GW of total installed capacity at the end of 2010, India ranks fifth in the world in terms of cumulative installed capacity. It accounts for nearly 70 per cent of the installed capacity in the RE sector in the country. The Indian wind market grew by almost 68 per cent on a year-on-year basis with 2,139 MW of new capacity installed between January and December 2010. This made India the third largest annual market after China and the USA for 2010. A long coastline and relatively low construction costs could make India a favoured destination for offshore wind power.

There also exists a vast untapped hydropower resource, only 23 per cent of which has been harnessed so far. The current hydropower deployment is about 40,000 MW which accounts for about 20 per cent of total power generation. The public sector has a predominant share of 97 per cent in total hydropower generation. The total hydro potential is estimated around 130,000 MW. Out of this, small hydropower plants of up to 25 MW capacities have the potential of 15,000 MW. The government has set the target for India's optimum power system mix at 40 per cent from hydropower and 60 per cent from other sources. However, there has been a lot of public protest in recent years due to increasing environmental and social impacts. Several projects in upper Ganges have been put to halt due to increasing protests from religious leaders, and public as they feel that river Ganges should not be disturbed by creating dams and storage reservoirs. Ecosystem degradation, land acquisition, resettlement, relocation and rehabilitation (R&R) are some of the critical issues associated with large hydro projects. Early attention to social and environmental aspects of projects is very important, in particular, mitigating the negative social and environmental impacts.

Table 1

Grid connected renewable energy (RE) potential in India

Energy source	*Capacity (MW)*	*Assumed Plant Load Factor (PLF)*	*Annual energy generation (billion kWh)*
Wind (onshore)	100,000	25	219
Small Hydro	15,000	45	46
Bagasse	5000	60	26.3
Biomass	16,881	60	88.72
Large Hydro (existing and future)	100,000	60	525.6
Large Hydro in Bhutan	16,000	60	84.1
Waste to Energy	5,000	60	26.28
Solar CSP based power generation*	200,000	35	613.2
Solar PV/CPV based power generation*	200,000	20	350.4
Geothermal	10,000	80	70.1
Total	662,881		2,049.70

Source: WISE, January 2011.

Note: Resource Potential of other RE Sources including offshore wind, tidal, biogas-based power not considered.

* In India a total suitable desert area of 208,110 square kilometres is available for solar power generation. At 10 per cent utilization of this area the stated CSP/PV potential can be achieved (assuming 20 MW/sq.km), CSP: Concentrated Solar Power; PV: Photovoltaic

IV

Experts' Opinion on India's Position on Low-Carbon Growth Strategies

The interview was designed to have an interactive format on a structured questionnaire. The survey questions covered a range of topics related to low-carbon growth in some of the critical sectors of India's economy and energy, including: issue priorities, policy preferences, risk perceptions and political and consumer beliefs and behaviour. Twenty-five key experts concerned with India's climate change policies have been consulted and interviewed during March–April 2012. Twelve per cent of

Figure 2

Perception of most serious problem India faces currently for achieving inclusive growth (relative ranking of issues in percentage)

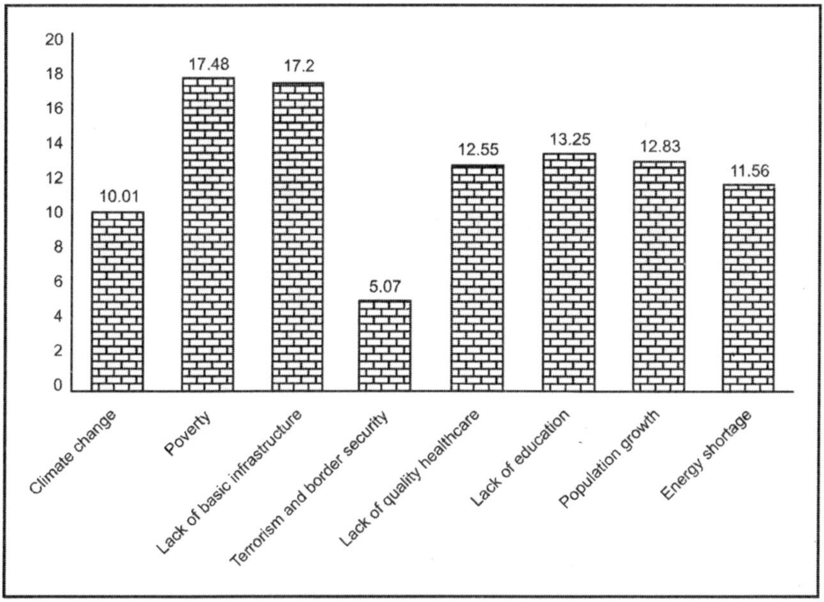

the experts came from the government, 36 per cent from research and policy think tanks, 32 per cent from academia and 20 per cent from the industries. The opinions of experts are enumerated under following headings: (i) need for inclusive growth and low-carbon development, (ii) domestic actions and capacity, (iii) barriers to low-carbon growth, (iv) international partnership and support, and (v) policy options for low-carbon growth. Expert elicitation is especially relevant for addressing issues like low-carbon growth as it can improve decision making in the face of high ambiguity when data are insufficient or unreliable. Based on our results, we argue that by eliciting experts' opinion on barriers to carbon growth and their relative importance, existing models and knowledge can be enhanced.

Need for Inclusive Growth and Low-Carbon Development

Perception of climate change with respect to other development problems (Figure 2): More than one third of the experts felt that poverty and lack of basic infrastructure are the top two most serious problems constraining India's growth. India's development strategy is focused on multiple goals – largely on strengthening the economy and alleviating poverty. Therefore, it is believed that development goals will have large climate co-benefits such as meeting large energy deficits, access to improved

Figure 3

Key sectors and resources for which the impacts of climate change would be most felt in India (relative ranking of expert's responses in percentage)

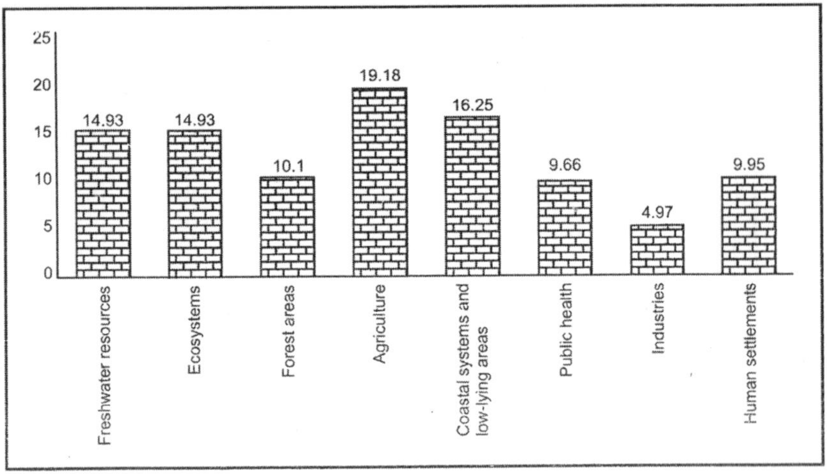

transport systems, improvements in local environmental quality, employment generation etc. The common consensus is that India will not halt development to avoid GHGs emissions, but will be proactively taking steps to mitigate and adapt to climate change. Some experts felt that it was difficult to distinguish adaptation to climate change from the process of development itself, as a lot of development measures are also measures that would strengthen the coping abilities (Patwardhan 2011).

Relative impact of climate change on key sectors: Agriculture, coastal systems and low-lying areas, freshwater resources and ecosystems are some of the important sectors for which the impacts of climate change would be most felt in India. Experts believe that impact of climate change would be maximum on agriculture, and it is important that policy initiatives target on potential adaptation measures. Since Indian agriculture is largely based on rainfall, changing rainfall patterns are likely to affect food security. Some experts expressed that India's ability to adapt to future climatic situation would depend on its state of development. As there exists wide developmental gaps,[3] even with known risks and coping skills, there would be limitation in adapting to a changed climate. India's large regional and socio-economic differences in its various states would affect local communities' ability to adapt significantly. Coastal systems and low-lying areas are also vulnerable to rise in sea levels as it could threaten coastal mangrove and wetland systems, and increase the flood risk faced by the quarter of India's population that lives on the coast (Figure 3).

Relative importance of key-sectors in achieving inclusive growth: Experts rank agriculture, power, healthcare, industry, transport and housing as

Figure 4

Relative importance of key sectors in achieving inclusive growth in India

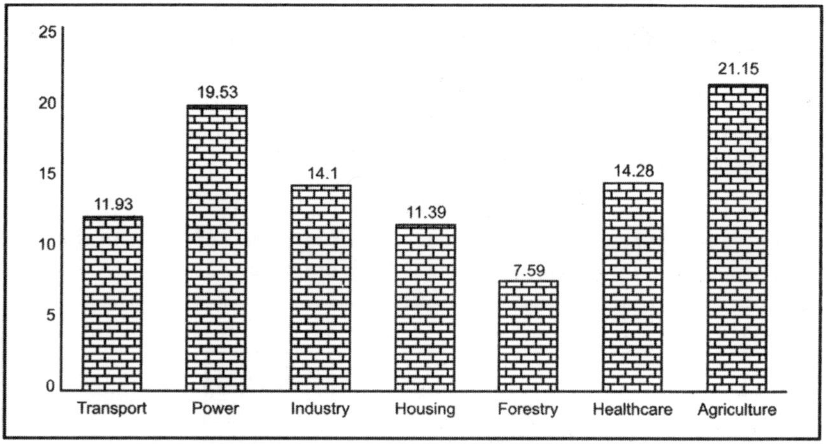

key-sectors for achieving inclusive growth India. The experts think that investment should focus on sectors which offer clear and existing comparative advantages and low-risk premium rather than divert significant resources to the new sources such as CCS. CCS is not considered as one of the options for reducing carbon intensity at the moment. This is largely due to the fact that it is yet to be proven at a commercial scale elsewhere. There are also uncertainties about the environmental risks (Figure 4).

Impact of investment in low-carbon growth on poverty reduction: High population density, increased natural resource depletion and wider socio-economic inequalities will add to climate anomalies in India. It is believed that investment in low-carbon initiatives will bring about poverty reduction. The experts are of the opinion that there is no inherent contradiction between poverty eradication and moving toward a low-carbon economy. However, it is not clear to what extent investment in low-carbon economy will bring about poverty reduction – about 21 per cent of the respondents think that investment will have more than marginal impact on poverty reduction while 37 per cent of the respondents think that investment will have only marginal effects on poverty reduction. One-fourth of the respondents are not sure about impact of investment in low-carbon growth on poverty reduction.

Inclusive growth versus low-carbon growth: A few experts feel that India must not inhibit economic development for the sake of lowering emissions and should pay more attention to economic development than climate change. Some experts also advocate keeping the goals of low-carbon economy and inclusive growth separate. There is also a widely shared perception among the policy community that low-carbon growth is less than the customary economic growth and costlier. This belief comes from the perceived notion that shifting to low carbon economy will mean high initial costs coming from adoption of energy efficient technologies, and could compromise development priorities and lead to an inflationary situation in the local economy. However, majority of experts feel that low-carbon growth priorities should be integrated with inclusive growth in India's 12th Five-Year Plan as they would lead to better development outcomes (Figures 5, 6, 7).

Figure 5

India should first focus on developmental needs regardless of GHG emissions as the development needs are huge

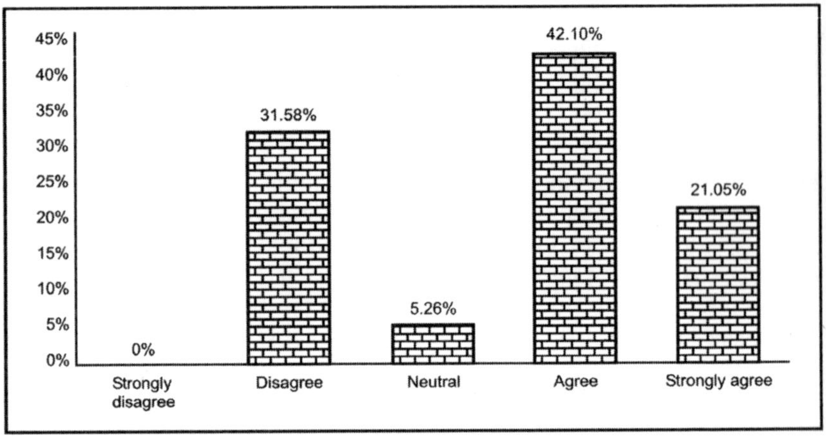

Figure 6

A determined effort to bring down carbon intensity will impose a high cost burden on India's economy and constrain growth

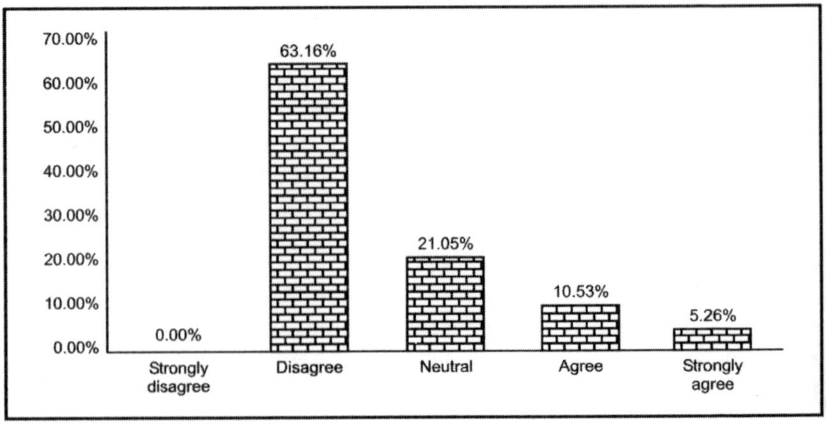

Figure 7

Approaching low-carbon growth to meet multiple objectives of inclusive growth is a complex task. Therefore, it is better to focus them separately rather than integrate into one objective

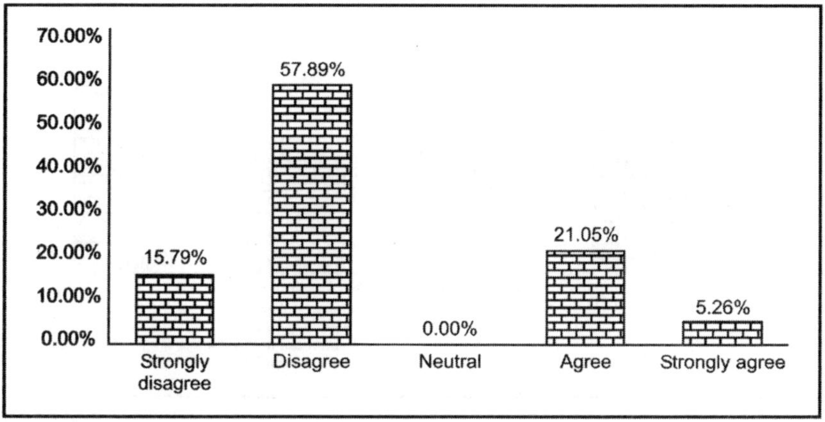

Figure 8

Low-carbon growth priorities should be integrated with inclusive growth in India's 12th Five-Year Plan as they would lead to better development outcomes

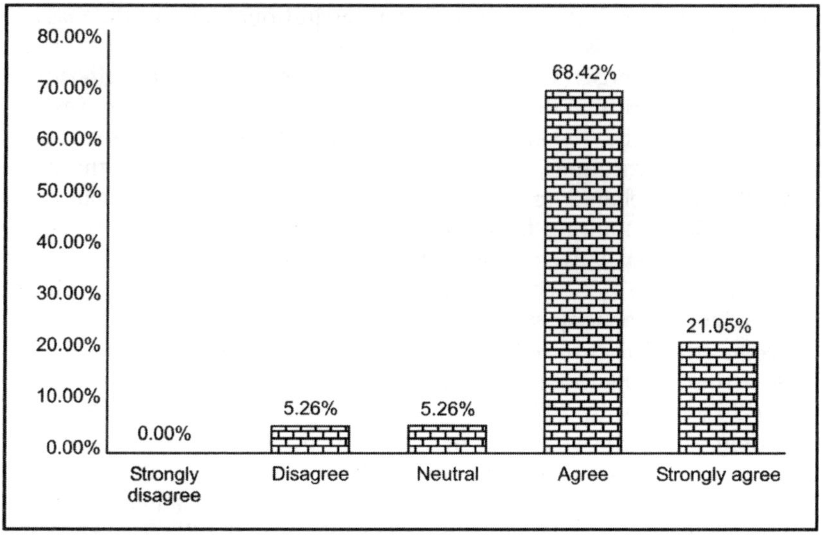

Domestic Actions and Capacity

Institutional set-up to deal with climate responsive strategies: Only a handful of experts think that India has a very organized institutional set-up to deal with climate responsive strategies, majority of them feel

Figure 9

Investment in low-carbon initiatives will bring about poverty reduction

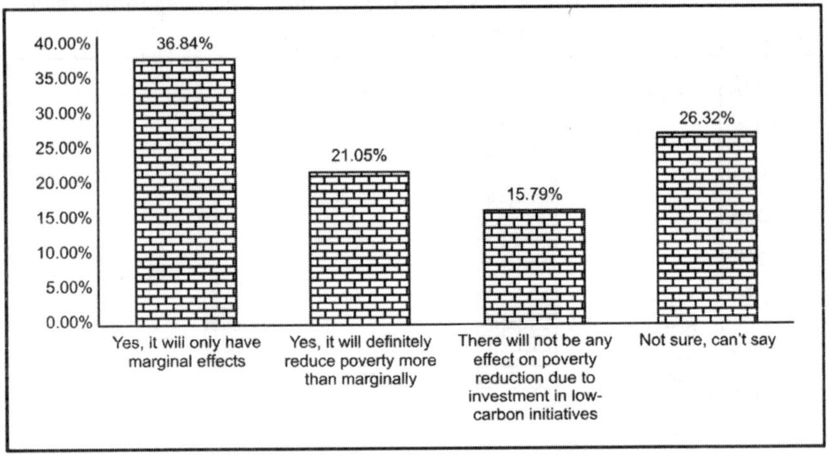

that India's institutional capacity is not geared enough to take the challenge of low-carbon growth. Only 5 per cent of the respondents observe that local bodies and city governments have adequate institutional set-up to deal with climate responsive strategies/actions. 58 per cent respondents felt there is insufficient institutional set up, while 32 per cent feel that there is no institutional set-up at all.

Institutional crisis will expose the climate governance offering a chance for retrofitting and renewal. They believe that climate crises will further expose the weaknesses of central, state and local governance systems and institutions, thereby simultaneously aggravating the crisis and offering a chance for retrofitting and renewal. Institutional effectiveness will be limiting factor towards implementing change from the business as usual.[4] Even though NGOs are working towards creating awareness, there is a need for orientation of grassroots NGOs in developing adaptive capacity within their areas of operation.

Domestic capacity to reduce emission intensity: A little over one-fourth of the experts think that India can reduce its emission intensity of its GDP by 20–25 per cent by 2020, about 32 per cent of respondents are poorly confident about the target, while about 16 per cent of the respondents think that the target is too ambitious and not possible to reach within the timeframe. More than 50 per cent of the respondents believe that a reduction of less than 15 per cent is feasible over the 2005 level by the year 2020, in view of the current emission intensity and development requirements.

Current levels of climate action: Almost 21 per cent of the respondents believe that State governments are taking appropriate climate actions,

Figure 10

India can reduce its emission intensity of its GDP by 20–25 per cent over the 2005 level by the year 2020

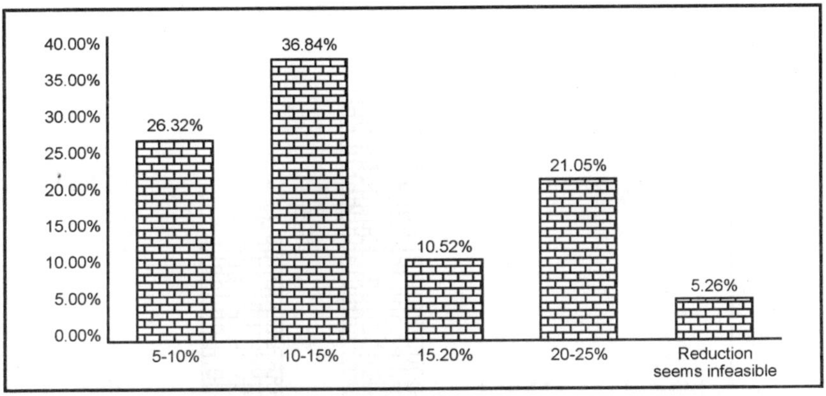

Figure 11

What level of reduction of emission intensity seems feasible by 2020 in view of current emissions and developmental requirements?

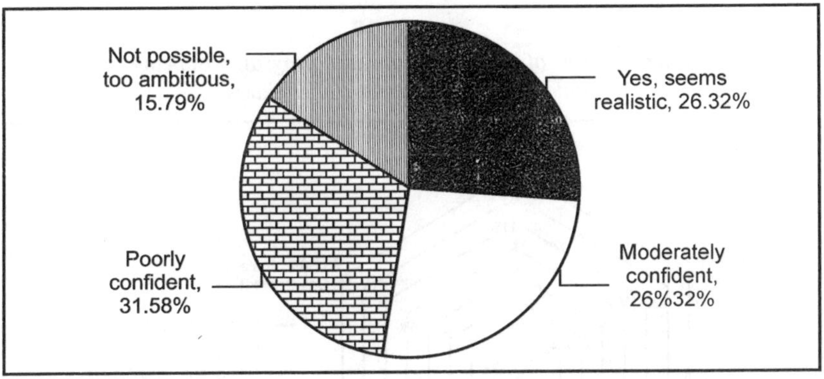

while a majority, about 74 per cent thinks that actions are not sufficient. Around 53 per cent of the respondents think that the Central government is more proactive in climate actions and it is doing about the right amount. Around 58 per cent of the respondents believe that private sectors are taking climate actions, but they are not sufficient. Almost 42 per cent of the respondents think that both municipalities and citizens are not doing at in terms of local-level climate initiatives. Experts also believe that there is a lack of an effective forum where state and central government, private companies, municipalities, NGOs and citizens share knowledge and drive meaningful and measurable action on climate change. Several states have made their climate change action plans laying

Figure 12

Do you think our local bodies and city governments have adequate institutional set-up to deal with climate responsive strategies/actions?

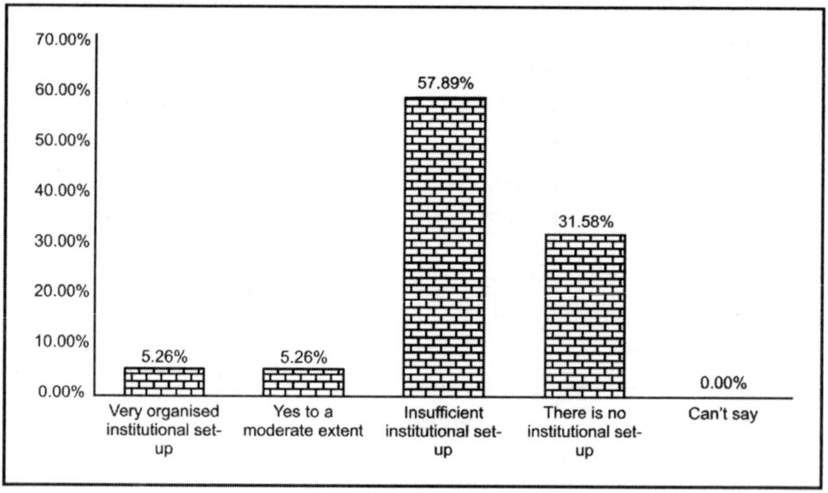

Figure 13

Current levels of climate actions and programmes being undertaken by different stakeholders

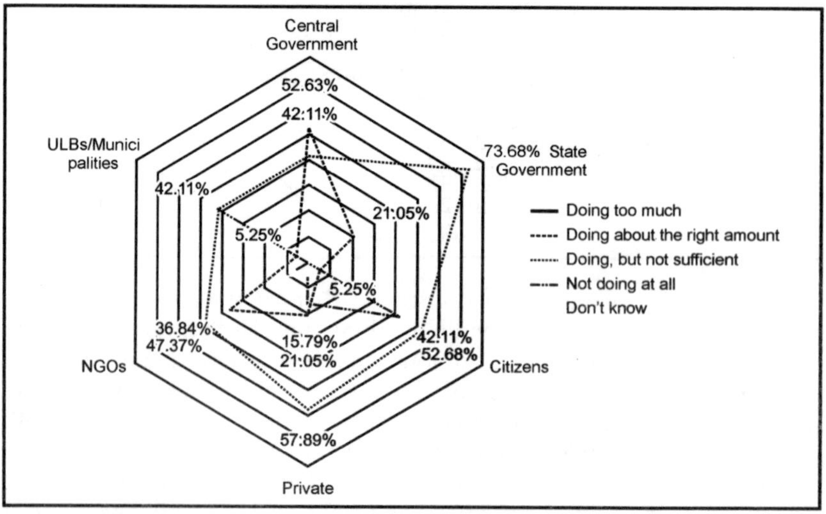

out a climate strategy, including specific policy recommendations, but there is resistance to structural changes mainly due to lack of supporting institutions and unwillingness to modify current institutional structures.[5]

Figure 14

Reasons that could potentially hinder the implementation of low-carbon strategies in India

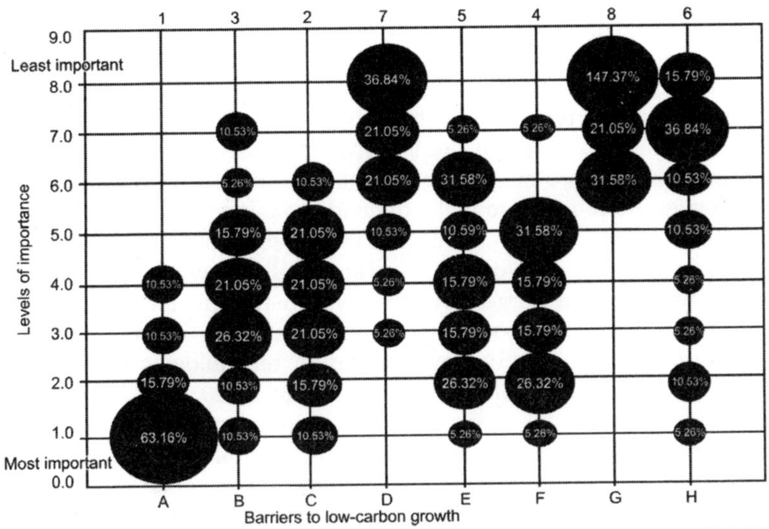

A	High cost of cleaner and new technology – capital and O&M
B	Limited access to private capital
C	Split incentives – cost/benefit mismatch
D	Cleaner and new technology has negative externalities
E	Lack of clear regulation
F	Inadequate institutional mechanisms
G	CDM institutions are overburdened
H	Lack of knowledge and awareness about the seriousness of the problem

Barriers to Low-Carbon Growth

High cost of cleaner and new technology – both capital and O&M, seems to be the most important barrier to low-carbon growth in India, followed by split incentives – cost/benefit mismatch, and limited access to private capital. New projects that offer radical reduction in carbon intensity are often front loaded with large capital requirements and stretched payback periods. Both private players and government can increase scope for accessing large capital and provide direct and indirect financial incentives such as venture capital.

Experts believe that markets alone may not automatically deliver low-carbon outcomes that promote long-term inclusive growth, a long-term regulatory certainty for investors and other economic agents is necessary for deployment and diffusion of low-carbon technology. Following barriers have been identified by the experts during the survey:

1. *High cost of cleaner and new technology – capital and O&M:* Despite an impressive return on investment and net present value, many clean technology projects do not go forward due to lack of available capital. The biggest pools of capital investment required to support clean technologies in India will be in the field of low-cost, massive-scale power, particularly RE. India and China have major growth markets for clean technology investments, with a focus on RE technologies. Many promising new technologies have been deployed because they have become more scalable in the last five years. So far, cleaner and new energy sources are more expensive than conventional ones, and large subsidies are required to make them affordable to the large masses.
2. *Limited access to private capital:* The incremental investment required to shift India's economy on a low-carbon pathway would be enormous; this will require large private investments in technologies that contribute to mitigation of GHG emission. The overall investment quantum required to deploy low-carbon technologies could require tens of billions of dollars annually out to 2030. The initial capital investments will be critical in large-scale deployment requiring a long-term and comprehensive plan across several sectors. According to the UK's BERR study (2009) the current market value of low-carbon and environmental goods and services (LCEGS) sector in India has been estimated at £191 billion. This is 6.3 per cent of the £3,046 billion global market which will grow by 45 per cent in the next few years, mostly in low-carbon renewables. According to TERI (2008) if India moves along the *business-as-usual* pathway, the cumulative investment required in the power generation technologies alone from 2001 to 2036

Figure 15

Has the strategy of low-carbon options for reducing GHG emissions (such as NAPCC, Energy Efficiency Mission), has been embedded within the existing policy (Integrated Energy Policy, Auto Fuel Policy, JNURM etc.)?

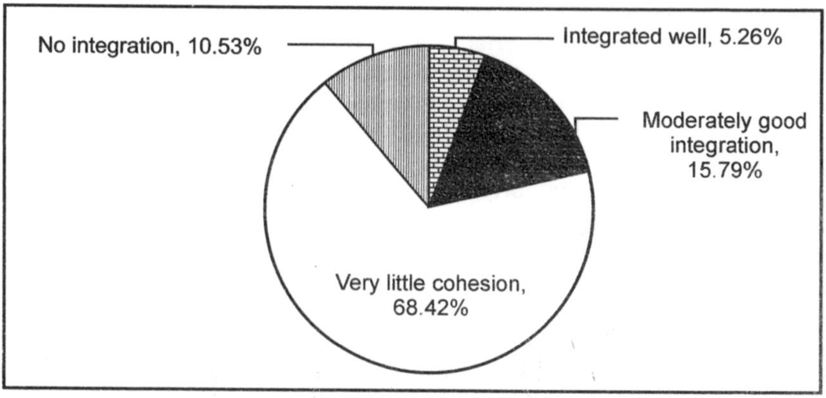

would be $1.4 billion. This investment requirement would increase by 4 to 7 times, if the transition were to follow a low-carbon model.

3. *Split incentives – cost/benefit mismatch:* The investment in low-carbon pathways reinforces the fundamental market problem of split incentives, i.e. costs and benefits do not always accrue to the investing party and hence savings cannot be fully recouped in the short and middle terms. Most of the clean energy projects require large upfront investments, followed by a long period of payback. Some experts suggest that fossil-fuel subsidy policies have further made renewable energy technologies less cost-effective. Most Indian power plants are fueled by coal and generate electricity at about ₹ 4 per kilowatt hour (7.5 cents) – less than half of solar's cost. However, in near future, the price of solar power will continue to decline, eventually approaching the cost of electricity generated through conventional methods.
4. *Cleaner and new technology has negative externalities:* Some of the new technologies may have unproven log-term impact on the environment, or they may compete with other resources. For example, biofuels compete with arable land for food crops and water, increasing share of nuclear energy posing safety risk associated with radiation leakages and radioactive waste. Technologies like CCS are not proven at this stage, and may entail risk of contamination of groundwater.
5. *Lack of clear regulation:* Lack of clear regulation and inconsistent policies make investment difficult. There is general consensus among the experts that the strategy of low-carbon options for reducing the GHG emissions has not been embedded effectively within the existing policy and there is very little cohesion between the policies.
6. *Inadequate institutional mechanisms:* Even though state governments have come up with the action plan on climate change, the present institutional set-up does not foster dialogue and thinking, on low carbon policies and mitigation strategies across sectors and beyond traditional boundaries.
7. *CDM institutions are overburdened:* CDM's Executive Board (EB) and the designated operational entities (DOEs) are overburdened due to lack of capacity and growing number of projects. DOEs are essentially accredited private certifiers who validate projects that claim reduction in GHG emissions and subsequently verify emission reduction targets. CDM's EB issues CERs according to the recommendations of the DOEs. India's designated national authority is vested with Ministry of Environment and Forests (MoEF) which is overburdened with other equally important tasks.
8. *Lack of knowledge and awareness about the seriousness of the problem:* There is currently no clear or agreed definition of what constitutes a green or low-carbon economy. Most of the SMEs have insufficient expertise and knowledge about the energy and carbon intensity of

their products/processes. There is also a strong belief that the main barriers to adoption of low-carbon technologies may have much more to do with the level of absorptive capacity, social and economic conditions than with the market barriers.

International Partnership and Support

Around 42 per cent of the respondents feel that international partnership in promoting low-carbon investment in India has been inadequate; however, 53 per cent respondents feel that it has been moderately successful. The international negotiations have found India largely on the defensive, defending India's *'right to development'*. The traditional Indian negotiating position is based on equity. There is a difference in placement of ideas about climate proofing among developed and developing countries – the developed countries see the need for climate actions through a 'low carbon' lens whereas the developing countries outlook is primarily based on 'right to develop' notion addressing food and livelihood security, social equity and poverty alleviation.

Most of the experts disagree on prospects of defining quantitative emission reduction targets that may bind national policies. Experts mention that climate negotiation should recognize India's development divide and its policy path, and be in tandem with the principles of 'common but differentiated responsibility' and 'equity'. Around 47 per cent of the respondents disagree on a treaty that requires India to legally cut its emissions at the end of the second commitment period, 32 per cent remain neutral while 21 per cent support on binding agreements at the end of Kyoto's 2nd commitment period. Some experts are of the opinion that India should appear flexible and constructive, not argumentative and polemical while setting the discussion for legally binding agreements on

Figure 16

Do you think India should sign an international treaty that requires it to legally cut its GHG emission at the end of the second commitment period of the Kyoto Protocol (2017)?

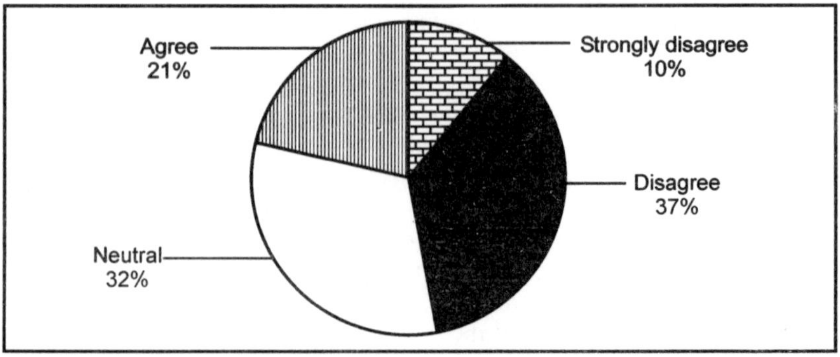

Figure 17

To what extent do you agree with India's position for getting financial support from developed countries on the basis of historical responsibility and equity?

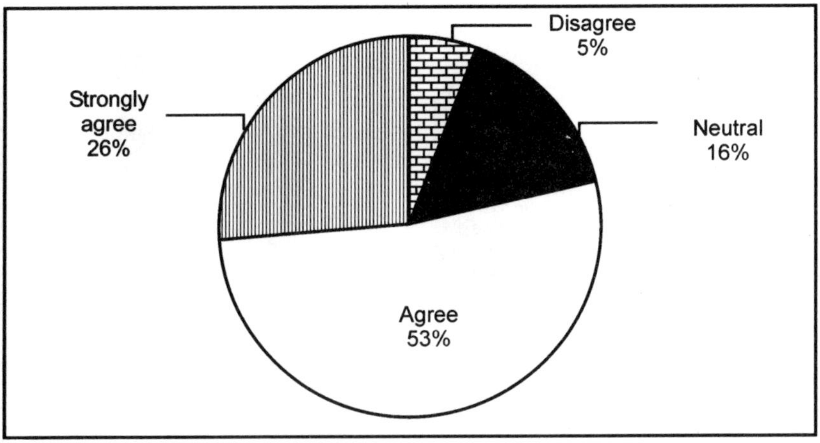

emission reduction. Around 80 per cent of respondents support India's position on getting financial and technical assistance from Annex 1 countries on the basis of historic climate burden and equity.

Climate Finance for Mitigation and Adaptation

Climate finance has a crucial role to play in helping India make the transition to more environmentally sustainable infrastructure, at the same time addressing widespread developmental priorities in ensuring basic services. The experts felt that there was a need for more effective architecture for equitable climate financing. The distribution of adaptation finance to the recipient countries remains uneven. The scale of finance is also not commensurate with the estimated needs. The experts felt that India has not received appropriate funding for adaptation from the donor agencies.

Climate funds for adaptation: Meeting the costs of adaptation to climate change in developing countries is a major challenge for the international community. The UNFCCC projects adaptation costs in the range of \$28 – 67 billion per year for developing countries by 2030. There has been a considerable increase in adaptation finance from dedicated climate financing instruments in 2011. Sub-Saharan Africa is the region with the highest volume of adaptation finance approved and disbursed and the highest number of adaptation projects with an average of \$2 million spent on each project. The Least Developed Countries Fund and the Special Climate Change Fund have disbursed the most finance for adaptation, but the Adaptation Fund, the Pilot Programme on Climate Resilience of the

Figure 18

Performance of international partnership and support in India's GHG mitigation efforts

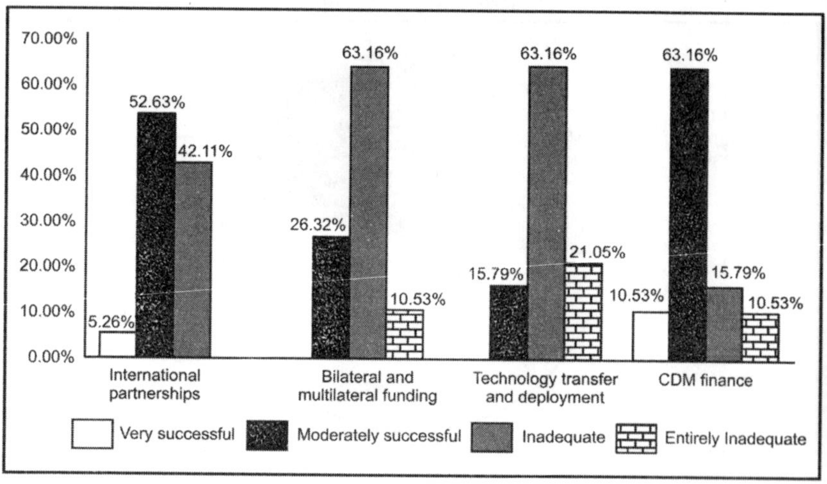

Figure 19

Climate funding to India from GEF and International Climate Initiative during 2007–2012

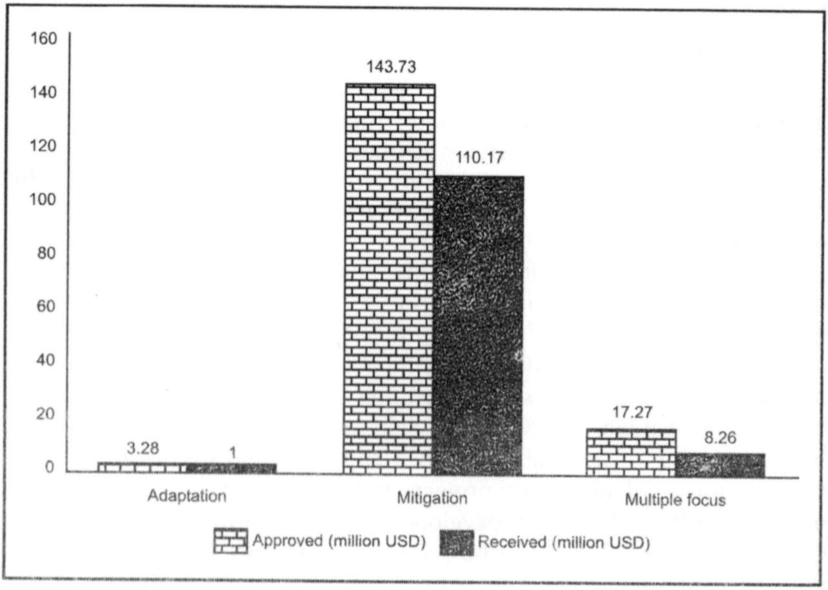

Figure 20

Adaptation finance approved and disbursed during 2007–2012

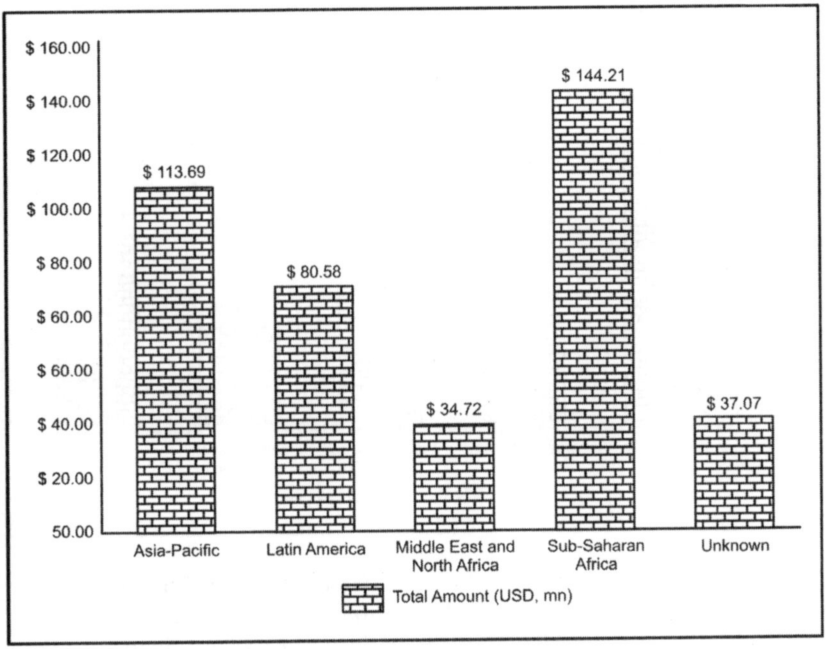

Figure 21

Climate funding to India during 2007–2012 by fund schemes

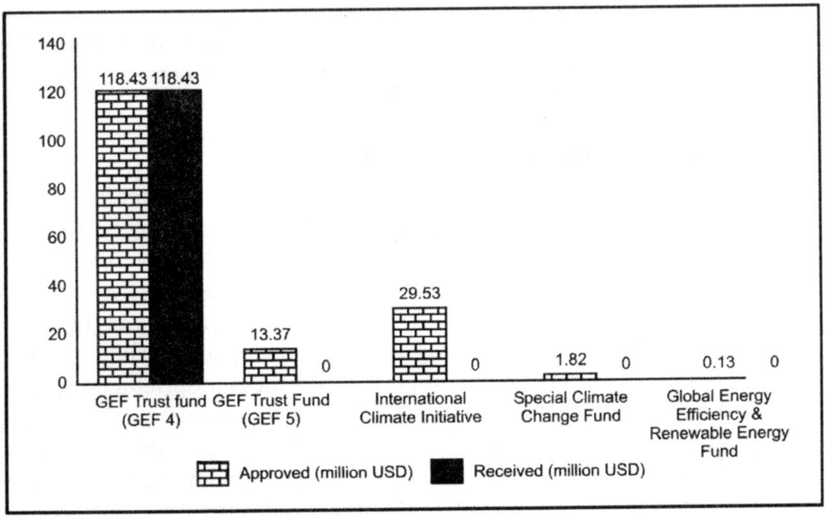

Figure 22

Mitigation finance approved and disbursed during 2007–2012

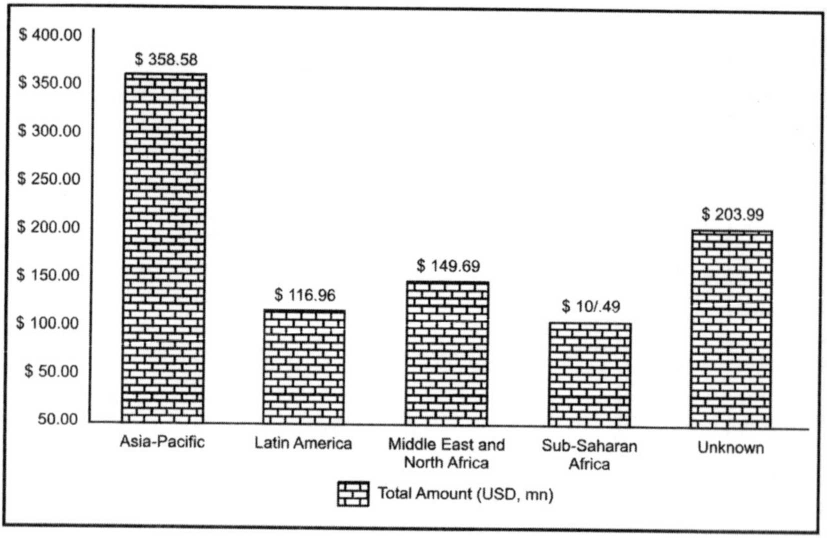

Climate Investment Funds, and the EU Global Climate Change Alliance have also substantially added to the volume of finance available.

Climate funds for mitigation: The majority of climate finance to date has been directed to mitigation activities, particularly in fast growing emerging economies where emissions are growing rapidly such as in China, South Africa, India, Turkey and Mexico. The GEF Trust Fund has disbursed the largest amount of financing for mitigation to date. The Clean Technology Fund administered by the World Bank in partnership with regional development banks has the largest amounts of pledged, deposited and approved amounts financing for mitigation.

Technology Transfer and Deployment

India is in a phase of massive infrastructure build up. Therefore, it is important that it deploys best available technology for creating new infrastructure or expansion of already existing infrastructure. Slow progress in technology transfer could lead to a lock-in in high-emissions systems for several years to come. R&D and technology transfer have a significant potential to contribute towards progress in achieving low-carbon growth in India. Many of the clean technologies are business proprietary and are protected by strong patent regimes. According to a survey of 70 developing nations conducted by UNFCCC in 2009 (UNFCCC 2010), it was revealed that the main barriers to clean technology transfer were economic and market barriers. Some potential measures that could address these barriers were identified as: individual country's policy regime to attract foreign investments; increased involvement of the private sector in technology transfer; removal of subsidies and price distortions; and improvement of collaborative R&D of clean technologies.

According to Seres et al. (2009) about 36 per cent of the total CDM projects accounting for 59 per cent of the annual emission reductions, claim to involve technology transfer. Four countries constitute about 75 per cent of the CDM projects: Brazil, China, India and Mexico. While 68 per cent of Mexican projects include an international transfer of technology, the rates are, 12 per cent, 40 per cent and 59 per cent for India, Brazil and China respectively. According to Dechezleprêtre et al. (2009) high rate of technology transfers to Mexico, Brazil and China are mainly due to the strong involvement of foreign partners and good technological capabilities. The rate of technology transfer for new projects in India has fallen significantly. The authors cite that low levels of technology transfer to India may be due to a better capability to diffuse domestic technologies. Experts feel that, in India the CDM has contributed to some individual project level technology transfer without creating more widespread policy support for technology transfer in energy systems.

Figure 23

Rate of technology transfer through CDM projects showing steep decline in India and China

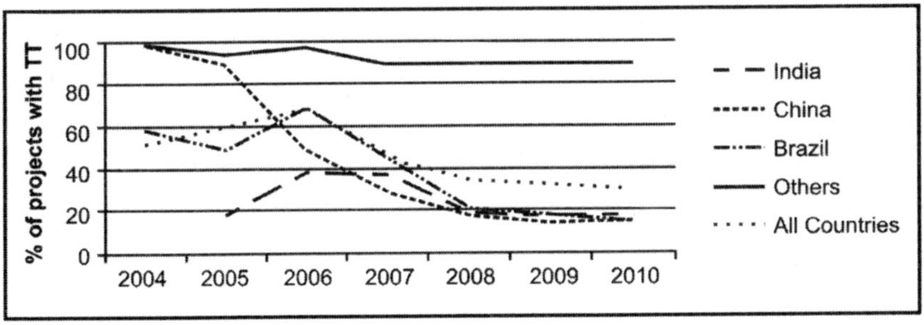

Source: UNFCCC 2010.

In a regression analysis by Dechezleprêtre et al. (2008), it was found that there is a significant relationship between the technological capabilities of a country and the types of technology transfer it attracts. Countries with higher technology capability are found to attract Technology Transfer (TT) in the energy sector and in the chemicals industry. However, in a recent analysis, it appears that countries with higher technology capability and numbers of CDM projects in specific project categories have a tendency to develop their own capacity for domestic technology in that specific sector in the long run irrespective of the level

Figure 24

Top low-carbon options that India should adopt considering inclusive growth targets and carbon space requirements in the future

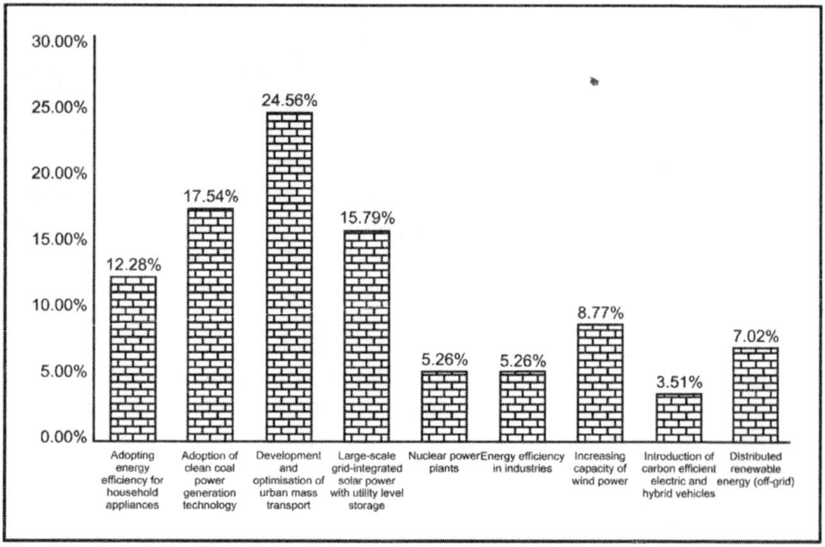

of sophistication (UNFCCC 2010). This is evident in the declining trend in technology transfer through the CDM in China, India and Brazil.

According to Kumar (2011) in developing countries, growth in productivity is led by innovations under strong disposability of emissions, but under weak disposability, it is function of both innovations and diffusion of technologies. Climate change lowers the productivity growth in developing counties through lowering the process of technological adoption. Low-income states in India have high relative share of population as well as high emissions due to concentration of polluting industries as well as less resources to invest in clean technology. Therefore, they will need technology transfer and financial support from the centre and richer states (Gupta and Kumar 2011). Therefore, for long-term benefits and sustainable growth, more focus should be on promoting low-carbon investment with greater priority in less-developed states, even if it has adverse growth impacts in the initial phases. It is also reported by the authors that better institutions help in lowering the impacts of climate change.

Only about 16 per cent respondents feel that technology transfer and deployment has been somewhat successful in India; 84 per cent respondents feel that there has been no progress. The rich nations remained non-committal on financing clean technology transfer to India and other developing countries in a recent ministerial level consultation on 'Green

Economy and Inclusive Growth' involving 54 countries held in India in October 2011 (Kapur 2012).

CDM Finance

Many experts in the survey felt that CDM offsets have strengthened the project economics for isolated projects only in the short to medium term, without delivering technology leapfrog or other long-term benefits. Some experts feel that CDM has become merely a CERs purchase tool without any integration of sustainable development to its core.[6] Technology transfer has become absent in new CDM projects without any substantial investment. However, a large number of experts feel that progress in CDM financing has been good as it has led to several new projects that target reduction in GHG emissions. They also believe that CDM cash flow increased the IRR of the project making it more interesting for the investors.

Indian market could capture a sizeable share of emerging competitive CDM markets resulting in large amount of CERs, if the institutional set-up in the Indian finance sector is suitably geared up. Experts felt that banks and financial institutions have not developed procedures for efficient financing of CDM projects. Mainstream financing is yet to formally engage in CDM transactions. Both conventional and new financing instruments such as project finance for large projects, Special Purpose Vehicles, CDM funds or CDM bonds could become attractive to banks. There is also a need to accelerate the integration of private sector perspective in CDM decision-making process. As CERs cannot be used as a dependable source of cash flow within the overall non-recourse project financing structure, improving equity returns to investors and sponsors using CER revenue stream is a better possibility.

Instead of project-based CDM, sectoral trading mechanisms (STM) can strengthen demand for emission reduction through a cap on emissions per unit of output, e.g. energy intensity caps (Chatterjee 2011). This can also motivate India to import advanced low-carbon technologies which can lead to greater emission reductions and generate revenues. STM are more likely to support large-scale technology transfer while at the same time creating an incentive to import and adapt new technologies (Chatterjee 2011; Baron and Aasrud 2009).

V

Policy Options for Low-Carbon Growth

According to the experts, the low-carbon interventions in transport, power and housing sectors have large socio-economic benefits, more than any directly climate related benefits. They will also advance India's competitive advantage in the global economy due to large-scale technological leapfrogging. A brief account of these three sectors are provided below:

Transport Sector

The Indian transport sector requires technology upgrades in setting up high speed and high capacity public transport systems in cities and along high-density traffic corridors. In metropolitan cities, there has been rapid growth in private transport due to unreliable public transport leading to excessive concentration of fuel-inefficient vehicles, increasing congestion and pollution. Around 15 per cent of the total urban population lives in 3 megacities of India, but contributes to 50 per cent of the total emission from urban transport. The public transport services are below the demand and involves long commutation period due to declining operating speed. In designing new suburban areas, corridor identification for Metro/BRTS/Rapid Rail and inter-modal integration should be developed from the beginning. Further gains have been identified in the transport sector through development and optimization of urban mass transport, improved vehicle fuel efficiency standards, reduced reliance on private transport, and improved transport planning. Among the five criteria for evaluation of transport policy, the experts placed their highest emphasis on developing public transport infrastructure followed by integrated transport and land use planning (ITLP), freight transport management (FTM), emissions standards and fuel efficiency (ESFE) and finally, tax, incentives and pricing regulation (TIPR).

In India, the transport sector is facing four main problems: (i) degrading bus service and increasing risk to pedestrians and cyclists (ii) declining use of non-motorized transport (NMT) and public transport system (iii) increasing use of personal motorized vehicles, and (iv) increasing emission levels. Cars and Motorized-Two-Wheelers (MTW) contribute 65 to 80 per cent of the total transport emissions, whereas bus is the least CO_2 emitting mode of transport according to data collected from three Indian cities, Delhi, Pune and Patna. Delhi has highest emission levels because of more motorized vehicles travelling longer distances (Jain and Tiwari 2011).

The best policy option for reducing emission and energy intensity in transport sector as well as promoting inclusive growth has been assessed as better and affordable provision of public transport systems such as metro rails and metro buses. Public transportation can be provided at a far lower cost by using buses. Bus Rapid Transit Systems have become popular in many cities. Indian cities are dense and this will make public transport profitable as transportation fuel consumption per capita declines by one-half to two-thirds as urban densities rise from four to 12 persons per acre (Kenworthy and Laube 1999). Higher density in the cities and towns will reduce the allocation of required infrastructure per person, thereby lowering the cost of infrastructure construction as well as operation and maintenance.

One of the reasons for unregulated urban growth and sprawl in India is the lack of integration of land use and transportation planning. A developing city's land use and transportation planning decisions directly influence whether people and businesses will have mobility choices that allow them to save time, energy and money. The second best policy to achieve this goal is the better integration of transport planning and land-use zoning, scheduling and space coordination. A better transport corridor within the urban city and non-urban inter-city transport is also very important in promoting inclusive growth and low-carbon development.

According to Kahn et al. (2007), in transport sector there are five emission reduction options which have clear link with development and have positive social, environmental and economic effects. These are: use of carbon-efficient fuels such as biofuels, adoption of energy efficient technology, encouraging public transport, non-motorized transport and urban planning. Promotion of non-motorized transport (NMT) has the large and consistent co-benefits of GHG reduction, air quality improvements and overall people's health benefits (Mohan and Tiwari 1999).

Improving public transport system infrastructure, as well as safe and secure use of NMT is seen as two major interventions for addressing these problems (Jain and Tiwari 2011). NMT is the dominant mode of transport in most of the Indian cities – ranging from 30 to 60 per cent of total transport needs, however, it is not given due attention by the transport planners (Jain and Tiwari 2011). Developing proper infrastructure for NMT for pedestrians and bicyclists is very important, especially in newer cities which offer scope for developing bicycle lanes. NMT infrastructure along with bus infrastructure is required to reduce emissions from mega and large cities. In properly planned cities, the central business district (CBD) areas are either non-motorized or rely on only public transport; therefore by creating suitable infrastructure and incentives, congestion can be eliminated and mobility increased. In bigger cities, suburbanites are more likely to drive longer distances relative to central city residents, thus proper planning of residential and business districts are very important in limiting the need for longer transportation (Mohan 2011).

The economic functioning of big cities like Mumbai, Delhi, Bengaluru are hampered by the immobility of its roads. Congestion pricing could be one of the means to create better policies towards movement in congested urban core within cities. It needs to be coupled with other approaches including developing other means of public transportation in such areas. Congestion charges provide substantial benefits for GHG reduction (Cifuentes et al. 2001; Cifuentes and Jorquera 2002). After introduction of congestion charges in the city of London in 2003, the investment in developing alternative public transport increased, and CO_2 emission in such zones decreased by 20 per cent after one year of the scheme. It contributed to a 30 per cent decrease of the traffic volume by the chargeable vehicles and less congestion. It also led to higher speed of private vehicles by 20 per cent and buses by 7 per cent, and more walking and bicycling (Transport for London 2006). Similar congestion-based traffic zoning is also in operation in Singapore since 1975 with similar benefits.

Power Sector

India's power sector is largely dependent upon coal which constitutes about 53 per cent of the total generation capacity. There is substantial scope for increasing the Plant Load Factor (PLF) of thermal power plants through renovation and modernization. According to the CEA's performance review of thermal power plants in 2008, the PLF of thermal power plants increased from 63 per cent in 1995–96 to 77 per cent in 2006–07. Another area for improvement is plugging transmission and distribution losses, which came down merely from 34 per cent to 30.4 per cent between 2002 and 2006 (Prayas 2009).

Adoption of high carbon-efficiency coal power generation technology: As conventional coal-based power plants will continue to emit substantial amount of GHG emissions, new coal power plants based on super-critical technology could be an important way of achieving low-carbon emissions if technological advancements are adopted. India's future carbon intensity will depend heavily on technology shifts in the coal towards super-critical and ultra super-critical thermal power plants.

Large-scale grid-integrated solar power with utility level storage: The efficient integration of power sourced from renewable energy sources into transmission and distribution networks is of high priority in India. In ensuring high diffusion of solar power, grid interconnection of photovoltaic (PV) solar systems is very important.

Energy efficiency in industries: Significant energy efficiency gains are possible in both large energy-intensive and small and medium-sized industries (SMEs).

Nuclear power as source of low-carbon energy: According to India's Central Electricity Authority data, nuclear power is the fourth-largest source of electricity after thermal, hydroelectric and renewable sources of electricity (CEA 2012). As of 2011, India has 20 nuclear reactors in operation in six nuclear power plants, generating 4780 MW while seven other reactors are under construction and are expected to generate an additional 5300 MW.

According to Nuclear Power Corporation Limited (NPCIL) estimates, India has drawn up an ambitious plan to reach a nuclear power capacity of 20,000 MW by 2020 and 63,000 MW in 2032 by setting up of 16 new indigenous Pressurized Heavy Water Reactors (PHWR), including ten based on reprocessed uranium.[7] Out of the total target of 63,000 MW, about 40,000 MW is planned to be generated through Light Water Reactors (LWR) with international cooperation. Currently, about 3 per cent of electricity comes from nuclear power, with a targeted capacity enhancement of about 25 per cent of electricity from nuclear power by

2050. India has a vision of becoming a world leader in nuclear technology due to its expertise in fast reactors and thorium fuel cycle. Due to lack of indigenous uranium, India has uniquely been developing a nuclear fuel cycle to exploit its reserves of thorium.

Table 2

Opinions of experts for various policy options targeting emission reduction

Policy Choice	*Strongly Oppose*	*Somewhat Oppose*	*Neutral*	*Support*	*Strongly Support*
Policies that would put a **price on carbon**	5.26%	15.79%	26.32%	42.11%	10.53%
Require automobile companies to increase the **fuel efficiency** even if it costs customers more to buy	5.26%			52.63%	42.11%
Provide **tax rebate** for people who purchase carbon-efficient vehicles or solar panels			10.53%	42.11%	47.37%
Establish a **special fund** to help make buildings more energy-efficient by imposing surcharge to the household's electricity bill		15.79%	21.05%	26.32%	36.84%
Increase taxes on petrol/diesel and return revenues to develop public transport systems	5.26%	10.53%	10.53%	47.37%	26.32%
Regulate GHGs through stricter Air pollution control **Acts/Rules**		5.26%	10.53%	57.89%	26.32%
Require electric utilities to produce at least 20% of their electricity from **renewable** sources, even if it costs extra to the average household	0.00%	10.53%	26.32%	42.11%	21.05%
Fund more research into renewable energy sources	0.00%	0.00%	0.00%	47.37%	52.63%
Create a **new national market** that allows companies to buy and sell the right to emit the GHGs	0.00%	26.32%	21.05%	31.58%	21.05%
Build more **nuclear power** plants	15.79%	21.05%	31.58%	26.32%	5.26%

Some people remained neutral while others raised questions about nuclear energy as a clean and safe alternative to other fossil fuels and renewable energy sources. The experts felt that the new nuclear power projects and expansion plans have to face key challenges such as continued efforts in achieving enhanced safety and reliability standards in the existing and upcoming nuclear power projects as well as gaining the acceptability and confidence of the public. People living around the nuclear plant have strong reservations to its presence and have voiced their opposition to it. Other critical challenges are: management of spent fuel and radioactive waste, maintaining confidence in nuclear non-proliferation and achieving effective and sustainable use of resources for the long term.

Several people are protesting against the commissioning of new nuclear power plants and there is also a public interest litigation filed in the Supreme Court to stop all ongoing projects till all satisfactory safety standards are put in place and cost-benefit analyses are completed by independent agencies. Majority of the experts felt that nuclear energy has to be extremely safe, particularly in light of the March 2011 accident in Fukushima in Japan. There have been mass protests against the French-backed 9900 MW Jaitapur Nuclear Power Project in Maharashtra and the 2000 MW Koodankulam Nuclear Power Plant in Tamil Nadu. The state government of West Bengal state has also refused permission to a proposed 6000 MW facility near the town of Haripur that intended to instal six Russian reactors.

'Without convincing local people, often in remote, undeveloped areas, of the merits and benefits of nuclear power, we will not be able to progress.' Experts also felt that current expansion of nuclear power is being done in non-transparent manner. 'We never talk about nuclear waste, never talk about decommissioning. It does not tell us the full story.'

Housing Sector

The vast majority of people in India still lack standard housing, and this sector is growing at a rapid pace given the high rate of urbanization and demand for better living in urban areas. There is estimated shortage of affordable urban housing to the tune of 26.5 million units in 2012 which will further rise to 29 million units by 2017, most of which is for the low and middle income households (Hingorani and Tiwari 2012). With the addition of more housing units there would be huge upsurge in their energy and resources demand with consequent increase in emissions. The current energy and resource utilization pattern in the Indian housing sector is quite intensive. The residential sector alone accounts for 21 per cent of the total electricity consumption which is rising by 8 per cent annually (CWF 2010). According to the BEE, ceiling fans (34%), lighting (28%) and refrigeration (13%) constitute 75 per cent of the total energy use in the residential sector, where substantial scope for energy saving exists. There is a huge scope for increasing efficiency for household appliances such as refrigerators, room air conditioners, fans, and lighting which could range from 12 to 60 per cent depending upon the product class (McNeil et al. 2008; Prayas 2009). This would save electricity consumption by 4.7 per cent by 2020 (McNeil et al. 2008; Ghosh 2010). In terms of CO_2 abatement potential in housing sector, the estimates varies from 140 to 250 million tonnes of CO_2 per year by 2030, if mitigation opportunities are tapped (Indian Green Building Council estimates).

There is an urgent need to tap this opportunity for enhancing inclusive growth through social housing schemes as well as promote low-carbon initiatives. This can be done through several interventions such as actively promoting energy efficient appliances with incentives to adopt energy-saving designs and materials, phasing out old and polluting technologies, and enhancing customer's awareness on energy savings through lifestyle and behavioural changes. Currently, National Building Code (NBC), Energy Conservation Building Codes (ECBC), and Bureau of Energy Efficiency (BEE) rating, and energy star ratings are modifying the construction industry and purchasing behaviour of households towards energy-efficient design and devices, but there is much more scope to diffuse better technologies from both demand and supply side.

Table 3

Present and future energy consumption of various appliances (kWh per year)

Appliance	*Average Appliance Sold in 2005*	*Highly Efficient Appliance*	*Super Efficient Appliance*
Refrigerator	418	314	104
Fan	71	55	38
TV	175	134	79
Air conditioner	1370	1072	831
Lighting Bulb (Watts)	60	15	13
Lighting Tube (Watts)	51	40	32

Source: Prayas Energy Group, 2011.

The 'National Housing and Habitat Policy', should explicitly include new buildings rating system with low-carbon model of housing for making statutory support effective. Currently, this policy does not integrate energy efficiency and savings with the design of new units as well as in the provision of other related basic infrastructure. Consequently, the housing schemes that emerge from this policy are not in sync with energy and resources efficiency. There is also a need for stricter supply side interventions for promoting sustainable building materials. The buildings code should be supported by manufacturing of devices and building materials – design that increases the daylight and reduce the need for daytime lighting, use of natural ventilation, and wall and roof insulation.

Conclusion

The paper relies on expert opinion to draw common consensus on India's strategy for low-carbon growth and attempts to underpin possible barriers to adopt such a path. The opinion explains India's unique situation of addressing inclusive growth and climate-resilient path and reflects preference for policy design. In view of the large demand-supply gap in key infrastructure sectors, the experts believe that economic growth objectives will continue to take precedence over any climate change agenda.

It appears that the current low-carbon intensity of economy in part reflects energy and infrastructure poverty. Development remains a major challenge for India, a large chunk of population still remains to be provided with basic 'quality of living' and simple survival is still a concern for a majority of the population. India has outlined strategies for low-carbon growth according to its national circumstances; the various options have large development co-benefits and are intended to reduce internal inequities.

The pertinent question is whether low-carbon growth is different from customary economic growth? Or framing this question in the Indian context could mean that, can the usual high-growth path taken by the economy be decoupled from carbon and resource use intensity and environmental degradation? In the current background of growth strategy and carbon intensity, the answer could be in the negative, even though the energy intensity of the Indian economy has been relatively low compared to other countries, and it continues to go down which reflects a reasonably controlled growth.

India has to allocate more resources for social sectors. Given the current development divide, there may be a clear trade-off between customary growth and low-carbon growth. Shifting to low-carbon pathways will require substantially higher costs in the beginning as greater capital has to be allocated to new and energy-saving technologies that may be costlier than long-established conventional technologies. Since current investment in infrastructure and technology will decide the future emissions due to higher locked-in emission pathways, the priority of investment is very important. With targeted allocation emphasizing on more social sector outlays such as transport, housing and power (and prioritized investment within these sectors), the growth can be more inclusive and at the same time carbon-friendly.

As India has a huge backlog of infrastructure needs, it has also a unique opportunity to redefine conventional growth paradigm, by adopting a low-carbon and climate-resilient path. According to the expert's opinion, there is a clear link between low-carbon growth and long-term prosperity. Therefore, with right kind of policies, investment and incentives, India can advance inclusive growth priorities while significantly reducing GHGs. India is at an advantageous position with respect to placing its objective of low-carbon economy with inclusive growth – it has relatively low levels of carbon footprint and steadily declining energy intensity. However, high absolute emissions coupled with constrained carbon space will mean that it has to make a balance between using a *'fair'* share of carbon space and focusing on ensuring access to appropriate technology that would enable transition to a low carbon model of development that is inclusive. Given the current development challenge, policy makers have to be extremely cautious in arriving at a framework that takes into consideration inclusive growth and the long-term carbon emission reduction strategies. The wide range of strategies would vary from implementing sector-specific technology to large-scale capital and in many situations addressing behavioural systems – if a meaningful shift from *business-as-usual* is to take place.

India's transition to low carbon economy will present tremendous growth opportunities for renewable energy industries as current capacities are insufficient to meet the growing demand. The mandatory target of meeting 15 per cent renewable energy by 2020 will call for substantial investments in renewable energy technologies, thereby creating plethora of opportunities along the entire manufacturing value chain. However, the current institutional structure, inadequate availability of resources, inefficient subsidy systems, limited opportunities for global partnership, and overreliance of regulatory approaches for reducing carbon emissions may limit the success of some ambitious policy measures. There is huge gap between the scale of current international finance mechanisms, technology transfer and deployment and the scale of transformative change needed to shift India's emission pathways towards cleaner options. India is aggressively required to implement NAPCC targets, in particular, public transportation in smaller and medium sized cities with scope for improved non-motorized mobility, off-grid low-carbon electricity with driving down costs of the solar technology with local manufacturing, and energy efficiency technology for deployment in both housing and S&M scale industries.

Notes

1. NAPCC targets 15 per cent renewable energy for India by 2020. However, a more aggressive effort is required to limit the GHG emissions and meet up energy shortages.
2. According to the KPMG's Solar Grid Parity Model, in India, grid parity is likely to happen in 2019–20 when the levelized tariffs from solar power would be comparable with the levelized tariffs of grid power.
3. There is significant disparity between and within Indian states with 65 per cent of India's poor located in just eight states.
4. In a recent study using India as case study, it is observed that better institutions help in better utilization of resources and improve productivity thus lowering the impacts of climate change and cost of abatement (Kumar 2011).

5. Some researchers argue that state action plans are a repackaging of existing policies under a climate banner (Dubash 2011). For example, in several state action plans, the existing schemes on agriculture and forests are packaged as climate adaptation measures, with no complementary effort to scrutinize and rethink existing schemes that may add to climate vulnerability.
6. CDM gave small negotiating returns, to provide relatively cheap short-term GHG offsetting options, and to kick-start a broad-based emissions trading system. Several researchers claim that CDM finance has led to increasing diffusion rates of technologies, reducing the payback period and improving the internal rate of return (IRR) of clean technology projects in developing countries (Ang 2009; Hansen 2008). However, some researchers think that CDM projects benefit short-term abatements and fail to incentivize innovation and structural changes within the sector (Pueyoa et al. 2012; De Sepibus 2009; van der Gaast et al. 2009). According to Dasgupta and Taneja (2011) CDM's role in transferring superior technology from developed countries to India has been limited.
7. The 12th Five-Year Plan for 2012–17 is targeting the addition of 100 GW over the period, three quarters of which would be coal – or lignite-fired, and only 3.4 GW nuclear, including two imported 1000 MW units at one site and two indigenous 700 MW units at another. By 2032, total installed capacity of 700 GW is planned to meet 7 to 9 per cent GDP growth, and this is to include 63 GW nuclear.

References

Ang, J.B. 2009. 'CO_2 Emissions, Research and Technology Transfer in China', *Ecological Economics*, 68 (10): 2658–2665.

Baron, R. and A. Aasrud. 2009. *Charting a Path to Sectoral Mechanism*, Paris: OECD and IEA.

BEE (Bureau of Energy Efficiency). 2008. *Accredited Energy Service Companies (ESCOs)*, New Delhi: Government of India, Ministry of Power, http://www.bee-dsm.in/docs%5caccreditedwscos.pdf

BERR. 2009. 'Low Carbon and Environmental Goods and Services: An Industry Analysis', Department for Business, Enterprise and Regulatory Reform, UK and Innovas, a research consultancy.

Biesbroek, R., J. Klostermann, C. Termeer and P. Kabat. 2011. 'Barriers to Climate Change Adaptation in the Netherlands', *Climate Law*, 2 (2): 181–199.

CEA. 2012. *Highlights of the Power Sector*, Delhi: Central Electricity Authority.

Chatterjee, B. 2011. 'Technology Transfer through the Clean Development Mechanism (CDM)', Briefing Paper, AEA Technology, ED 56638, Issue Number 1, http://ec.europaipv4.eu/clima/policies/ets/linking/docs/technology_transfer_en.pdf, accessed on 25 May 2012.

Cifuentes, L. and H. Jorquera. 2002. 'IES Developments in Chile', October 2002, http://www.epa.gov/ies/documents/chile/cifuentespres.pdf, accessed on 30 May 2007.

Cifuentes, L., H. Jorquera, E. Sauma and F. Soto. 2001. *International CO-Controls Analysis Program*, Final Report, The Catholic University of Chile, December, 65.

CWF. 2010. Reducing GHG Emissions in the Building Sector in India: A Strategy Paper, Climate Works Foundation, India.

Dasgupta, P. and N. Taneja. 2011. 'Low Carbon Growth: An Indian Perspective on Sustainability and Technology Transfer', *Problems of Sustainable Development*, 6 (1): 65–74.

De Sepibus, J. 2009. 'Reforming the Clean Development Mechanism to Accelerate Technology Transfer', NCCR Trade Working Paper No. 2009/42, NCCR Trade Regulation, Swiss National Centre of Competence in Research,Switzerland,http://phase1.nccrtrade.org/images/stories/publications/ip6/cdm%20and%20technology%20transfer%2020%20october%20finalfinal.pdf, accessed on 25 May 2012.

Dechezleprêtre, A., M. Glachant and Y. Ménière. 2009. 'Technology Transfer by CDM Projects: A Comparison of Brazil, China, India and Mexico', *Energy Policy*, 37: 703–711.

Dechezleprêtre, A., M. Glachant, I. Hascic, N. Johnstone and Y. Ménière. 2008. 'Invention and Transfer of Climate Change Mitigation Technologies on a Global Scale: A Study Drawing on Patent Data', CERNA Research Programme on Technology Transfer and Climate Change, Paris, France, December.

Doriaa, Miguel de Franca, E. Boydb, L.T. Emma and W.N. Adgerd. 2009. 'Using Expert Elicitation to Define Successful Adaptation to Climate Change', *Environmental Science and Policy*, 12 (7): 810–819.

Dubash, N.K. 2011. *Handbook of Climate Change and India: Development, Politics and Governance*, New Delhi: Oxford University Press, 400.

Ghosh, S. 2010. 'Examining Carbon Emissions Economic Growth Nexus for India: A Multivariate Cointegration Approach', *Energy Policy*, 38 (6): 3008–3014.

Gupta, P. and S. Kumar. 2011. 'Tracking the Anthropogenic Drivers of CO_2 Emissions Growth in India: A State-level Analysis', http://cseindia.org/userfiles/parul_gupta.pdf, accessed on 15 June 2012.

Hansen, U. 2008. 'Technology and Knowledge Transfer from Annex I Countries to Non Annex I Countries under the Kyoto Protocol's Clean Development Mechanism (CDM): An Empirical Case Study of CDM Projects Implemented in Malaysia', Master's Thesis, University of Copenhagen, Working Paper 5, CD4CDM.

Hingorani, P. and P. Tiwari. 2012. 'Housing and Basic Infrastructure Services for All: A Conceptual Framework for Urban India', paper presented at the 13th GDN Annual Conference, Budapest, 16–18 June, http://depot.gdnet.org/newkb/submissions/pritika_hingorani.pdf, accessed on 17 June 2012.

IEA. 2008. *World Energy Outlook*, France; International Energy Agency.

——. 2009. *World Energy Outlook*, France: International Energy Agency.

——. 2010. *Energy Technology Perspectives 2010: Scenarios and Strategies to 2050*, France: International Energy Agency.

IMF. 2012. *World Economic Outlook (WEO): Growth Resuming, Dangers Remain, World Economic and Financial Surveys*, US: International Monetary Fund, April.

Integrated Energy Policy. 2006. Planning Commission, Government of India, August 2006.

Jain, D. and G. Tiwari. 2011. 'Impact of Improving Public Transport and NMT Facilities on CO_2 Emissions in Indian Cities', 2nd National Research Conference on Climate Change, TRIPP/CED, Indian Institute of Technology, Delhi, 5–6 November.

Kahn, R.S., S. Kobayashi, M. Beuthe, J. Gasca, D. Greene, D.S. Lee, Y. Muromachi, P.J. Newton, S. Plotkin, D. Sperling, R. Wit and P.J. Zhou. 2007. 'Transport and its Infrastructure', in B. Metz, O.R. Davidson, P.R. Bosch, R. Dave and L.A. Meyer (eds), *Climate Change 2007: Mitigation*, Contribution of Working Group III to the Fourth Assessment Report of the Intergovernmental Panel on Climate Change, Cambridge University Press, Cambridge, UK and New York, USA.

Kapur, S. 2012. 'Building a Green Economy', *The Financial Express*, 5 June.

Kenworthy, J. and F. Laube. 1999. *Patterns of Automobile Dependence in Cities: An International Overview of Key Physical and Economic Dimensions with Some Implications for Urban Policy*, Institute for Science and Technology Policy, Mudoch University, Perth, Australia.

Kumar, S. 2011. 'Carbon Sensitive Productivity Growth, Climate Change and Institutions', paper presented at the Second National Research Conference on Climate Change, IIT, Delhi, 5–6 November, http://cseindia.org/userfiles/surender_kumar.pdf, accessed on 7 June 2012.

Markantonis, V. and K. Bithas. 2010. 'The Application of the Contingent Valuation Method in Estimating the Climate Change Mitigation and Adaptation Policies in Greece: An Expert-based Approach', *Environment, Development and Sustainability*, 12 (5): 807–824.

Mathew, S., A. Henderson-Sellers and R. Taplin. 2011. 'Prioritizing Climatic Change Adaptation Investment at Local Government Levels', Macquarie University Research Online.

McDaniels, T., T. Mills, R. Gregory and D. Ohlson. 2012. 'Using Expert Judgments to Explore Robust Alternatives for Forest Management under Climate Change', *Risk Analysis*, 32 (12), December: 2098–2112.

McKinsey. 2009. *Environmental and Energy Sustainability: An Approach for India*, Mumbai, India.

McNeil, M.A., M. Iyer, S. Meyers, V.E. Letschert and J.E. McMahon. 2008. 'Potential Benefits from Improved Energy Efficiency of Key Electrical Products: The Case of India', *Energy Policy*, 36 (9): 3467–3476.

Michaelowaa, K. and A. Michaelowaa. 2012. 'India as an Emerging Power in International Climate Negotiations', *Climate Policy*, 12 (5): 575–590.

Ministry of Power. 2012. 'Power Sector at a Glance', Government of India, http://www.powermin.nic.in/jsp_servlets/internal.jsp, accessed on 10 June 2012.

MoEF. 2004. India's Initial National Communication to the UNFCC, Ministry of Environment and Forests, Government of India.

——. 2009. India's GHG Emissions Profile: Results of Five Climate Modelling Studies, Climate Modelling Forum, India, for the Ministry of Environment and Forests, Government of India.

——. 2010. India's Greenhouse Gas Emissions 2007, Indian Network of Climate Change Assessment (INCCA), Ministry of Environment and Forests, Government of India.

——. 2011. 'Clean Technology Fund Investment Plan for India', Ministry of Environment and Forests, Government of India, http://moef.nic.in/downloads/public-information/ip-ctf-2011.pdf, accessed on 28 May 2012.

——. 2012. Second National Communication to the United Nations Framework Convention on Climate Change, Ministry of Environment and Forests, Government of India.

Mohan, D. and D. Tiwari. 1999. 'Sustainable Transport Systems, Linkages between Environmental Issues: Public Transport, Non-motorised Transport and Safety', *Economic and Political Weekly*, XXXIV (25), 19 June: 1580–1596.

Mohan, D. 2011. 'Speed, Distance and Social Space: What Do We Make of Our Cities?', 2nd National Research Conference on Climate Change, IIT, Delhi, 5–6 November.

Morgan, M.G. and D.W. Keith. 1995. 'Subjective Judgments by Climate Experts', *Environ. Sci. Technol.*, 29 (10): 468–476.

National Action Plan on Climate Change (NAPCC). 2008. Government of India.

Parikh, K. 2012. 'Sustainable Development and Low Carbon Growth Strategy for India', *Energy*, 40 (2012): 31–38.

Parikh, J. and K. Parikh. 2011. 'India's Energy Needs and Low Carbon Options', *Energy*, 36 (2011): 3650–3658.

Patwardhan, A. 2011. 'Co-Benefits Approaches to Climate Change Mitigation and Adaptation', presentation at the side-event, UNFCCC COP 17/CMP 7 at Durban, 1 December.

Planning Commission. 2011. *Low Carbon Strategies for Inclusive Growth: An Interim Report*, Government of India.

Prayas. 2009. *An Overview of Indian Energy Trends: Low Carbon Growth and Development Challenges*, Pune: Prayas Energy Group.

Prayas Energy Group. 2011. Potential Savings from Selected Super-Efficient Electric Appliances in India: A Discussion Paper, June 2011.

Pueyoa, A., M. Mendiluceb, M.S. Naranjoa and J. Lumbrerasa. 2012. 'How to Increase Technology Transfers to Developing Countries: A Synthesis of the Evidence', *Climate Policy*, 12 (3): 320–340.

Sankar, U. 2010. 'India's Low Carbon Inclusive Growth Strategy', presented at SANDEE Conference on Environment and Development, Kathmandu, Nepal, 6–7 December.

Seres, S., E. Haites and K. Murphy. 2009. 'Analysis of Technology Transfer in CDM Projects: An Update', *Energy Policy*, 37: 4919–4926.

TERI. 2008. *Mitigation Options for India: The Role of the International Community*, New Delhi: The Energy and Resources Institute.

——. 2009. *Reducing Greenhouse Gas Emissions in India: Financial Mechanisms and Opportunities for EU-India Collaboration*, Report for the Swedish Ministry of Environment, October.

The Economist. 2012. 'Solar Power in India: Waiting for the Sun – Is the Sun the Answer to India's Energy Problems?, April.

Transport for London. 2006. *Central London Congestion Charging: Fourth Annual Report*, 215.

UNFCCC. 2010. 'The Contribution of the Clean Development Mechanism under the Kyoto Protocol to Technology Transfer', http://cdm.unfccc.int/about/cdm_tt/index.html

van der Gaast, W., K. Begg and A. Flamos. 2009. 'Promoting Sustainable Energy Technology Transfers to Developing Countries through the CDM', *Applied Energy*, 86: 230–236.

WISE. 2011. World Institute of Sustainable Energy, Pune.

World Bank. 2009. *Low Carbon Growth Country Studies – Getting Started: Experience from Six Countries*, Energy Sector Management Assistance Program, Washington, D.C.

——. 2012. *An Update to the World Bank's Estimates of Consumption Poverty in the Developing World*, Washington, D.C.

WRI CAIT. 2011. Online Database of Emissions, World Resources Institute Climate Analysis Indicators Tool, http://cait.wri.org

WRI. 2012. 'Building an Inclusive Green Economy for All', A Poverty-Environment Partnership Joint Agency Paper, World Resources Institute, Washington, D.C., June.

8

Developmentality, Multimedia Technologies and Growing Ecological Concerns in Global Era

A Critical Appraisal from Rural India

Ashish Saxena and Vijaylaxmi Saxena

It is true that technological innovation is essential for human progress. People all over the world have high hopes that the new technologies will lead to better quality of life in terms of greater social freedom, increased knowledge and livelihoods that are more productive. Today's technological transformations are thus intertwined with globalization and creating a new paradigm: the network age. These transformations expand opportunities and increase the social and economic rewards of creating and using technology. This deserves special attention because digital breakthroughs are pushing forward the frontiers of how people can use technology to eradicate poverty. These breakthroughs are creating new possibilities for improving health and nutrition, expanding knowledge, stimulating economic growth and empowering people to participate in their communities. However, there is, at the same time, a great fear of the unknown. Many people fear that these technologies may be of little use to the developing world – or that they might actually widen the already savage inequality between North and South, rich and poor. Without innovative public policy, these technologies could become a source of exclusion, not a tool of progress. Within this framework, the ultimate significance of the network age is that it can empower people by enabling them to use and contribute to the world's collective knowledge. Moreover, the great challenge of the new century is to ensure that the entire human race is so empowered – not just a lucky few.

Since, the prospect for a reduction in inequality within orthodox capitalist development would appear to be quiet weak, so contemporarily alternative models of development have tended to opt for the antithesis of the orthodox approach. Thus, an urban and industrial bias is replaced with an emphasis on the rural and the agricultural; the top-down directionality and centralized character of development policy is challenged by decentralized, devolved and bottom-up initiatives; small-scale and particularistic development is seen as preferable to large-scale and universalistic approaches; and so on. Initial efforts of 'growth centre strategies' were the successful case of the same. Hence, it was argued that a self-reliant form of development may ensue which could deliver basic needs, social equality and a solid foundation for future growth and progress. The strategy is centred around seemingly contradictory concepts such as 'rural urbanization' and 'rural industrialization' with absolute priority being given to agriculture and small-scale artisanal activities, rural settlement, labour intensive activities and appropriate technology, local democracy and popular participation in the decision-making process. In corollary to this, the notion of 'bottom-up development' goes a stage further by fundamentally challenging the directionality of development decision-making. It argues that, what is required is another antithetical development, which is localized, and contextually rooted, small in scale, flexible, culturally sensitive, democratic and participatory, and which centres on the empowerment of the poor (Michael 2000: 115; Saxena and Rao 2005).

Putting the discussion in the framework of developing country like India, we find that transforming rural India is a challenge that should focus the best of Indian minds – it is perhaps the single biggest barrier to making India a developed country. Initially, the government efforts for rural development could not make a revolutionary change in rural India therefore the Indian villages need disruptive innovations to make the giant leap forward. Observing the trajectory of independent India, one finds that it has sought rapid economic growth through 'planned development'. This has entailed large-scale public investments involving land acquisition. Unfortunately, large numbers of people, especially marginalized sections, have been displaced from their original habitats to make way for these development projects. Such projects have also permanently changed the patterns of land use and natural resources that previously prevailed in these areas. Often, such acquisition of land leads to displacement of people, depriving them of their livelihood and shelter, restricting access to their traditional resource base, and uprooting them from their socio-cultural environment. These have traumatic, psychological and socio-cultural consequences for the affected population, especially 'subalterns'. This thus results into double marginalization, i.e. dislocation of indigenous people from their native inhabitants and also poor compensation in the form of resettlement and rehabilitation. Generally, they become passive owing to their underdog position; however, in many instances they show resistance and adopt violent protests. Glaring exemplifications of contemporary resistances are the

the 1950s, the vision of development articulated by the Indian government led by Nehru and the Congress Party, was not only socialist and nationalist but also technocratic and paternalistic. Sunil Khilnani, for example, claims that Nehru believed that only a national state, centrally responsible for directing economic development, could safeguard India's future progress and independence. There were also massive public investments in capital-intensive rural projects, such as the 'Green Revolution' in the Punjab. Since land reforms have been largely ineffective because of the influence of the landed interests in the coalition of ruling classes, the entire planning process has been an exercise in trying to promote industrialization without radical agricultural transformation (Ghosh 2001).

A technocratic strategy dubbed the 'green revolution' was introduced in the Indian countryside in the mid-1960s. Since then, its impact on the political economy of rural India has been the subject of enduring scholarly debate. Some analysts have claimed that the effects of the 'green revolution' have been positive in terms of promoting agricultural development. Others have argued the opposite, claiming that its spread has significantly undermined long-term agricultural sustainability, exacerbated rural poverty and hunger (despite growing national food-stocks), and intensified class and caste conflict in the countryside (Bardhan 1985; Lipton 1989; Shiva 1989).

A survey of the literature reveals that the technocratic strategy was 'top-down', a clear case of 'development from above'. For example, according to Frankel (1978), their orientation of the agricultural strategy represented a monumental 'attack' on the principles of Nehruvian socialism and a 'retreat from the social goals of planning. Hence, although much of the 'green revolution' technology was relatively scale-neutral in its application, its availability was (and is) generally restricted by inequitable distribution of land, income, and credit. In fact, the profitability of commercial agriculture has released an entrepreneurial dynamism among the wealthy landowners.

Emphasis on industry developed into a coherent ideological articulation of the post-colonial nation-state by the Second Five-Year Plan. By this time, Nehru was committed solely to developing a heavy-industry sector. The basic assumptions of the plan demonstrated ideological consistency with the narrative of modernization which held that 'backward' nations like India occupied a lower but clearly recognizable place on the ladder to modernity, a modernity that was represented by industrialized societies. Thus, Nehru could put almost the entirety of national savings into industrialization and still maintain without contradiction that the 'keystone' of development was agriculture: 'It is from agriculture and from the increasing production on land that we can build up our surpluses for future growth'.

II

Green Revolution to LPG – Arian Shifts in Rural India

Green Revolution technologies placed in the hands of the state reproduced the vision of the state as social engineer, but in a manner that placed the countryside at the centre of that engineering. In this context, the development of HYVs within national laboratories was hit upon as a solution out of dependency. Yet high-yield varieties (HYVs) were efficient only in areas already well endowed with water, and with low person-to-land ratios that would lend itself to intensive, widespread commercial farming. This exacerbated already extant inequalities between rural areas as well as inequalities within the farming population: Those farmers with resources that were capable of taking advantage of widespread chemical farming were placed in an advantageous position from the outset. Consequently, at the same time that the Green Revolution underscored the role of agriculture within nationalist ideology, it also produced a system of large-scale commercial farming that gave rise to a class of surplus-producing farmers.

This history of the process by which large-scale, state-sponsored agricultural production rose to occupy a central place in the making of the modern Indian nation-state suggests that sovereignty, the driving political imperative of the post-colonial nation, is inevitably linked with the issue of food security.

Mies and Shiva (1993) suggests that the militant cultures of nationalism that have emerged in the last fifteen years as essential moments of public identification in India are a reaction to the homogenizing processes of global capital, on the one hand, and to a perceived decline in the power of the Indian state on the other. Conversely, the ethnicized territory of peasant nationalism is – unambivalently – a site of collective resistance against the 'West's paradigm of development' and the homogenizing cultural forces of a capitalism represented by a few multinationals and the superpowers that support them.

Thus, the social dislocations induced by development (as when populations are moved to make way fŏr dams, roadways, plants, and so on) are construed as 'uprootings' that displace 'the rhythms and patterns of the universe.' Mies and Shiva (1993) are right to draw attention the dislocations of local communities. Indeed, it is precisely such dislocations that need to be highlighted as sites of injustice and as sites of resistance against national and transnational governmentalitiy. For her, development is associated with the death of the 'sacred'. They write, 'Development has meant the ecological and cultural rupture of bonds with nature, and within society, it has meant the transformation of organic communities into groups of uprooted and alienated individuals searching for abstract identities.' They make a similar point a bit later: 'In effect, the process of development leads to turning away from the soil as a source of meaning and survival, and turning to the state and its resources for both. The destruction of organic links with the soil also leads to the destruction of organic links within society.'

Movements of landless, unorganized labour in rural and urban areas, Adivasis, Dalits, displaced people, peasants, urban poor, small entrepreneurs and unemployed youth took up the issues of livelihood, opportunities, dignity and development. Moreover, the process of development itself resulted in large-scale displacement, destitution, centralization and destruction of resources. As rightly pointed out by Sangvai (2007: 112) 'These processes and their ideology have intrinsically been against both the 'modernist' baggage of development and at the same time also against post-modernist tendencies to postulate each struggle as fragmented from another and the predominance of the 'politics of the local'. The struggles are different, issue based and localized, yet their context and purpose is overarching and interrelated.

With the 'New Economic Policy' and concomitant liberalization, privatization and globalization (LPG) policies, various existing organizations and new forums like Azadi Bachao Andolan have launched a campaign against the onslaught of national and multinational capital. The united struggle by local communities, trade unions and various organizations under the National Alliance of People's Movements (NAPM) against the infamous Enron Company in Konkan and more recently against special economic zones (SEZ) are examples of major struggles against LPG. The struggle against the World Trade Organization (WTO) and free trade by thousands of peasants in Karnataka under the banner of the Karnataka Rajya Raitha Sangha (KRRS), or against the World Bank-sponsored forestry project by Madhya Pradesh organizations are some more illustrative examples. The struggles against Enron in the mid-1990s and against the Coca-Cola Company in early 2000 also brought in sharp focus the collusion between the powerful political elite in India with MNCs (Sangvai 2007).

In addition to the elitist development, globalization of the Indian economy has dramatically influenced social life in the country and the countryside, especially the so-called marginals (SCs/STs and OBCs) of rural India. The state has mounted policies of economic liberalization, claiming that the policies will ultimately benefit the entire population but a number of social scientists have argued that marginalized groups will be adversely affected. In particular, they have emphasized problems such as uneven regional development, wages disparities and increased level of inequality. Even the large sectors of the 'backward classes and castes remain extremely hard-hit by inflation, contradiction in secure rural employment, the virtual impossibility of entry into lucrative export markets, continual poor levels of literacy, health and living standards, and overall social and political marginality. It is realized that in the present era of privatization, the State's indirect withdrawal from 'welfare' of service sector people, particularly Dalits, became vulnerable. Further the shift of decision-making power into the hands of trans-national institutions like the World Trade Organization, International Monetary Fund, World Banks and others has severely reduced the sovereignty of national government and resulted in a very serious drift. In this regard we find that the Dalits are at the lowest end of the economic and social ladder and are quite obviously at the receiving end of the elitist culture promoted by urban world and by globalization.

For Mies and Shiva (1993), GATT and its enforcement through the WTO establish a neo-colonial platform through which the capitalist dominates nature. In their narrative, the new agents of agriculture – agribusinesses – have a very different relationship to the earth than the romanticized indigenous farmer. Rather than sustainably and harmoniously cultivating the soil, agribusiness is seen as a predator that tends to exploit and exhaust it: on Shiva's account, in addition to affecting the freedom of small farmers to steward the land, free trade's preference for corporate agriculture degrades the environment. They write, 'Local control over natural resources is an essential precondition for farmers' freedom. But free trade which, as we have seen, implies a relocation of control over natural resources for farmers and Third World governments to global institutions has serious environmental consequences.'

As rightly pointed out by Chand et al. (2011: 26–27) between 1970–71 and 2005–06, the total number of operational holdings in India increased from 71.01 million to 128.89 million and operational holdings area declined from 162.18 million ha to 156.62 million ha. This resulted in the reduction of average farm size from 2.28 ha to 1.21 ha. In the same period, the share of small and marginal holdings in operated area doubled. Smallholders now cultivate 42 per cent of operated land and constitute 83 per cent of total landholdings.

To substantiate further, some social scientists have considered the implications of economic liberalization and forecast that marginalized groups will be adversely affected (Acharya 1995). In particular, they have argued that economic liberalization and marketization will reinforce asymmetrical gender relations. Yet, these are predictions, based on experiences of the poor elsewhere in the developing world, rather than actual case studies in India. The vulnerability of labour has increased considerably under globalization because the volatility of the global market does not ensure stable jobs to workers, and increased competition in the domestic labour market results in closures, down-sizing and restructuring of units. As the International Labour Organization (ILO) (2004) has pointed out, at present society in developing countries is 'a risk society' because most groups, particularly those at the bottom, are exposed to uncertainty and risks that emanate from new policy regimes. Their insecurity is multidimensional – related to employment, income, work, job, and skills.

In short, sociological discussions of economic liberalization in India focus on class polarization in general terms, rather than on empirical evidence. For example, it is argued that: '... market forces will polarize classes in both agricultural and industrial sectors. The growing global demand for vegetables, fruits and flowers as Indian agriculture links up with international agri-business and the consequent shrinkage of the area under food crops is inimical to the interests of the poor who will be made to pay higher prices for their daily food requirements. The poor will become poorer, whereas the rich farmers will attain prosperity because of their linkages to the global market' (Panini 1995: 54).

III

Growing Agrarian Concerns and Multiple Crises in Rural India

No doubt, the green revolution technologies have helped in the reduction of poverty but have resulted in widespread regional disparities as dry lands are not covered. It took a long time for the technology to spread to these areas and even then rainfed farmers did not benefit to the same extent as those in well-endowed regions. This is because of the exclusive emphasis on increasing production without keeping the objective of equity in view.

Sociological discussions of economic liberalization in India focus on class polarization in general terms. For example, it is argued that: ... market forces will polarize classes in both agricultural and industrial sectors. The growing global demand for vegetables, fruits and flowers as Indian agriculture links up with international agri-business and the consequent shrinkage of the area under food crops is inimical to the interests of the poor who will be made to pay higher prices for their daily food requirements. The poor will become poorer, whereas the rich farmers will attain prosperity because of their linkages to the global market.

Shiva (1998) too rightly said that 'the present process of globalization threatens to undermine aspects of agricultural policy. The transnational corporations (TNC) are not guided by any philosophy of welfare of the marginalized. Liberalization emphasizes competition, efficiency and providing commodity for the market. Such an approach goes against the prevalent practice of poor farmers who produce for their own use and their methods are based on traditional wisdom.

Talking about agrarian economy, we find that the agricultural growth during 1990–93 to 2003–06 reflects the impact of economic reforms on agriculture performance. The most important feature of this period is that agricultural growth decelerated sharply at the all-India level and in all regions. At the all-India level, the output growth decelerated to 1.74 per cent pa during 1990–93 to 2003–06 compared with a growth rate of 3.37 per cent per annum (pa) during 1980–83 to 1990–93. At the regional level, during the same period, the growth rate of agricultural output decelerated highest from 3.55 per cent to 1.58 per cent pa in the north-western region … . The main reason for the deceleration of growth during the post-reform period was a visible deceleration in investment in irrigation and other rural infrastructure. In India, there are competing demands on area available for cultivation from increase in rural habitation, forestation, urbanization and industrialization. Consequently, net sown area in the country has registered a rapid deceleration in its growth over time. During 1962–65 to 1980–83, net area sown rose at a rate of 0.15 per cent pa at the all-India level – its growth rate decelerated to 0.11 per cent pa during 1980–83 to 1990–93 and further to – 0.05 per cent pa during 1990–93 to 2003–06. Even, the economic reforms initiated during the early 1990s were expected to hasten the process of crop diversification from low value food grains to high value non-food grain crops (Bhalla and Singh 2009; Natarajan 1999).

Economic globalization in India has had a severe ecological impact, with adverse effects on several hundred million people who depend directly on nature for their subsistence and livelihood. Liberalization of trade (exports and imports) has had two consequences: rapid increase in exploitation of natural resources to earn foreign exchange, and a massive inflow of consumer goods and waste into India. This has created serious disposal and health problems, and impacted traditional livelihood in forestry, fisheries, pastoralism, agriculture, health, and handicrafts. Environmental standards and regulations have been relaxed, or allowed to be ignored, in the bid to make the investment climate 'friendlier' to both domestic and foreign corporations. Governments are sacrificing more and more natural habitats and prime food-growing land to make way for commercial enterprises. The need for infrastructure or raw materials or commercial energy is determined not by the imperatives of human welfare and equity, but by economic growth rate targets, even where the growth rates may have no necessary correlation with human welfare. The last couple of decades have, therefore, seen a massive increase in new infrastructure creation (highways, airports, urban infrastructure and power stations). This has meant increasing diversion of land, mostly natural ecosystem like forests or farms and pastures.

As an alarming situation, it is rightly pointed out by V. Shiva (2010: 23) – Two aspects of the food and agrarian crisis in India are the epidemic of farmer's suicide and increase in hunger. Per capita calorie intake has fallen from 1830 Kcal in 1989 to 1600 Kcal in 1998. In 1999–2000 almost 77 per cent of the rural population consumed less than the poverty line calorie requirement of 2400 calories.

Farmer's suicide and increase in malnutrition are related processes – both are the result of ten years of rapid, unthinking trade liberalization. Globalization has led to:

1. Increase in cost of production.
2. Falling prices of farm produce.
3. Rising costs of food as a result of the dismantling of the PDS system and weakening of the Essential Commodities Act.

Main impact of liberalization of industrial policy has been a shift from developmentalism based on the strategy of growth with justice towards an elite-oriented model of growth. The green revolution model, and now the globalization model of agriculture are based on the destruction of diversity, uprooting of small peasants through indebtedness, chemical intensification, capital intensification, centralization and long distance transport. They are based on intensification of external inputs, and integration with global markets.

A very large section of India's population is going through severe and multiple crises: food insecurity, water shortages, inadequate fuel availability, and dislocation of livelihoods with limited alternative options. Finally, there is the crisis of livelihoods. As ecosystem disruption and land/water degradation intensifies, or as access to natural resources and traditional consumer declines, communities who have been traditionally self-employed (as farmers, pastoralist, fishermen, hunter-gatherers) are increasingly impacted.

The debate on agricultural biotechnology is focused mainly on the environmental impact, bio-safety issues, and intellectual property rights. The State will have to play a proactive role in priority setting for pro-poor agricultural development and accessing tools and technologies from the private sector by forging creative public-private partnerships to focus on crops and problems of importance to small farmers. It is to be stressed that ecologically, patent related to biological resources have major implications for the conservation of biodiversity and its sustainable use. By creating 'property' in life through patents, the economically powerful corporations can become the new 'life-lords', just as earlier we had landlords or *zamindars*.

IV

Multimedia Aspects in Global Era and its Efficacy in India's Rural Development

Apart from the above pessimistic interpretation of the globalization on rural India, we should also observe the strengths of globalization and multimedia in transforming agrarian India. We find that global aspects characterize the post-traditional society, by new forms of interdependence. Individuals are linked in local societies to change at the global level and actions are done without personal contacts via the global net of mass media, phone, computer etc. The structure of local communities is thus constructed no longer at the local places, but is constructed organization far away. Thus, today the community networking is an idea that has caught the global imagination. Communities worldwide are creating tele-village, tele-cottages, tele-service centres, community technology skills centres and more. Community networking in the past has involved face-to-face meetings, chance encounters, and various technologies that aided our sharing the newspaper, radio, TV and the telephone. Today, additional means of sharing, in the form of modern telecommunications technology, help us continue to build our communities in even more powerful ways. One of the Castells' (1996) core thesis is that networks, which have shared information at their core, and which result in new social and economic relations, increasingly dominate the world. We thus find that multimedia can play a significant role in managing the local information. It can make it more effective, speedy and extensive.

It is said that technology innovation, technology diffusion and technology acceptance are the pillars of technical promotion and enhancement. In the Indian context the technologies do not reach every villager, not every villager can access technologies, and hence there is limited flow of information among the rural community. To make information technologies and electronic communication network more accessible to rural population, to collect and disseminate information beyond national boundaries, more emphasis should be given in the development of information technologies, especially in the remote rural areas. The main rationale for establishing multimedia tele-centre is to enable rural people to have information access for rural development. It includes determining information needs of different communities and designing systems for linking the groups to relevant information and training resources. The approach taken to achieve this broad objective is a modular concept using Multi-media to develop and disseminate Multi-disciplinary information and knowledge from Multiple-sources to Multiple-users with built in user needs assessment and feedback mechanisms. This modular approach may support an evolving pluralistic knowledge-based rural extension system consisting of the following elements:

1. Accelerates the transition process – recognizing the importance of diverse information and knowledge user needs.
2. Increasing importance of non-farm activities to supplement farm incomes especially for the emerging small private farmer.
3. Evolution of diversified service providers consisting of contractual arrangements, public-private sector partnerships involving NGOs, producer organizations, association of private farmers, water user associations etc.

4. Creation of a decentralized and localized extension programme management and delivery where all forms of media – from traditional to modern will play a role.
5. Creation of strong linkages between education, research, and various forms of farming systems and works in coordination with other providers of rural information and knowledge – such as health, education, micro and small enterprise credit, environment, eco-tourism etc.

In India, a number of innovative experiments already under way indicate that achieving global digital access may not be as difficult as many think. More than fifty grassroot projects in India are using modern Information and Communications Technologies (ICTs) for the benefit of urban and rural citizen consumers. To justify the utility of multimedia channels in rural development, following initiatives are worth quoting; some of the database applications developed by NIC (Dhingra 2001) for the Ministry of Rural Development includes.

1. Financial Release software for various poverty alleviation schemes, besides, Below Poverty Line (BPL) Survey Computerization has been done for various District Rural Development Agencies (DRDAs) in the country.
2. Management Information System (MIS) to process data related to poverty alleviation schemes known by the product name of RuralSoft2000, has also been designed, developed and distributed to all State RDs and NIC State Units
3. Web Sites for Ministry of Rural Development (www.rural.nic.in), Council for Advancement of People's Action and Rural Technology (CAPART) (http://capart.nic.in)
4. Web Sites have also been developed for Rural Development Departments of various States like Gujarat, Uttar Pradesh etc.
5. Programme Status Report (PSR) providing physical and financial status of rural development schemes have been designed and developed for MoRD.
6. *Daily* is a dynamic web based application focusing day-to-day activities and requirements of the Ministry of Rural Development (MoRD) officials like file monitoring, lodging infrastructure related complaints etc.
7. A Geographic Information System (GIS) based analysis has been carried out for some of the watersheds in East District and North District (Sikkim), Raipur, Etah, Udaipur, Salem, and Pune Districts.
8. A cadastral level spatial database has been developed for Tintek block, East District, Sikkim, which has been suitably integrated with attribute data to evolve a Management Information System.
9. Several multimedia applications have been developed to cater to the training and monitoring needs of Rural Development Institutes, which include; Ministry at a Glance, Convergence of Services and Hill Area Development with Specific Reference to Uttarakhand.

V

Few Successful Case Studies of Multimedia Usage in Rural India

Samadhan Kendras (Soochnalaya Gumtis)

In Madhya Pradesh, a regional network connects 21 rural cyber cafes called *Soochanalayas*. Each *Soochanalaya* provides services to about 10 to 15 Gram Panchayat, 20 to 30 villages, and aggregate population of between 20,000 to 30,000 people. The network covers 5 of 13 Blocks and 3 of 7 Tahsils in the district. *Soochnalayas* are located at Block headquarters, haat bazaars, villages and bus depot centres.

Bhoomi Project

Bhoomi is the Karnataka government's one of the prestigious e-governance initiatives through which the entire agricultural land records in the State (20 million land records of 6.7 million landowners in 177 taluks of Karnataka) have been digitized. This has been developed with the comprehensive software called 'Bhoomi' designed fully in-house by NIC, Bangalore. Either the name of the farmer or the assigned plot number can be used to access information (Tadasad and Mahesharappa 2003).

Gyandoot

Gyandoot means 'messenger of information' and is the name of an intranet launched by the Government of India in January 2000 in the tribal dominated Dhar district of the state of Madhya Pradesh. It is a community owned, self-sustainable and a low cost rural project developed to bring the benefits of ICTs to the doorsteps of villagers who are dependent on farming and agricultural industries. It is a unique form of Government to Citizen (G2C) e-commerce activity, wherein the District Panchayat is enabling over half a million rural tribal citizens affordable access to various government and market-related needs through IT (Asian Vendors 2001).

Information Village Experiment

Swaminathan Research Foundation (MSSRF) launched it in January 1998 in Pondicherry. It acts as a hub or Value Addition Centre (VAC) of a local area network set up over 10 villages, serving a population of about 21,500. The villages act as Village Knowledge Centres (VKC) or 'Information Shops' and are connected by a hybrid wired and wireless network (http://www.mssrf.org).

TARAhaat

TARAhaat means 'star market place'. It is a private sector initiative whose goal is to bring information and marketing services using e-business to rural North India. Development Alternatives in two rural regions launched it in September 2000: Bundelkhand (one of the poorest in Uttar Pradesh) and Bhatinda (one of the richest in Punjab). It acts as a gateway that connects the village user to information, social services, and healthy entertainment and to all kinds of markets, through a network of 20–odd franchised *TARAkendras* or cyber centres, customized in the language of their choice (English, Punjabi and Hindi) (Babu 2002).

Akashganga

Akashganga (meaning 'the milky way') was launched in 1996 at Anand, Gujarat and implemented by Shree Kamadhenu Electronics Pvt. Ltd, with the objective of using appropriate IT to facilitate the timely collection of milk and thereby generate higher profits for the rural milk producers. *Akashganga* is being used at the Dairy Cooperative Society (DCS) a farmer-owned, grassroots unit in the cooperative structure. The basic milk collection transaction comprises measuring weight of milk with electronic weighing scale; fat testing using Milko Tester; capture of unique member ID by the PC software; and printing of a pay slip with all this data and the amount to be paid.

E-Choupal

ITC's International Business Division, one of India's largest exporters of agricultural commodities, has launched 'e-Choupal', in June 2000 (http://www.itcportal). It serves more than 3.1 million farmers growing a range of crops in over 31,000 villages through 5,050 kiosks across six states (Madhya Pradesh, Karnataka, Andhra Pradesh, Uttar Pradesh, Maharashtra and Rajasthan). And it enables the agricultural community access ready information in their local language on the weather and market prices, disseminate knowledge on scientific farm practices and risk management, facilitate the sale of farm inputs (now with embedded knowledge) and purchase farm produce from the farmers' doorsteps (decision making is now information-based).

Viswa Gram

The Viswa Gram project module 'e-Gram Panchayat Monitoring System' is for maintaining record of village information of all families' data and issue of necessary certificates to citizens. This module is designed and developed by National Informatics Centre. In 32 villages of Bhavnagar District, the Viswa Gram project is implemented. The Viswa Gram is based on survey form of each individual family of the village, which contains the detailed information of each member of family. This is used to provide many certificates such as Income, Caste, Domicile, Character, and Farmer etc. to citizens. Also, the database of property of each family is maintained to provide necessary certificate and to use in panchayat tax collection property-wise. Birth, Death and BPL Beneficiaries data is also maintained to provide respective certificates.

VI

Barriers in Implementation

Most of the Indian villages are deprived of communication facilities. Among numerous other problems are specified political conditions and rigid or ineffective bureaucracies, high rate of illiteracy, social discrimination and others. However, with the advent of global information society, new communication technologies are increasingly being adopted as an effective tool for reaching rural audiences. Yet the benefits of the information revolution are still much debated, particularly, in the case of developing countries like India. There is serious concern that the gap between the information 'haves and have-nots' will continue to grow unless the developing countries acquire the infrastructure and resources to access these new technologies. The situation is more serious for remote rural communities where basic communication infrastructure such as newspaper, television, radio and telephone is lacking. If the benefits of new technologies are to reach rural areas in developing countries, it is essential to disseminate information in local languages with simplified version and ensure that it is relevant to local development needs. Thus, it appears that a complete system is required for technology transfer; this system must accommodate the movement from a physical transaction to indigenous technological endowment to technology adoption and development in a specific setting.

Concluding Remarks

The above arguments reflect the critical analysis of developmental model and the level of inequitable inclusion of the rural population and the marginals. At the onslaught of liberalization and globalization, there is gradual decline in the nation-state policy to tackle the countryside and the 'social concern' (welfare goals) for marginal peasantry is emphasized. It is manifested that the new regime promotes more inequality, starvation, indebtedness, suicides, deprivation and de-peasantization vis-à-vis agriculture and consolidation of their holdings in favour of corporate farming. In brief, there is a serious concern for holistic evaluation of the purpose and necessity of adverse policies against the countryside at the cost of the affected population. It is suggestive that for such resolution there is a need of multi-specialists having sensitivity towards social, economic, psychological and ecological impacts of acquisition. No doubt, there is a need for such an alternative paradigm of development, i.e. participatory and down-top approach which takes into cognizance that the necessary land is available for 'genuine' public purposes and in the process the land losers are not impoverished. Apart from such societal and democratic concerns the approach of the decision-making body should be pro-poor, pro-rural and pro-sustainable development, i.e. the policies should bridge the gaps and not create conflicts and inequalities. Land is still a vital asset more so in India with its burgeoning populace growth along with high population density. The judicious use of land is very necessary keeping in mind the interests of the subalterns and economically weaker sections, which constitute majority of the Indian population. The developemental model and rural policies needs to be reworked beyond elitist sections, i.e. government, political parties and bureaucracy and should involve the intervention of civil societies, representatives of affected population, members of local bodies, human rights activists and others before arriving at a judicious decision. Few constructive efforts can be:

- Need for reviving the synergy between Nehruvian and Gandhian model of development with its emphasis on rural development and technological decentralization.
- A new era is being constructed in which state, people, and the environment are forced to adjust to the market, and it is the markets, not people, that require freedom. Today, we have to look beyond the state and the market place with 'people' as the driving force and to protect the right of the majority of Indians – the rural communities. The 'commons' cannot be monopolized by the economically powerful citizens or corporations or by the politically powerful state.
- Any large-scale bio-fuel programme has to ensure that it does not compromise with the nation's food and nutritional security. Further, utilizing unirrigated, barren and wasteland for growing non-food bio-fuel crops would prove fruitful with multiple benefits like greening up of unused lands, creation of employment opportunities, greater people's participation, and so on.
- Policy shifts and reforms-like RTI Act 2005, NREGA 2006; Scheduled Tribe and Other Traditional Forest Dwellers (Recognition of Forest Rights) Act 2006 should be encouraged for more participatory and assertive development.

At the end, for rural India specific, there is a need of disruptive innovations to make the giant leap forward and in this context, harnessing the tools of the Space Age is needed for rural transformation through quality ICT/multimedia. Here the intervention through multimedia can be made successful by retaining the taste of traditionality, i.e. the hybrid of tradition and modernity in the development process. The successful database applications developed by NIC and other rural multimedia projects like E-choupal, Taarahat, Gyandoot, Akashganga, Viswa Gram, Bhoomi project etc. are few attempts in this direction to boost up the future transformation of rural India. Ultimately, there is also an increasing need to create summative local resources accessible anytime via Internet to keep rural citizens current on the local status of surroundings and other threats to health, safety, economy, pursuit of life, liberty and happiness to bring improvement in quality of life of rural India in particular and rest of India in general.

References

Acharya, S.K. 1995. 'Spatial Implications of the New Economic Policy: Reflections on Some Issues for North East India with Special Reference to Women', *Man and Development*, 17 (1): 56–64.

Asian Vendors. 2001. 'Gyandoot: The Rural Intranet Project', http://www.asianvendors.org/trade/in/mag/200112/Infrastructure200112.html

Babu, M. 2002. 'Raising the Pyramid to Help People Help Themselves'.

Bardhan, P. 1985. 'Poverty and Trickle-down in Rural India: A Quantitative Analysis', in J. Mellor and G.N. Desai (eds), *Agricultural Change and Rural Poverty*, Baltimore: John Hopkins University Press.

Bhalla, G.S. and Gurmail Singh. 2009. 'Economic Liberalisation and Indian Agriculture: A Statewise Analysis', *Economic and Political Weekly*, XLIV (52), 26 December: 34–44.

Castells, M. 1996. *The Rise of the Network Society*, Oxford: Blackwell Publishers.

Chand, Ramesh, P.A. Lakshmi Prasanna and Aruna Singh. 2011. 'Farm Size and Productivity: Understanding the Strengths of Smallholders and Improving Their Livelihoods', *Economic and Political Weekly*, XLVI (26–27), 25 June.

Dhingra, Anjali. 2001. 'Rural Informatics Network for E-Governance in Rural Development Sector', *Journal of Rural Development*, October–December, http://ruralinformatics. nic.in/files/4_9_0_0.pdf-41

Frankel, F. 1978. *India's Political Economy 1947–77*, Princeton: Princeton University Press.

Ghosh, Devleena. 2001. 'Water Out of Fire: Novel Women, National Fictions and the Legacy of Nehruvian Developmentalism in India', *Third World Quarterly*, 22 (6): 951–967.

International Labour Organization (ILO) 2004. *Economic Survey for a Better World: ILO Report*, ILO Socio-economic Security Programme, Geneva.

Lipton, M. 1989. *New Seeds and Poor People*, Baltimore: John Hopkins University Press.

Michael, J.G. Parmwell. 2000. 'Agropolitan and Bottom-up Development', in Vandana Desai and R.B. Potter (eds), *The Companion to Development Studies*, New York: Oxford University Press, 115.

Mies, Maria and Vandana Shiva. 1993. *Eco-feminism*, London: Zed Books, 9–98.

Natarajan, I. 1999. *India Market Demographics Report*, New Delhi: National Council of Applied Economic Research.

Panini, M. 1995. 'The Social Logic of Economic Liberalization', *Sociological Bulletin*, 44 (1): 54.

Radhakrishna, R. 1993. 'Indian Agriculture: Growth, Employment and Equity', in Tapas Majumdar (ed.), *Nature, Man and the Indian Economy*, New Delhi: Oxford University Press.

Sangvai, Sanjay. 2007. 'New People's Movement in India', *Economic and Political Weekly*, 15 December: 112.

Saxena, Ashish and Shivarama Rao. 2005. 'Efficacy of Multimedia Technology in Indigenous Knowledge Management', *Convention Journal of Lucknow Management Association* (IIM, Lucknow), 1 (1): 248–258, on the theme 'Transforming Villages into Vibrant and Prosperous Entities: Innovative Strategies', Lucknow Management Association Publication, Lucknow.

Saxena, Ashish. 2011. 'Land Acquisition Bill', *The Radical Humanist*, 75 (6), September.

Sen, Amartya. 1999. *Development as Freedom*, New Delhi: Oxford University Press.

Shanin, T. 1986. 'Chayanov's Message: Illuminations, Miscomprehensions, and the Contemporary Development Theory', in D. Thorner, B. Kerblay and R.E.F. Smith (eds), *The Theory of Peasant Economy*, Madison, WI: University of Wisconsin Press, 1–24.

Shanin, T. (ed.) 1971. *Peasant and Peasant Societies*, Harmondsworth: Penguin.

Shiva, V. 1989. *The Violence of the Green Revolution: Ecological Degradation and Political Conflict*, London: Zed Press.

——. 1991. *The Violence of the Green Revolution: Third World Agriculture, Ecology, and Politics*, Penang, Malaysia: Third World Network.

——. 1998. *Globalization of Agriculture, Food Security and Sustainability*, New Delhi: Research Foundation for Science and Technology, June.

——. 2010. 'The Social Costs of Economic Globalization', in P.J. George (ed.), *Reflections on Vital Issues*, Hyderabad: Orient Blackswan, 23.

Tadasad, P.G. and S. Mahesharappa. 2003. 'Electronic Information to Farmers: The Government of Karnataka's Initiatives', in Akhtar Parvez and others (eds), *Electronic Information Environment and Library Services: A Contemporary Paradigm*, Bangalore: ILA, 486–489.

9

Perils of a Fragile Planet

Paradigms and Possibilities

Tapan R. Mohanty

Global environmental issues are a growing concern among the people and communities of various regions today. As it is the changing ecological base which will ultimately affect them as they seek to exploit the available resources in order to raise their standard of living. The challenge is not so much that of improving the quality of life of the population at the cost of their standard of living – a dilemma faced in particular by the developed countries – but rather to improve the standard of living in an environmentally sustainable manner. This involves recognition of the fact that alternation in the human and natural environment are underlining factors in the development process so that the topic of the environment cannot be absent from the minds of those who are responsible for guiding that process.

The rational arrangement of the natural resources is a necessary precondition for achieving economic growth and sustainable improvements in the standard of living of the populace. The natural capital is of fundamental importance in achieving changing production patterns with equity. The nation has entered upon a stage in which the existing resources will soon threaten the process of development. India now is on the threshold of a number of environmental problems, which, if not solved, will adversely affect the productive capacity of the economy. India as an agrarian country producing agricultural goods, heavily depends on the management of natural resources such as soil, water, vegetation and the climate. These elements are already beginning to suffer from considerable strains, which are having a negative effect on the quality and quantity of agricultural products. The relation between growth, equity and environmental sustainability are extremely complex. On the one hand, the transformation of natural resources into goods is essential for growth and the raising of living standards. Thus, for example, the expansion of the water supply is what makes it possible for growth of human settlements. On the other hand, however, there is the danger of development processes affecting the which are crucial to flowering of the above areas to draw on the interest of the wider society.

Mohapatra (1994) argues that in the age of rapid economic growth and technological advancement culture has taken a back seat and constraint to make adjustment. This path of adjustment has taken all leanings and curves resulting in dilution, distortion and death of cultural forms.

Dove (1994) also echoes the same sentiment in his studies of Indonesian society. According to him 'traditional cultures and lifestyles are regarded as clear signs of underdevelopment and as formidable obstacles to the necessary socio-economic advancement.

The dimension of gender too is neglected in the paradigm of development. Female subordination of all interpersonal relations is deeply ingrained in the consciousness of both sexes. This creates a fissure among the units of society and the recent movements like eco-feminism is the logical outcome of genuine outburst against a repressive social order.

Another important aspect, which has been bypassed in the drive to development, is the shabby treatment meted out to nature in general and environment in particular. In the name of industrialization, forests were destroyed, mining and quarrying went unabated causing enormous damage to mountains and springs, toxic and harmful effluents were drained to the water bodies without giving a second thought. Huge dams and massive reservoirs were constructed in order to tame the rivers and generate hydroelectric power. This has not only altered the ecological balance of the area but also changed the socio-economic bases of the population besides causing large-scale displacement. The design of development has created major problems not only in terms of environmental degradation but also in terms of human suffering.

II

Development Alternatives: The Quest for Best

This has, in fact, given rise to the quest of searching for an alternative design of development, which will not only improve the standard of living and life chances but also arrest the by-products of contemporary development strategy. Then there is the question of alternatives, its meaning and importance. 'The alternative is something new and better, both in the qualitative and quantitative aspect, than the preceding and existing conditions and state of affairs (Guha and Vivekananda 1985: 41).

The alternative to development was not an accident but the product of continuous search for a strategy that will harmoniously combine nature and culture. The capitalist and socialist developmental models are designed to perpetuate inequality and imperialism in the garb of an ideology and their threat to nature is menacing. This was in fact, a late realization on the part of the victims of the Third World countries. The failure of various models of development in fulfilling the objectives has only hastened the process of searching an alternative. Present slogan for 'sustainable development' is in fact a reaction against the existing models of development and deeply rooted in an environmental awareness.

Man's use and reuse of environmental resources has not only confined itself to food, shelter and clothing as the basic needs of life, but also goes much beyond it. The present form of use of nature can be defined in terms of exploitation plunder. Today, there is hardly any pocket of land, island, polar region or space left virgin from the scars of human interference in a bid to realize and translate man's aspirations and dreams into reality.

Realizing the imminent danger of ignoring the environmental issues there has been some kind of new awakening and the New World Order addressed itself to protect the environmental cause. There has been a plethora of environmental issues that need to be debated and focused. Environmental issues do not lie in technical and scientific arena alone but is enmeshed with social, economic and political fabric, besides in the ethos of humanity.

It was only in the early seventies that the world made an extraordinary rapid transition in terms of social awareness from capitalist economy to environmental concern particularly, the mutilation of natural resources and subsequent ecological imbalances. Administration, monitoring, abatement and maintenance of environment is linked with certain decision of political nature and legal sanctions. Here exists an urgency to examine ecological improvements and maintenance in the context of local condition and native population. It was this awareness on conservation of nature that led to the Stockholm conference in 1972, which not only recognized the crisis that environment was facing but also emphasized that solving these problem is a political issue and could not be left in the hands of a few environmentalist alone.

The effect of the conference is evident today with the change in the attitude and perception of Third World countries towards environmental problems in these nations where it had been unnoticed, uncared and unemphasized. Issues relating to environment were then considered as the domain of the industrialized world.

The catastrophe of Bhopal in 1984 and followed by Chernobyl in 1986 provided a glimpse of the destructive potential of global environmental hazards. Six years later at Rio de Janerio in 1992 the earth summit presented a programme for action to avert global environmental disaster based on the underlying principle of sustainable development. There has been the ending of the of the cold war, the break up of the former Soviet empire, the war in the Gulf, which along with the emergence of new-nation states provides an unique opportunity to give a fresh look at the immediate environmental problem. The uneasy stability of super powers has been replaced by an increasingly unpredictable world, but it certainly has delayed the imminent nuclear threat and opened up the possibility of diverting resources towards more humane concerns than in the horse race of global domination.

In the advanced western countries, public interest in the environment is mounting. In Eastern Europe and former USSR the emerging evidence of grossly degraded and polluted areas underlined the urgency of the problem. In the developing countries problem of deforestation, desertification and resource depletion was also attracting grater attention. And the long-term but apparently inevitable threats of global warming and ozone depletion also underline the global interest in environmental protection. At regional, national and global level environmental policy has achieved a higher priority and underwent vigorous development, the leitmotif of the Bruntland Report in 1987, has become the accepted goal of policy at Rio five years later.

It soon became clear that optimism must necessarily be qualified. Interests in the environment, though sustained, were not always translated into action in the face of more immediate and pressing economic concerns. The crumbling of Soviet empire, persistence of poverty in the Third World and the growing recession in the West, emphasized the competing priorities and inherent conflicts in the development theory. The eponymous process of Balkanization in areas such as former Yugoslavia threatened to entangle countries in a serious regional conflict. As the Gulf War has shown, such conflicts could do serious harm to sensitive ecosystem (Barnaby 1991). Overall was the threat of nuclear proliferation, more alarming as the Soviet nuclear complex was dispersed and which, if not restrained, would pose a far more dangerous threat to world survival than the more published global warming. However, some steps have been taken in order to check global pollution, i.e. the international agreements in the Vienna, Montreal and London in 1985, 1987 and 1990 respectively. These achievements have led to the banning of Chlorofluorocarbons (CFCs) to protect the ozone layer and various other agreements designed to restrict the trade in hazardous waste (Basel convention-1989, Bamko convention-1991 and related environment commission directives).

III

The Paradigm of Nature Culture Synergy

This process of global negotiations and agreements on environmental problems culminated at Rio summit which saw the adoption of 27 principles constituting a declaration on environment and development, the first of which was 'human beings are at the centre of concerns for sustainable development'. Over 150 nations signed a framework convention on climate change to tackle the problem of global warming and also a convention on biological diversity. By consensus, the summit endorsed agenda 21, an 800-page action programme of follow-up (United Nations). Although the summit had been preceded by conflict especially between North and South over the responsibility and resources, Rio produced a number of positive features. It was a process engaging more governments than even before and the participation of 6,500 non-governmental organizations and 15,000 participants at their global forum. The involvement of NGOs had begun at the Bergen conference on sustainable development in 1990. Despite the failure to reach agreements on certain issues, notably forests, and USA's reluctance to enter into binding targets or to sign Biodiversity Convention, Rio was a global endeavour at cooperation in the interest of mutual security. 'It defined new international values of equity and environment, linked them inseparably, and dramatized how powerfully they affect North-South relations' (Speth 1994: 146–61). Rio should also be seen as stage in a sustained process of developing policy and securing implementation.

IV

Sustainable Development: The Beauty of the Buzzword

Sustainable development has now passed a threshold of public and political concern that makes it fully and permanently established as an issue of high priority on the national and international policy making agenda. Although the environment is on the agenda, environmental policy making is fraught with uncertainty and conflict, particularly between North and South.

Sustainable development is the acknowledged subject of most recent development thinking but little headway appears to have made in terms of a rigorous definition of the concept. Therefore, not surprisingly, efforts to 'operationalize' sustainable development and to show how it can be integrated into practical decision-making have been few and generally unpersuasive. The use of the term 'development' rather than 'economic growth' implies acceptance of the limitations of the indexes such as Gross National Product (GNP) to measure the well-being of nations. Instead, development embraces wider concerns such as the quality of life, educational attainment, nutritional status, and access to basic freedom and spiritual welfare (Pearce et al. 1990).

The emphasis on sustainability suggests that what is needed is a policy effort aimed at making these developmental achievements last, well into the future. According to Winepenny, 'sustainable development is that which leaves our total patrimony, including natural environment assets, intact over a particular period. We should be able to bequeath to future generations at the same capital, embodying opportunities for potential welfare, that we currently enjoy' (Winepenny 1990).

Perhaps Repetto gives the best definition of sustainable development. For him, 'sustainable development is a development strategy that manages all assets, natural resources, and human resources, as well as financial and physical assets, for increasing long-term wealth and well-being. Sustainable development as a goal rejects policies and practices that support current living standards by depleting the productive base, including natural resources and leaves future generations with poorer prospects and greater risks than our own.'

The analysis of this definition depicts that sustainability appears to be accepted as the mediating term designed to bridge the gulf between 'developers' and 'environmentalists'. Its beguiling simplicity and apparently self-evident meaning have obscured its inherent ambiguity.

The notion of sustainability appears most convenient to the replenishable use of renewable resources. The aim is to benefit from the advantage provided by such resources to the point where the rate of 'take' equals to the rate of renewal, restoration and replenishment. So, in agriculture, the farmer derives fertility from soil equal to the ability of the soil to supply nutrition. Similarly, the woodsman removes trees or the products at a rate equal to the regeneration. The fisherman catches marine resources in amounts that are equivalent to their refurbishment. This begs the question of whether inherent rates of renewability can be enhanced through scientific management.

The relation between growth, equity and environmental sustainability are extremely complex. On the one hand, the transformation of natural resources into goods is essential for growth and raising of living standards. Thus, for example, the expansion of the water supply is what makes it possible for growth of human settlements. On the other hand, however, economic development process affects the quality of the environment. For example, air and water pollution reduces the capacity of ecosystems to provide the community with vital goods and services.

Though the paradigm of sustainable development is riddled with contradictions and conflicts at the conceptual as well as operational level, the much-propagated conflict between economic development and environmental concerns seem to be less, in the light of socio-ecological attributes like awareness and attitudes. These characteristics are more crucial in the context of agricultural sustainability in India which is known for its inter and intra-regional and cultural variation. Lack of comprehensive or integral understanding of these aspects both at the conceptual and operational levels make the question of agricultural sustainability ambiguous. In this context Reddy in an article argues that the 'models of sustainable development (i.e. agricultural) which focus on the people as their primary concern, should not stop at providing them with livelihood alone. The problems of environment and sustainable agriculture are not limited to fragile resource regions or poor regions, though the nature and degree of problems may vary between agriculturally developed and backward regions. Unless people's awareness, attitude and perceptions towards environment are changed, sustainable agricultural practices as conceived in the present form appear to be a distant dream. The factors that could change people's perception are literacy, market forces, technologies and institutional changes. Therefore, to recapitulate, human resource development should occupy the centre stage in the overall developmental planning, in order to achieve the objective of sustainable agricultural development. And this should be fostered by state powers by integrating environmental concerns with markets and promoting efforts towards developing appropriate technologies which are sustainable economically as well as environmentally and removing interferences in agrarian structure, i.e. land labour and capital markets.

The policies of globalization and liberalization have ushered new changes in Indian society, particularly in the economic and environmental dimension, through the structural adjustment programme. Structural adjustment programme is an implicit acknowledgement of unsustainable growth policy of 1980s, if not in ecological terms but certainly in terms of fiscal and foreign exchange. However, an alternative was offered to structural adjustment programme known as needs-oriented Economic and Ecological Development. Mohan Rao in an article has vigorously defended NEED. According to him, the growth strategy currently in place is built on faith in trickle down and what may be termed long-run Malthusian optimism. It does little to address the problem of transition and sustainable economy. While the prospect of accelerated growth under the strategy is, to say the least, uncertain, it is all but certain that both inequality and environmental degradation will be of concern over the years. Major policy thrusts and institutional restructuring under NEED are designed to promote equitable outcome and to exploit complimentarily between income gains, employment growth and environment refurbishment. In the medium terms, equitable growth is green growth; this only enables us to a faster and less uncertain transition to a sustainable economy (Rao 1995).

The threat to environment and ecology due to population, poverty, pollution and indiscriminate industrialization is a hard reality in India. People have begun to realize the dangers to ecosystem inherent in the development design that India has chosen at the behest of capitalist forces. Alternative to this kind of development has come as a blessing to fill the gap. There have been some fundamental problems endemic to sustainable development like transfer of environmental technology; the need for social policy etc. and India has to decide its future in a development policy, which will encapsulate equity, equality and excellence. The sooner we find the alternative, the better for the nation as well as for the world (Reddy 1995). Sustainable development as a goal rejects policies and practices that support current living standards by depleting the productive base, including natural resources and leaves future generations with poorer prospects and greater risks than our own (Pearce et al. 1990).

V

Principles of Scientific Management

1. Knowability: the amount, rate and other characteristics of renewability are knowable and calculable.

2. Homeostasis: renewable resource systems operate broadly around equilibrium or can be manipulated to approximate following human intervention. Homeostasis is a preferential state of nature.
3. Internal bioethics: the act of thinking upon a renewable resource even below some threshold has implications only for the tightly confined ecosystems that is the resource.
4. External bioethics: utilizing a renewable resource up to the point of sustainable yield is morally justifiable even though that resource, below in threshold of optimal 'take', may have other ecological values and function. Sustainable growth is primarily a technical concept, bounded by formalistic rules of efficiency and administration. Sustainability is a much broader phenomenon embracing ethical norms pertaining to the survival of living matter, to the rights of future generation and to institutions responsible for ensuring that such rights are fully taken into account while formulating polices and actions. The first two of the four premises of sustainability outlined above, are pertaining to renewability and homeostasis, apply to the concept of sustainable utilization. The latter two, which embrace a more bioethical perspective with implications for a great variety of rights and obligations, impinge more directly on the notion of sustainability. Sustainable utilization is a prior condition for sustainability, but not a sufficient one. The analysis which follows asserts that sustainable utilization is manageable and politically acceptable because it is safely ambiguous. Sustainability on the other hand, is politically treacherous since it challenges the status quo. Paradoxically, the objectives of sustainable utilization cannot be met without incorporating the principles of sustainability, hence the confusion of misunderstanding that has grown up around the sustainability debate should be removed.

It was the Brundtland Commission's publication 'Our Common Future' in 1987 that really put the concept of sustainable development on the international agenda and highlighted its applicability to the environmental problems in the developing world. The report recognizes that in most developing countries there is a greater dependence on natural resources and environment as an input of production and economic growth. Essentially, development and the environment are complementary, it is not economic growth per se that is to be rejected but there is a need to search for alternative development strategies and technologies based on sustaining and expanding the environmental resource base in the word of the Brundtland Commission.

There has been a growing realization among national governments and multilateral institutions that it is impossible to separate economic development issues from environmental issues; many forms of development erode the environmental resources upon which they must be based, and environmental degradation can undermine economic development (WCED 1987: 3).

The report reflects the importance of economic efficiency in achieving the goals of sustainable development but also stresses that the benefits of development must be distributed equitably. Social equity, both within and across generations, is a fundamental goal and prerequisite to achieve sustainable development. In this context, poverty is seen as a major cause and effect of global environmental problems, and attempts to deal with environmental problems will be thwarted unless a broader perspective that encompasses the factors underlying world poverty and international inequality is adopted. Brundtland Commission perceived 'sustainable development as a strategy to ensure the (development) needs of the present without compromising the ability of future generations to meet their needs' (WCED 1987: 8).

Markandya has written extensively on the topic of sustainable development. In a paper titled as 'Criteria for sustainable agricultural development', he suggests a set of working rules on targets covering equity, resilience and efficiency as a first step in making the concept operational in the agricultural sector. In working towards these sustainability targets, three broad areas of action are identified – those of valuation, regulation and monitoring.

Another area is environmental monitoring which is also fundamental to the pursuit of policy of sustainable development. Markandya and others stress the need to develop a set of sustainability indicators that can be used to evaluate performance. However in broad terms the objectives and premises of sustainable development can be summarized in following points.

1. Survival of human beings.
2. Survival of all other life forms.
3. Satisfaction of basic human needs.
4. Maintenance of biophysical productivity.
5. Economic efficiency and growth.
6. Preservation of environmental quality and ecosystem.
7. Inter and intra-generational equity.
8. Social justice.
9. Self-reliance and people's participation.
10. Stabilization of human population.
11. Promotion of values and ethics.

The emergence of sustainable development in the discourse of development has raised wide-ranging debate on the issues of environment equity and redistribution of natural resources. The result is a growing realization about … . Excessive stresses on biophysical environment for short-term gain is ultimately counterproductive, so there is a limit up to which productivity can be raised.

Human survival can only be assured by preserving the basic resource base. Even the renewable resources are not perpetual unless prudent management praxis is followed. Conventional economics does not necessarily indicate the actual status of resource utilization.

Social justice and equality are prime objectives of any form of development. Quite naturally, they are encapsulated in the premise of sustainable development. Sustainable development at the central level can be assumed as a process to maximize three goals – biophysical, economic and social. From another angle, sustainable development can be depicted as a process of development that operates within a framework confined by the biophysical, economic, and social coordinates. To operate within such condition, it is essential to resort to a process of trade-off, primarily to recognize that there are certain limits within which development has to be functional. These limits can be broadly identified as biophysical and ethico-social limits. The biophysical limit, which actually controls the resource availability, resources use and economic viability is due to (i) finitudes, (ii) entropy, (iii) complex ecological interdependence and (iv) resilient-self ameliorating capacity.

The ethico-social limit has more to do with the philosophical back-drop of a society including its legal, executive and policy making authorities. Bio-physical conditions differ regionally as well as locally, hence it requires utmost attention in adopting management practices. Another point to note is that, when the biophysical limit is exceeded then it can be detected from the degraded conditions of land; information is quite inadequate to anticipate the triggering of degradation. This is one of the serious problems in natural resource management.

VI

Operationalizing Sustainable Development: The Road Map

On the one hand, intervention leads to determination and on the other hand intervention is required because the local productivity is perceptibly low to support today's population. Thus, the new development strategy should be sustained amidst various pressures and the resource management has to be viewed in its totality considering all the available uses and focusing not on the sustainability of any individual project on sector but on the entire production system. Analysts and planners have become increasingly concerned about the futility of maximization of growth rate in ensuring economic equity and social justice.

Growth with justice also can constitute the necessary and sufficient conditions for the fulfilment of the basic needs of the poor. Therefore, growth-oriented development planning needs to be replaced by planning for sustainable development. Recent ecological critiques of development have focused on the impact of development in degrading ecosystems, and degrading consequently in poverty and deprivation. Because the poor in developing countries have to depend directly on natural resources like land, water, forest products and marine life; any exploitation and commercialization of these resources tend to threaten the very survival of the poor. This is more so in countries which have predominantly agrarian social structure. In this context, the basic need approaches of sustainable development open new vistas to social life.

As science and technology go on enlarging the space for human creativity, simultaneously, nature and ecology impose restrictions on the extent to which this creativity could be purposefully utilized to satisfy human desires without jeopardizing their own future. Since consumption standards and perceptions of 'better life' are culturally defined, 'sustainable development' then becomes an ideological construct, in the sense, that it requires promotion of values that either encourage or discourage kinds of needs within the bounds of ecological possibilities. Some staunch Marxist economists may believe that the socio-cultural systems are themselves propelled by a set of forces that have their origins in prices and markets and in other institutions such as the State. In other words, such an epiphenomenal view could tend to treat culturally defined uses of ecological resources as basically anchored in political economy. 'But sustainability or unsustainability of development process is essentially a consequence of the interaction between socio-economic systems and ecological systems. Both are co-evolutionary systems which suggest not only that they reinforce each other but also that each encounter constraints and impasses that originate in the other.' The notion of sustainable development suggests that the process of production which functions within a given ecological system must determine the optimum level at which we should operate.

Some scholars optimistically argue, that ecosystem is autonomous and its residence provides a safety margin, so that it has capacity to withstand and absorb crisis resulting from imbalances. Certain modes of environmental management do suggest that every ecosystem has a certain waste accumulative capacity. Only when the accumulation of effluence goes beyond this limit then it tends to erode ecological balance. However, just because a certain type of production or industrial material pollutes the environment that does not render the entire development process 'unsustainable'. The operational concept of 'Sustainable Development' would therefore, differ in a synchronic and in a diachronic context. In the short run, it refers to the balance between ecosystem and production system that exploits it at given point of time. In the long run, it is just to be achieved by a given society where its production process sustains the needs of present without jeopardizing the future.

Resources of the ecosystem are both renewable and non-renewable. And, water, forests, fish stocks etc. are renewable and do not get depleted by their exploitation provided it is within certain limits. The rate at which these resources are replenished periodically is the limiting factor. The non-renewable resources include fossil fuels and minerals. Their indiscrete and wanton exploitation by the present generation can lead to depletion of resources that are not easily sustainable. The coal and fuel crisis in Europe, some centuries ago, is a pointer to the critical dangers inherent in indiscriminate exploitation of non-renewable resources of the eco-system. Sustainable development then is a process in which an equilibrium is maintained between utilization of natural resources and environment, the direction of investment in technological development and institutional change should be in harmony with one another in a bid to enhance the present as well as the future capabilities of human societies to meet their needs and aspirations as well as to enhance their quality of life (Dhanagare 1995).

The successful operation of sustainable development depends upon interdependent principles aimed at meeting human needs, maintaining ecological integrity, attaining social self-determination and establishing social equality. Conventional interpretations of development either pay too little attention to equality consideration or else, assume inequalities will be explained through economic growth on a scale that would threaten global ecological integrity and jeopardize future generations. Sustainable development depends upon a better understanding and satisfaction of non-material human needs in the developed nations and a reorientation of social and economic policy. Human needs can be divided into material and non-material needs. Material on basic needs include physical necessities of life such as adequate food, water and shelter. One of the primary goals of sustainable development is to raise the standard of living of the world's poor to a level where these basic needs are met (Reddy 1995). Non-material needs include 'quality of life' requirements such as health, political and spiritual freedom, human rights, clean, healthy and accessible natural environments and meaningful work. These needs also include our needs for personal development and self-fulfilment. Maintaining ecological integrity can also be divided into two parts. One is characterized by principles of consumption, stewardship and appropriate land uses and is pertinent to all interpretation of sustainable development. The second, reducing the inequitable consumption of energy and materials by the industrial houses is somehow downplayed in conventional interpretation of sustainable development and must be considered an integral part of sustainable development.

Social Self-Determination rests on certain prerequisites, not the least of which is place. Militarism and overconsumption by the affluent section of the society causes massive destruction both to natural resources and to the ability of many of the world's people to achieve self-governance. Place and equity require community self-reliance, in both developed and under-developed communities. Other important elements of social self-determination include people's participation in decision-making, human scale communities, participatory governance, decentralized management, grassroot activity and cultural integrity.

Equity means, both future equity and current equity. Proponents of conventional sustainable development have embraced intergenerational equity in principle; it depends largely on maintaining ecological integrity. Sustainable development gives close attention to issues of current inequalities which in turn, requires not only maintaining ecological integrity but also meeting the entire range of human needs and achieving self-determination. Steady state economy is defined as an economy in which the total population and the total stock of wealth are maintained at constant at some desired levels by a minimal rate of maintenance throughout. The need for a steady economy is based on the fact that the resources are finite. The human economy is a subsystem of the steady state ecosystem. Therefore, at some level, over some time period, the subsystem must become a steady state.

VII

Sustainable Development in India

India's tryst with destiny is fraught with plethora of challenges and innumerable complexities. Its history, culture and polity reflect a wide range of diversity and multiplicity, quite unparalleled in human civilizations. It is this ubiquity of uniqueness that has brought the nation into a perpetual process of realignment and assimilation. In the event of the approaching millennium and subsequent change in global order in the form of globalization, rise of international terrorism, unbridled consumerism, climatic change and emergence of unipolar world order, India's challenge has gone manifold. The dilemma is more pronounced in the economic and social front, particularly in its effort to achieve social harmony and the need to ensure sustainable management of natural resources and the environment. The difficulty in attaining social harmony in the face of multiculturalism and a predominance of pluralist worldviews has been encapsulated in the historicity of its tradition. The second, however, has gradually been incorporated into the core of the development debate in recent years.

India's quest for economic growth, scientific achievement and social development has been aptly reflected in the Nehruvian policy of post-independent India. But a critical analysis of India's progress after half a century, depicts a sad reality. Poverty, unemployment, illiteracy and poor health care facilities continue to haunt the nation. These facts question the validity of our development model and its effect. The recent emphasis on sustainable development is a product of our search for alternatives in pursuing human happiness and welfare.

A conservative estimate of environmental damage in India puts the figure at more than $10 billion a year or 4.5 per cent of GDP in 1992. That is, urban air pollution costs India $1.3 billion a year. Water degradation leads to health costs amounting to $5.7b every year, nearly three-fifths of the total environmental costs. Soil erosion affects 83–163 million hectares of land every year. Deforestation, which proceeded at the rate of 0.6 per cent a year between 1981 and 1990, leads to annual costs of $214 million.

Taking an example of a development process of industrialization and urbanization and its impact on the environment, it is noted that of the 3 million premature deaths in the world that occur each year due to outdoor and indoor air pollution, the highest number are assessed to occur in India. According to the World Health Organization, the capital city of New Delhi is one of the top ten most polluted cities in the world. Surveys indicate that in New Delhi the incidence of respiratory diseases due to air pollution is about 12 times the national average. According to another study, while India's gross domestic product has increased 2.5 times over the past two decades, vehicular pollution has increased eight times, while pollution from industries has quadrupled.

These disastrous revelations came at a stage when controlling such pollution was a Herculean task. The development of India can be divided into phases. In the phase lasting almost till 1972, development essentially meant usage of natural resource to gain material wealth or rather conversion of natural resource to market commodities. There was absolutely no consciousness of natural resource depletion or other issues of environment degradation. The 1980s continued with speedy conversion of the green tree cover into the green moolah cover in industrial houses and world state economies. It was somewhere in the late 1980s and early 1990s that the alarm bells rang and the concepts of sustainable development etc. sprung up. Thereafter, the global community has become conscious of the possible results of the present rate of natural resource depletion, world over.

Several enactments world over came into force; many of them based on general environmental protection and quite a few on specific aspects of environment protection. The crucial point, however, is not in the framing of exclusive enactment but in their enforcement. This is of special relevance in a developing country like India.

A developing country like India is on its way to achieve the standards set by its developed counterparts like the United States of America; while on to this arduous task, the international community imposes on it an obligation to take care of its environment. The situation, which arises, is one of prioritization; which in itself is the most challenging decision for the state authorities. It is a position in the middle of the valley and the well.

The issues for consideration are:

The socio-economic cost for environment protection.
Environmental cost for industrial development.
Retarded development for environment protection.

For a nation, which has yet to provide its citizen with the bare minimum necessities; is it viable for it to take steps towards environmental protection which could use up funds spared for development activities? Must the state give preference to activities of national development or the larger issue of environmental protection? Also in a nation, where people do not have a day's meal; would they be in a position to resort to environment friendly products which are relatively more expensive than ordinary commodities?

Sometimes, these developing nations facing the multitude questions are also victims of choiceless situations. For instance, a harmful substance-producing factory is to establish by a foreign corporation in the developing country for the sake of economy. The developing country looking for all avenues for development might or may be forced to permit this establishment for fear of international sanctions and also for revenue purposes. The solution suggested for balancing developmental needs and environmental protection has been sustainable development. There has been fiery debate on this issue and much needs to be resolved on this aspect.

The phenomenon of development also highlights another situation, where the impact of developmental policies show a direct bearing on aspects of national life and may affect one segment of the society more strongly than another segment. As much as a segment of the global community moves towards urbanization, mechanization, industrialization, another segment of the community would, in all probability, be paying for it in relative terms.

VIII

Environmental Impact Assessment

In view of the Bhopal gas tragedy, future projects in a developing country like in India must apply Environmental Impact Assessment (EIA) to fulfil the following objectives:

- to identify adverse environmental problems that may be expected to occur;
- to incorporate into the development action appropriate mitigation measures;
- to identify the environmental benefits and drawbacks of the project, as well as its economic and environmental acceptability of the community;
- to identify critical environmental problems which require further studies and/or monitoring;
  to examine and select the optimal alternative from the various relevant options available;
- to involve the public in the decision-making process related to the environment; and
- to assist all parties involved in development and environmental affairs to understand their roles, responsibilities and overall relationships with one another.

Environmental Impact Assessment is essentially a preventive process, which avoids costly mistakes in planning and development. Therefore, it is necessary to carry out EIA during the feasibility study stage of the planning process. India must carry out EIA for its development projects in order to ensure economic development. It should ensure rational geographic distribution of its development plans and try to avert adverse environmental impacts in the future. Training of skilled manpower and an easy access to other countries having somewhat similar physical and socio-economic conditions are important factors which could determine the quality of environmental assessment to be made.

Developing countries must carry out EIAs to the best of their national capability. Therefore, it is urgently necessary to train our own experts in EIA. Foreign experts are no substitute for well-trained local experts. The involvement of local expertise will not only ensure that EIAs are carried out more relevant to local needs, but will also ensure a significant reduction in EIA costs when compared with those conducted by foreign experts.

The need to take environmental considerations into account to ensure successful economic development is increasingly recognized throughout the world. Laws of many countries provide that reports on environment impacts must be submitted for public review. Environmental Impact Assessment provides an important methodology for acquiring a clear understanding of the relationship between economy, society and environment and offers positive measures for better harmonizing the relationship between economic development and environmental protection as an effective means for strengthening environmental planning and management. The implementation of this system is of great significance to both economic development and environmental protection, especially in a developing country like India, with its historical Bhopal gas tragedy.

IX

The Supreme Court on Sustainable Development

The Indian Supreme Court has earned itself an excellent record in the creation of environmental rights in India, and their contribution to the jurisprudence of environmentalism has provided an example for environmental lawyers all over. The court has also applied the principle of sustainable development in several cases before it. The Indian experience serves as a good example to look at how sustainable development theory has developed in a given legal context.

The case regarded as a landmark for the application of sustainable development in the Indian context is the decision in the *Calcutta Wetlands* case,[1] where a Public Interest Litigation was filed against the rapid dredging and filling of marshes in the Salt Lake area of Calcutta. The High Court recognized that the environmental concerns had to be addressed at the same time as developmental concerns, and that development must progress at a rate that took into account the interests of 'posterity' as well.

The High Court struck down the decision to dredge and fill the marshes in Salt Lake on the grounds that India was a party to the RAMSAR Convention, and had thereby signified its intent towards the protection of wetlands. Further, the High Court also recognized the fact that the industries could be built in other areas with less environmental damage.

The first concrete application of sustainable development theory arose in *State of Himachal Pradesh v. Ganesh Wood Products,*[2] where the Supreme Court held that the government could not allow the approval of these industries in the State where the industries sought to use the forest products collected by mechanized units as the units had a deleterious effect on forest wealth, ecology and the environment. The Supreme Court also held such approval to be contrary to established considerations of sustainable development and inter-generational equity. However, in the striking down of the approval of the forest industries that posed environmental threats, the Supreme Court did not go into an analysis of the application of the principle of sustainable development to the facts of that case, nor did it explain the principle itself. Hence, the value of the case as a precedent for the application of sustainable development in the Indian context may be regarded as slightly doubtful.

The actual leap from the principle of absolute liability laid down in the *Oleum Gas Leak*[3] case occurred in the case of *Indian Council for Enviro-Legal Action v. Union of India*[4] where the court held that the manufacturers of products discharging toxic effluents into the region would be liable for the payment of compensation as per the absolute liability rule laid down in the *Oleum Gas Leak* case, as well as the precautionary principle and sustainable development principles. This represented the first instance in which the Supreme Court used sustainable development theory as a basis for calculation of damages for environmental wrongs.

The decision in *Vellore Citizen's welfare Forum v. Union of India*[5] was the first decision in which the Supreme Court discussed the principle of sustainable development from a theoretical standpoint. In that case, the Court agitated its mind upon the matter of polluting tanneries in Tamil Nadu. The court held that the polluting industries had to prove that they were non-polluting, as they had failed to do so; they were liable to be shut down and pay compensation for the ecological damage caused by them.

The trend in the Supreme Court till recently was to apply the principle of sustainable development towards ecological harmful industrial activity, resulting in the closure or relocalization of industries, as most of them were found to be violative of the delicate sustainable development balance principle. However, of late, the trend has seen a few deviations. Recent Supreme Court decisions have tended to support incursions on the forest cover and the environment in the interests of 'development'. Good examples are as follows:

In a similar context the petitioner filed Public Interest Litigation against the alteration of boundaries of the Chinkara Sanctuary for carrying out mining activities inside the sanctuary. The Court ordered that the mining be allowed for a period of five years, after which it would examine if any deleterious impact of the mining could be noticed.[6] In *Live Oak Resort v. Panchgani Hill Station Municipal Council*[7] the petitioners sought to prevent the construction of a Five-star luxury hotel in the Panchgani Hill Station on the ground that it might affect the environment of the area. Court in this case allowed the construction on the grounds that:

> In a developing economy there cannot be only development or only ecology, but both must exist and thus a balance will have to be struck else otherwise society will perish in the absence of either of the elements.

In the *Narmada Bachao Andolan v. Union of India*[8] the legality of the proposal to raise the height of the Sardar Sarovar dam, which would directly impact people in parts of the states of Madhya Pradesh and Gujarat was challenged. When discussing the environmental impact of the dam, the court did not apply to shift the onus of proof upon the polluting industries as was laid down in *MV Nayudu's* case.[9] As dams were not so inherently hazardous, this shift in the onus of proof was unnecessary. The Court held:

> Merely because there will be a change is no reason to presume that There will be ecological disaster. It is when the effect of the project is Known then the principle of sustainable development would come into play, which will ensure that mitigative are and can be taken to preserve the ecological balance. Sustainable development means what type or extent of development can take place, which can be sustained by nature and ecology, with or without mitigation.

Having looked at the decision of the Supreme Court seeking to apply sustainable development theory, the correctness or authenticity of their application must also be analysed. In the initial phase, it cannot be denied that the Supreme Court was on a strong theoretical footing; seeking to apply sustainable development by means of its handmaidens – the polluter pays principle and the precautionary principle. The application of sustainable development in the Indian context may not satisfy conventional sustainable development theorists; however, as some of the later decisions conflict with this trend and seem to have the following common characteristics:

The reliance upon the precautionary principle, which pervaded earlier judicial decisions, seems to have been diluted in these decisions. The precautionary principle takes a very tricky approach to the enforcement of protective measures for the environment. Hence, any doubts that may reasonably arise may be justified on the basis of this principle. There also appears to be a shift away from the reconciliatory approach of sustainable development. The concept of sustainable development is reconciliation between environment and economy. Decisions of the Supreme Court seem to take a confrontational approach to the crisis of ecology and economy, by adopting to follow either a purely ecological trend, or a purely economic perspective, depending upon which has a more persuasive case.

Finally, it must be submitted that sustainable development is not purely an abstract concept. It must be clearly enunciated, and its tools and methods must be clearly specified. However, while prescribing the trend of sustainable development, the Supreme Court has failed to lay down a roadmap for sustainable development in its orders. While it is clear that the courts in India have not adhered to a rigid interpretation of sustainable development theory in India, the importance of the theory lies in its potential for use by the courts as a means of ensuring environmental compliance. Hence, in order to comply substantially with the requirements of sustainable development theory, the courts should seek to incorporate the suggestions made above.

Conclusion

To end the discussion it can be said that the threat to environment and ecology due to pollution, poverty, population and indiscriminate industrialization is a hard reality in India. People have begun to realize the dangers to ecosystem, inherent in the development design India has chosen at the behest of market forces. Alternatives to this kind of development are still being searched and sustainable development is one of them. There have been some fundamental problems endemic to sustainable development like transfer of environmental technology, the need for a social policy etc. But it appears in the present context, a better alternative. Therefore, it is a crucial decision on the part of the nation as it has decided its future in a development policy, which will encapsulate equity, equality and excellence. The sooner we find the alternative the better for the native as well as the world. But to our understanding and observation, we can achieve our objective of economic growth and equitable distribution of wealth and resources if we can restructure our present development policy and alter our priorities a bit. In fact, crucial to development practice is the notion of commitment and accountability. If our political decision makers and executives of development and planning could be accountable to the masses for their action and a little bit committed to the ethos of development, then we could see significant change in near future.

Notes

1. People United for Better Living in Calcutta v. State of West Bengal, AIR 1993 Cal. 215.
2. Himachal Pradesh v. Ganesh Wood Products, AIR 1996 SC 149.

3. M.C. Mehta and Others v. Shriram Food and Fertilizer Industries and Union of India (Oleum Gas Leak Case-I) AIR 1987 SC 965.
4. Indian Council for Enviro-Legal Action v. Union of India, 1996 AIR 1446.
5. Vellore Citizen's Welfare Forum v. Union of India, AIR 1996 SC 2715.
6. Consumer Education and Research Centre v. Union of India, 1995 3 SCC 42.
7. Live Oak Resort P. Ltd. and another v. Panchgani Hill Station Municipal Council and another, AIR 2001 SC 3478.
8. Narmada Bachao Andolan v. Union of India, AIR 2000 SC 3751.
9. A.P. Pollution Control Board v. Prof. M.V. Nayudu, AIR 1999 SC 812.

References

Agarwal, Anil and S. Narain. 1992. *Towards a Green World*, New Delhi: Centre for Science and Environment.

Alavi, H. 1972. 'The State in Post-Colonial Societies: Pakistan and Bangladesh', *New Left Review*, 74, July–August: 59–81.

Altman, I. and M. Chemers. 1984. *Culture and Environment*, Cambridge: Cambridge University Press.

Amin, S. 1976. *Unequal Development*, Hassocks: Harvester.

Apter, David. 1965. *The Politics of Modernization*, Chicago: Chicago University Press.

Ascher, W. and R. Healy. 1990. *Natural Resource Policymaking in Developing Countries*, Durham: Duke University Press.

Ashby, E. 1978. *Man with Environment*, London: Oxford University Press.

Attfield, R. 1990. *Ethics of Environmental Concern*, London: Basil Blackwell.

Barbier, E.B. (ed.) 1993. *Economics and Ecology: New Frontiers of Sustainable Development*, London: Chapman & Hall.

Barnaby, F. 1991. 'The Environmental Impact of the Gulf War', *The Ecologists*, 21 (4), July/August: 166–172.

Berry, J. (ed.) 1993. *Environmental Dilemmas: Ethics and Decisions*, London: Chapman and Hall.

Chakravarty, S. 1991. 'Sustainable Development', *The European Journal of Development Research*, 3 (1), June: 67–77.

Daly, H.E. and K. Townsend (eds) 1993. *Valuing the Earth: Economics, Ecology and Ethics*, Massachusetts: The MIT Press.

Dhanagare, D.N. 1995. 'Sustainable Development: Environment and Social Science Research in India' in Savur, M. (eds) *Contradictions in Indian Society*, Jaipur: Rawat Publications.

Dove, M.R. 1994. *The Real and Imagined Role of Culture in Development*, Honolulu: University of Hawaii Press.

Fanon, F. 1967. *The Wretched of the Earth*, Harmondsworth: Penguin.

Gadgil, M. and R.C. Guha. 1993. *The Fissured Land: An Ecological History of India*, Delhi: Oxford University Press.

——. 1995. *Ecology and Equity*, New Delhi: Penguin Books India (P) Ltd.

Government of India. 1992. *Environment and Development: India Approach*, New Delhi: Ministry of Environment and Forest.

Guha, A. and F. Vivekananda. 1985. *Development Alternative*, Stockholm: Bethany Books.

Guha, R.C. 1988. 'Ideological Trends in Indian Environmentalism', *Economic and Political Weekly*, 23 (49), 3 December: 2578–2581.

Haq, M.U. 1979. *The Poverty Curtain: Choices for the Third World*, New York: Columbia University Press.

Hayward, Tim. 1994. *Ecological Thought: An Introduction*, Cambridge: Polity Press.

Kenneth, E.F. et al. 1977. *The Unsteady State: Environmental Problems, Growth and Culture*, Hawaii: University Press.

Khoshoo, T.N. 1986. *Environmental Priorities in India and Sustainable Development*, New Delhi: ISCA.

Kothari, S. 1985. 'Ecology is Development: The Struggle for Survival', *Social Action*, 35 (4), October–December: 379–392.

Makenzie, R. 1967. 'The Ecological Approach to the Study of the Human Community', in R. Park and E. Burgess (eds), *The City*, Chicago: University of Chicago Press.

Mohapatra, S. 1994. 'Culture as Centre Piece', *The Times of India*, New Delhi, 5 August.

Pearce, D.W. et al. 1990. *Sustainable Development,* London: Edward Elgar.

——. 1993. Mexico Forestry and Conservation Sector Review: Sub-study of Economic Evaluation of Forests. Report to Latin American Technical Department, World Bank: Washington, DC.

Prah, K.K. 1993. 'African Languages: The Key to African Development', in M.V. Troil (ed.), *Changing Paradigms in Development: South, East and West*, Uppsala: The Scandinavian Institute of African Studies.

Rao, J. Mohan. 1995. 'Economic Reforms and Ecological Refurbishment: A Strategy for India', *Economic and Political Weekly*, XXX (28), July: 1759–1760.

Redclift, M. 1984. *Development and the Environmental Crisis: Red or Green Alternatives?*, London: Methuen.

——. 1987. *Sustainable Development: Exploring the Contradictions*, New York: Routledge.

Reddy, V.R. 1995. 'Environment and Sustainable Agricultural Development', *Economic and Political Weekly*, XXV (12), 25 March.

Repetto, R. 1986. *World Enough and Time: Successful Strategies for Resource Management,* New Haven: Yale University Press.

Savur, M. and M. Munshi (eds) 1995. *Contribution in Indian Society*, Jaipur: Rawat Publications.

Shiva, V. 1988. *Staying Alive: Women, Ecology and Survival in India*, New Delhi: Kali for Women.

Speth, J. 1994. 'A Post Rio Impact', Foreign Policy No. 88. Honolulu, East-West Centre.

Tisdell, Glen. 1988. 'Sustainable Development: Differing Perspectives of Ecologists and Economists, and Relevance to LDCs', *World Development*, 16 (3): 373–384.

Winpenny, J.T. 1991a. (eds) *Development Research: The Environmental Challenge,* London: Overseas Development Institute.

Winpenny, J.T. 1991b. *Values for the Environment: A Guide to Economic Appraisal*, HMSO Books for the Overseas Development Institute.

World Commission on Environment and Development (WECD) 1987. *Our Common Future.* Oxford: OUP.

Part III

Social Dimensions

10

Impact of Globalization on Women

Shweta Prasad

Starting with the first principles, the word 'globalization' was derived from the word 'globe'. When the different phenomena started transcending nation-states' boundaries, people started terming it global. This spread was in every field of societal activity.

Of late, this concept was almost rigidly being used for the economic globalization, as economy today dominates and decides the fate of nations, societies and individuals.

This economic globalization took place when different world economies attempted to integrate for obtaining a better economic model for their businesses. Here, the capital, technologies, procedures, processes and deliverables came primarily from the developed nations and the services; raw materials and consumers came from the developing world.

Globalization, for Bhagwati, is a process of 'integration of national economies into the international economy through trade, direct foreign investment, short-term capital flows, international flows of workers and humanity generally and flows of technology' (Bhagwati 2008: 3), while for Stiglitz fundamentally, it is the 'closer integration of the countries and people of the world which has been brought about by the enormous reduction of costs of transportation and communication, and the breaking down of the artificial barriers to the flows of goods, services, capital, knowledge and (to a lesser extent) people across borders' (Stiglitz 2003: 9).

Hence, it may be noted that globalization by itself is cross-cultural, multinational, cross-societal and comprises of highly specialized professions that have unique phraseologies, ethos, practices and culture. This, thus, makes training and mentoring for those idioms compulsory. That in turn provides yet another key for control to the proprietors of the global enterprises.

Thus, one finds an increase in global interconnectedness that allows for the interchange of the economic, social, cultural, political and technological attributes between different societies. This interchange of attributes facilitated by the globalization has brought about some desirable as well as undesirable changes for the different countries depending on their relative position in the global economy, local needs and the preparedness.

Opening up to international trade helped many countries grow far more quickly than they would have done otherwise. The rapid growth of the Asian Tigers (Singapore, Hongkong, Taiwan and South Korea) offers its concrete example. People there, are now living longer, have better paid jobs and the improved quality of life. It has reduced the sense of isolation in the developing world. Foreign aid has brought benefits to teeming millions by increasing the incomes of the farmers through the irrigation projects, induction of new agro-technologies and increasing literacy level through the educational projects.

At the same time, globalization and the introduction of a market economy has not produced the promised results in many other countries. It has neither succeeded in reducing poverty nor in ensuring stability. Crises in Asia and Latin America and the transition of the erstwhile USSR from the communism to the market economy brought unprecedented poverty amongst the masses. These could be ascribed to the undelivered claims of the globalization.

The West-driven agenda of globalization went against the interests of the poor countries and they were actually made worse off. Damages to the environment, massive unemployment, exploitation in the name of intellectual property rights, increase of poverty, erosion of union and labour rights, harming women and a threat to the indigenous culture were some of the fears accompanying the globalization that came true. Thus, the phenomenon called 'globalization' has been subjected to an unending controversy where both the proponents as well as the critics of globalization have their own views, substantiated by the facts at the global level.

Hence, it may be observed that riding on the crest of matured wave of information technology, contemporary globalization generated hyper-aspirations, the world over. For some countries, it worked as panacea for few people. For others, it generated utter frustration and huge disillusionment.

I

Globalization: The Claims

The UN Millennium Summit, 2000 embraced poverty alleviation as one of the goals. It was resolved during the Summit that by the year 2015, the proportion of the world's people with the income of less than one USD a day and the proportion of people suffering from hunger would be halved.

To achieve the target, an active, pull-up strategy was adopted to accelerate growth through varieties of policies. The examples of various countries like India, China, Vietnam, Mexico, and the 'Asian Tigers' had proved that the countries which adopted an outward-looking strategy, showed better performance on the economic front, whereas the non-globalizing countries experienced a slow down in the growth rate.

The proponents of globalization had claimed that the free trade would lead to prosperity in various countries of the world. Trade has been regarded as an 'engine of growth'. This, in turn, would result in reduced poverty and inequality the world over.

II

Globalization: The Reality

Globalization has been one of the most controversial phenomenon in the recent times. Riots and protests against the policies and actions of the institutions of globalization are hardly new. Since quite sometime, almost every meeting of the International Monetary Fund (IMF), the World Bank and the World Trade Organisation (WTO) has seen scenes of conflict and turmoil.

There are some thinkers (Bhagwati 2008) who have opposed the very idea of going global.

- French sociologist, Pierre Bourdieu has underlined the vices of globalization.
- Brazilian sociologist, Cardoso argued that the poor countries would be relegated to a dependent status in the international economy. However, later, on becoming the President of Brazil, he opened the Brazilian economy for the globalization.
- Raul Prebisch, the Argentinean economist, cautioned about the dangers in the international interactions between the 'core' and the 'periphery'.
- Osvaldo Sunkel, the Chilean sociologist argued that the integration into the international economy would lead to the disintegration of the national economy.

Why are so many people against globalization? What are their concerns? Are these concerns well founded?

There are two main groups of the critics of globalization.

The first group comprises of hardcore protesters who have deep-seated antipathy to the very concept of globalization. They come from the different intellectual and ideological directions.

The second group comprises of those whose discontents are well within the parameters of the mainstream dissent and discourse. They argue that the globalization is the cause of several social ills today ... that it harms and does not advance social agenda ... and ... that it needs a human face (Bhagwati 2008).

Most of the anti-globalizers are concerned about the following issues:

1. Increase in the poverty in the Third World countries: where the actual number of people living in poverty has increased by 100 million, at the same time when total income of the world actually increased by 2.5 per cent annually.
   In 1990, 2.718 billion people were living on less than 2 dollar per day which increased to 2.801 billion in 1998 (World Bank 2000).
2. Failure to provide stability to the world economy: as indicated by the Asian crises which almost destabilized the entire world in 1997 and 1998.
3. Perpetuation and accentuation of the problem of child labour: though it may not be a cause of child labour. According to ILO, 100 to 200 million children under the age of 15 are working, of which 95 per cent are in poor countries and half of them are in Asia. Further, an estimated 100 million of them are out of school (International Labour Organization 2002).
4. Downplaying the labour rights and the trade union related rights of the workers: In EPZs and SEZs, unionization is prohibited which has harmed the interests of the workers. However, India has retained the union rights for EPZs with the provision that wildcat strikes are not allowed in EPZs. Also, globalization has negatively influenced the wage and the labour standards in the rich countries of the world.
5. Environmentalists see globalization as a threat to the environment: They argue that the global trade and investment that results in growing incomes will result in environmental deterioration.
6. Multinationals have been blamed for exploiting the host countries and their workers: by imposing their terms of trade on the host countries on the one hand and getting several concessions in the form of tax exemption, rent-free use of public land etc. on the other. They are accused of paying low wages to the workers and violating their labour rights. Workers are required to work long hours especially in EPZs which is violation of their human rights.
7. The institutions of globalization, i.e. the WTO, the World Bank and the IMF have been charged with hypocrisy, double standards and unfair trade.
8. It is perceived as a threat to the indigenous culture of the host countries.
9. Women's organizations term it anti-women in nature as it has the potential to harm the well-being and the welfare of women.

III

Globalization and Women

Women constitute almost half the population of the world. Hence, no study can afford to ignore them while assessing the impact of globalization on the countries and the societies. On the one hand, globalization created much job opportunities for women but on the other, in the absence of proper regulations, left them extremely prone to exploitation. Development of global care chain that thrives on migration of women from poor countries to work as domestic help in the developed countries has led to the exploitation and abuse of these women. Women working in EPZs and SEZs have been found to be relatively more prone to exploitation than their male counterparts.

The WTO has been blamed for not paying requisite attention to the gender issues and the micro-level impact of its decisions on the women. Trade agreements failed to recognize the competing demands of the women and violated the rights of the women workers.

The 'conditionalities' imposed by the international funding agencies like the World Bank and the IMF for loans were quite tilted against the interests of the women.

Globalization and Domestic Work

Migrations across national borders are seen to be of three types: between poor and rich countries, flows of skilled and unskilled migrants and legal and illegal flows of human beings between different countries. Expectation of improved standard of living, better educational and other opportunities for children, and the growth prospects for the skilled workers are the major drivers of emigration. Moreover, due to the aging of the population, there is an increased demand of immigrants in the rich countries to supplement the domestic labour force. The demand for the skilled and semi-skilled labour force is also rising. The rich countries want skilled immigrants, whereas the poor countries are interested in having their unskilled citizens migrate. As a result, in the unskilled category of workforce, supply of immigrants exceeds demand in the rich countries causing illegal immigration. Migration of domestic workers is one such type of immigration in the rich countries having serious implications for both – the sending as well as the receiving countries.

Globalization has encouraged cross-country migration of human beings from one country to another in search of jobs. Women migrate in large scale for domestic and/or care work. The national level estimate of Philippines, Indonesia and Sri Lanka indicate that more women than men migrate to work abroad (Varia 2005). As these immigrants enter the host country mostly illegally, getting reliable data on domestic workers crossing borders is difficult. Further, being illegal they are never registered in the receiving country also. Being people with 'missing identity', tracing them and tracking their well-being is just not possible. This makes them face the brunt of worst category of exploitation. The problem gets aggravated because being illegal they are destined to bear the exploitation silently without approaching any authority for redressal.

Earlier, the developed countries like US, UK, Australia, Canada and the Sweden (UN Statistical Division 2000) were major 'receiving' countries for these women workers. However, evidence suggests that now countries like Singapore, Kuwait, Hong Kong and Saudi Arabia have become equally important destinations for domestic workers mainly after the economic growth of these countries.

As per an estimates, 22 million Asians work outside their own country (ILO 2004). In 2009, 215 million people had migrated. In Singapore, in 2005, the population of the migrant domestic workers was 1, 50,000 (Varia 2005: 2). In Spain, there were 3, 93,500 homes with a domestic employee in 2002. Most of these domestic workers were women (Albarracin 2002). According to the UK careers service, there were 9,28,000 private household workers in 1998 at UK (Career Website, UK). According to the Centre for the Migration Studies, at least 10 per cent, of the 3 million illegal immigrants of US, worked in childcare in 1993, with roughly another 10 per cent working in other private household occupations.

Ever since the women started working outside their home for wages, they needed someone to look after their household responsibilities like cleaning, cooking, care of elder and children in the family and these responsibilities were, then, shared by the domestic workers who were not that skilled to join the highly paid jobs but had to work in order to make their own families going. These domestic workers are often migrant women who travel to the developed countries, often illegally.

These women are often subjected to abuse and exploitation by their employers. Living away from their families in the house of the employer ensures that their working hours are very long. These migrant domestic workers leave their children behind who are looked after by the female relatives. As a result, they feel very lonely and depressed and to escape the situation, they transfer their love to the children of their employers.

The phenomenon of migration of women from poor countries who have children back at home looked after by the female relatives while the migrant women look after the children of other women in the developed countries has been termed as 'global care chain' by the sociologist Hochschild which puts every women at every point in the chain at a disadvantage. According to Hochschild, these women develop a 'hypothetical self', suffer emotional deficit and distress as the love they would have given to their own children is transferred to the children of the employers. Thus, 'the first world extracts the love from the third world' (Ehrenreich and Hochschild 2003). However, she concludes that the global care chain has liberating impact on these migrant women resulting in an increase in their self-respect and dignity (Bhagwati 2008).

Migration Related Remittances

The domestic workers send regular money to their families back at home. This constitutes a regular and good source of foreign exchange reserve for the sending country. According to Gina Mission, 34 to 54 per cent of the Filipino population is sustained by remittances (Mission 1998). A World

Bank Report indicated that 'officially recorded remittances worldwide exceeded USD 232 billion in 2005'. The report estimated that remittances sent through informal channels could add at least 50 per cent to this official tally, making them the largest source of foreign capital in many developing countries.

Remittances to the developing countries were USD 406 billion in 2012 and are projected to grow by 7.9 per cent in 2013, 10.1 per cent in 2014 and 10.7 per cent in 2015 to reach USD 534 billion in 2015. However, worldwide remittances, including those to high-income countries, were reported to total USD 512 billion in 2012 and projected to grow to USD 685 billion in 2015, according to the issue of the World Bank's Migration and Development Brief.

Despite the growth in remittance flows overall to developing countries, the continuing global economic crisis is dampening remittance flows to some regions, with Europe and Central Asia and Sub-Saharan Africa especially affected, while South Asia and the MENA countries (Middle East and North Africa) are expected to fare much better than previously estimated.

The top recipients of officially recorded remittances for 2012 are India (USD 70 billion), China (USD 66 billion), the Philippines and Mexico (USD 24 billion each), and Nigeria (USD 21 billion). Other large recipients include Egypt, Pakistan, Bangladesh, Vietnam, and Lebanon.

As a percentage of GDP, the top recipients of remittances, in 2011, were Tajikistan (47%), Liberia (31%), Kyrgyz Republic (29%), Lesotho (27%), Moldova (23%), Nepal (22%), and Samoa (21%).

Although migrant workers are, to a large extent, adversely affected by the slow growth in the global economy, remittance volumes have remained remarkably resilient, providing a vital lifeline to not only poor families but a steady and reliable source of foreign currency to many poor remittances recipient countries (World Bank Report 2012).

Thus, these domestic workers are not only supporting their own families and the families where they work but in the process they are supporting others in the sending and receiving countries like money traders, employment agencies and also the pimps.

International Funding Agencies and Women

The IMF and the World Bank, both originated in World War II as a result of the UN Monetary and Financial Conference at Bretton Woods in 1944. Both the institutions were part of a concerted effort to finance the rebuilding of Europe after its devastation of World War II and to save the world from future economic depression. During the depression, dawned the realization that markets often did not work well causing massive unemployment. In such situations, markets also fail to make requisite funds available to the countries for restoring their economies. Hence, the

IMF was created and assigned the task of ensuring global financial stability and preventing another depression.

In the 1980s, when US and UK started pushing forward the ideology of the free market economy in their countries, the IMF and the World Bank became their agents pressurizing the reluctant poor countries seeking loans and grants from these institutions.

In the 1980s, the World Bank went beyond just lending for projects. It started providing broad-based support in the form of structural adjustment loans with the approval of IMF. For this the IMF imposed conditions on the countries that included among others, commitment to reduce the budget deficit or tightening monetary policy.

The 'conditionalities' imposed by the IMF went against the interests of women in the Third World countries. Some women's groups believe that in the process of meeting the conditions imposed by these financial institutions, the governments of the Third World countries took steps like cutting down of expenditures on health, education, employment generation and poverty reduction programmes. The resultant unemployment hit the women most as they were the ones who lost their jobs before the men. Again, cutting down expenditure on the health and education implied that women left the workforce to provide these facilities to their family members. As she lost her direct earning ability, her position was lowered.

However, Bhagwati terms such charges as 'misdirected'. He argues that the IMF comes in scene only when a country is facing stabilization crisis and requires more revenue and less expenditure, meaning that the country, in any case will cut the expenditure whether IMF imposes 'conditions' or not. The IMF loans, in fact, support the countries in making transition to a better macro-economic situation easier. Thus, by doing so, it helps the women and not harms them.

WTO and Women

The World Trade Organization, which came into existence on 1 January 1995 replacing GATT, is an international organization setting out the global rules of trade between nations. While the GATT was a bilateral agreement, the decisions of the WTO are applicable to all the member nations. It was established to ensure smooth flow of the international trade by administering trade agreements, providing negotiation forum, handling trade disputes, monitoring national trade policies, providing technical assistance and training to developing countries and cooperating with other international organizations. WTO had 159 members on 2 March 2013 covering more than 97 per cent of the world trade (World Trade Organization Fact Files).

Some important international women's groups like the National Organization for Women, Feminist Majority and the International Gender and Trade Network are apprehensive of the policies of the WTO. They claim that gender issues and the micro-level gender impact are ignored by the WTO. They have argued that:

'Current international trade agreements, like NAFTA, violate the rights of women workers. Women workers in many factories, located in Export Processing Zones (EPZs) have reported physical abuse, sexual harassment, violence and an inhuman and demeaning mandatory pregnancy testing as a pre-requisite for employment. Women workers in EPZs are forced to work long hours for extraordinarily low wages under poor working conditions. In Mexico, over 200 women had been murdered, many of them on their way to and from their work in EPZs (Senser 2002: 1).

Here, it may be argued that this is the failure of the state to provide security to women in Mexico as the women have not been murdered at their place of work but on the roads outside the work premises. But the logic is flawed. It is the point of time and the locations of EPZs that made them vulnerable to violence. In India also, several incidences of rape and murder of women call centre employees have been reported while commuting in odd hours at lonely places. Even if we accept that these are more of a national issue rather than a WTO policy requiring active intervention from the governments of the respective countries, the glaring fact remains that in the EPZs, women are frequently subjected to exploitations.

There is one more annoyance for the women at WTO. Some women's organizations have blamed it for overlooking the unpaid housework done by women. This affects the efficiency of resource allocation within an economy. On converting unpaid housework into monetary terms, the UN has estimated its value to be equal to USD 11 trillion per year.

It has been argued that it is not feasible to analyse the impact of each and every policy of WTO from women's angle. Hence, it is prudent to assess the impact of the policies of WTO in totality for its ability to alleviate the distress to women.

Multinational Corporations and Women

One of the important criticisms labelled against multinational corporations (MNCs) is that they exploit the host countries and their workers. The MNCs reached the poor countries of the Third World in search of cheap labour and other economic benefits to reduce the cost of production and enhance their profits. At the same time, the Third World countries are in a race to lure MNCs by providing tax concessions and rent free use of public land. It has been observed that in this race, poor countries have generally been the losers.

Globalization has marginalized the poor in general and poor women in particular creating a class of new industrial proletariat ... the Third World women (Mason 2004).

Around 80–90 per cent of the low skilled assembly jobs in the Third World are performed by women as they are regarded as more submissive than men. In many countries, the MNCs prefer hiring women for assembly production as they are ready to work for lower wages than men. Further, they are also ready to work in conditions that trade unions would not permit (Moghadam 1999).

An Oxfam report, based on the experiences of 12 countries, highlighted the plight of women working in garment, food production, and supply goods to major Western retail companies. The report stated that the big retail companies based in the West have been able to benefit from the globalization of production, mainly at the expense of women workers.

The MNCs bring with them many job opportunities for the host countries. Why then, are they charged for exploitation of the workers? The most obvious reason is the low pay-package offered by the MNCs to the workers in the Third World countries. It has been found that the minimum wage in most East Asian and Latin American countries come nowhere near to covering basic living costs (Fuentes and Ehrenreich 1998). They are also blamed for running 'sweat-shops' in developing countries and for violating labour rights. In all these cases, labour mainly comprised of women workers. Hence, they are the worst sufferers.

Linda Lim, Graham and Paul Glewwe, the economists have reported that the MNCs, on an average, pay 10 per cent or in some cases even higher wages than the local companies. The wages may be 'low' by the standards of developed countries but in any case, their workers get actually higher wages than those available in alternative jobs in the host countries. This perception of good salary is relative to what people of similar skills, qualifications and experience would get in other industries, but it is much lower than the wages given to the women employed on the same job in the developed countries. For example, take the case of India where BPO industry has grown phenomenally over the years providing jobs to millions. The industry has been termed as the 'sunshine industry' and the young people – both men and women – vie for the job as they get handsome pay-package. The BPO jobs have come as a boon for Indian women liberating them from the age-old shackles of patriarchy. The BPO industry prefers hiring women, and women constitute nearly 31 per cent of the workforce in BPOs in India (The National Association of Software and Services Company 2012). This has changed the traditional nature of the (women) workforce and women are now making their presence felt in public spaces at night, challenging the systems of patriarchal social control.

MNCs are also accused of violating the labour laws related to the safety of the workers and other working conditions that negatively influences both men and women workers. But, the violation of the security related provisions especially harm the women workers. For example, lack of proper security to women employees of the call centres in India during late night movements have resulted in their rape and murder. The lack of protective tariffs allows them to take multifarious advantage of the under privileged and unprotected female workers. Besides paying them low wages, women are also subjected to hazardous working conditions, making them prone to occupational health problems.

Thus, these MNCs have become new means of reinforcing women's subordinate position within the society.

Special Economic Zones and Women

Special Economic Zones (SEZs) are specifically demarcated zones, holding special economic and other privileges, where production of trade and services take place. It receives very special and preferential treatment from the national governments and thereby is termed as SEZ. In other words, an SEZ is a specially demarcated geographical area within the territory of a country where economic activities of certain kinds are promoted by a set of policy instruments offering special incentives that are not generally applicable to the rest of the country (Reigstad 2007). Free Trade Zones and Export Processing Zones denote the earlier variants of SEZ.

According to a report, the number of EPZs has gone up from 79 in 1975 to 3,000 in 2002 and the number of countries with one or more EPZs has gone up from 25 in 1975 to 116 in 2002. Today, more than 43 million workers are employed in EPZs the world over, most of them working in China's ever growing 2000 SEZs (International Confederation of Free Trade Unions 2003).

The usual objections against the SEZs pertain to the huge amount of revenue loss due to tax and duty concessions, labour rights largely stand circumscribed in SEZs as law of the land does not apply to the SEZs aggravating the exploitations of the poor and the marginalized, acquisition of large tracts of land lead to massive eviction and displacement of rural folks and loss of livelihood. SEZs are also accused of the use of agricultural lands for non-agricultural purposes.

The workers in SEZs are paid lower wages and often they end up working overtime without any extra payment. They cannot even raise their voice against such practices. According to an ILO study, workers in Asian SEZs were found working 10–12 hour shifts that could go up to 16 hours during peak periods. In China, average working hours in SEZs vary from 54 to 77 hours per week and overtime is a rule. In Vietnam EPZs, workers are forced to work overtime often without any extra payment. Health standards in these zones are often very poor. Safety regulations were not followed adequately and fires were quite common in different Asian SEZs claiming hundreds of lives (Aggarwal 2007).

Women are an important and integral part of the SEZs. They dominate the workforce in SEZs in most developing countries. For example, in Philippines, 74 per cent of the workers in EPZs were women in 1994. In Bangladesh, more than 95 per cent of the workers in the garment industries in EPZs were women. Though the employment opportunities in SEZs have empowered these women but the high incidences of abuse of the women SEZ workers is a matter of growing concern. The contractors discriminate against these female workers by paying them less than their male counterparts for the same type of work. They are also subjected to gender violence in these SEZs (Halim 2011).

A study of the Falta SEZ in West Bengal, India reported that nearly 60 per cent of the workforce consisted of women who were employed on contract and on a daily basis. It is a practice in the Falta SEZ to pay ₹ 10 less to the female workers than the male workers. There is an atmosphere of job insecurity in this SEZ as even the permanent workers are not retained for a long period. Trade unions are prohibited in the SEZ. Many of the companies in the SEZ have hazardous working conditions posing serious risks to the health of the workers.

Accidents are rampant and medical help is hardly available. Working hours are totally unregulated. The horrible working conditions, force the women to leave the job but they are compelled to continue to generate income for their family. Sexual exploitation of female workers by the contractors is a regular feature in this SEZ. But all these are tolerated quietly by the female workers for the sake of their family (Ray and Sanhati 2009).

Trafficking of Women and Girls in a Globalized Era

During 1960s and 1970s, tourism flourished at global level as an industry. To attract the foreign capital, the IMF, the World Bank and agencies like USAID encouraged countries to exploit their natural resources to develop themselves as a tourist destination. With the increase in tourist related activities, flourished the sex industry which is regarded part and parcel of tourism.

Growth in the global sex market has led to the tremendous increase in the volume of human trafficking. It is estimated that 2.5 million people are in forced labour (including sexual exploitation) at any given time as a result of trafficking, of these 56 per cent are in Asia and Pacific region. Around 79 per cent of all global trafficking is for sexual exploitation (ILO 2007).

Also 161 countries are reported to be affected by human trafficking. People are reported to be trafficked from 127 countries to be exploited in 137 countries, affecting every continent and every type of economy (UN Office on Drugs and Crime 2006). Around 43 per cent of the victims are used for forced commercial exploitation, of which 98 per cent are women and girls (ILO 2007). This unparalleled increase in trafficking which has made it the third largest crime of the world and a multimillion dollar business has been assisted by the global flows of humanity, legally and illegally.

According to the ILO, 0.25 and 1.5 per cent of the total female population of Malaysia, Philippines, Thailand and Indonesia is engaged in prostitution and the sex industry. Further, it accounts for about 2 per cent to 14 per cent of the countries' GDP (Lim 1998). The sex industry has grown in Southeast Asia due, in part, to the surge in recent years in the number of women in Asia's migrant force, where they now outnumber male migrants (Kaban 1998).

The phenomenon is related to globalization which poses threats for the well-being of women. This illustrates how the opportunity to earn more can also turn into a disadvantage for women, if they are left unprotected.

Conclusion

The globalization has had significant impact on the participant countries. Global agencies like the ILO, while spearheading the globalization had expected it to eliminate poverty from the Third World countries and to decrease inequalities in the world. Unfortunately, it has had the reverse effect of increasing the poverty and inequality across the world. Whether this situation occurred because of inferior policies, poorer implementation or weaker governance of the host countries or the error in the concept design of globalization cannot be clearly stated.

However, one thing is clear that more and more countries are opening their market for the world economy but the corresponding socio-political institutions of various countries are not properly geared up to permit a smooth transition. Therefore, to ensure a smooth transition, there should be proper institutional mechanisms and policies to cope with the adverse outcomes of globalization as and when they arise.

Globalization is a heterogeneous and multifaceted phenomenon having different consequences for the societies experiencing it, depending on its absorption level, strength, preparedness and ability to govern and regulate associated activities.

Globalization has overall been very demanding and challenging for women. At its worst, it has hit her badly taking her life and dignity away. Extended hours of work in unsafe zones that are not provided with good occupational facilities have been a problem for her.

It is true that globalization on the one hand, has benefitted the women of the world by creating job opportunities for them, inside and outside their countries of residence and has helped them through the process of empowerment.

At the same time, it has reinforced the exploitation and subordination of women at global level. Therefore, looking at the consequences of globalization for the societies in general and for women in particular, a proper way to move ahead with the process of globalization would be to protect the vulnerable sections of the society from its harmful consequences through policy changes and their effective execution, providing sufficient measures for their safety and growth as per potential.

References

Aggarwal, Aradhna. 2007. 'Impact of Special Economic Zones on Employment, Poverty and Human Development', Working Paper No. 194, Indian Council for Research on International Economic Relations, May, http://www.icrier.org/pdf/working_paper_ 194.pdf, accessed on 16 January 2014.

Albarracin, Daniel. 2002. 'Domestic Work Examined', European Industrial Relations Observatory Online, http://www.eiro.eurofound.europa.eu/eiro/2002/05, accessed on 16 January 2014 and http://analysemycareer.co.uk/, accessed on 16 January 2014.

Bhagwati, Jagdish. 2008. *In Defense of Globalization*, New York: Oxford University Press.

Ehrenreich, Barbara and Arlie Russell Hochschild (eds) 2003. *Global Woman*, London: Granta Books.

Fuentes, A. and Barbara Ehrenreich. 1998. *Women and the Global Factory*, Boston: South End Press.

Halim, Ujjaini. 2011. 'Untold Agonies from Asia's SEZs: Lessons for PNG', http://nancysullivan.typepad.com/, accessed on 16 January 2014.

International Confederation of Free Trade Unions. 2003. 'Export Processing Zones: Symbols of Exploitation and a Development Dead-end', http://www.icftu.org/www/pdf/wtoepz, accessed on 16 January 2014.

International Labour Organization. 2002. International Programme on the Eradication of Child Labour, ILO, Geneva.

International Labour Organization. 2004. *Towards a Fair Deal for Migrant Workers in the Global Economy*, International Labour Office, Geneva.

——. 2007. *Forced Labour Statistics Factsheet.*

Kaban, Elif. 1998. *UN Labour Body Urges Recognition of Sex Industry*, ILO, Reuters.

Lim, L.L. 1998. *The Sex Sector: The Economic and Social Bases of Prostitution in Southeast Asia*, ILO, Geneva.

Mason, Barry. 2004. *Women Workers Face Super-exploitation by Global Corporations: An Oxfam Report*, http://www.wsws.org/en/articles/2004/02/wome-f25.html, accessed on 16 January 2014.

Mission, Gina. 1998. 'The Breadwinners: Female Migrant Workers', *WIN: Women's International Net Issue*.

Moghadam, V.M. 1999. 'Gender and Globalization: Female Labour and Women's Mobilization', *Journal of World Systems Research*, 5 (2): 301–314.

NASSOM Report. 2009. Gender Inclusivity in India: Building Empowered Organizations, NASSCOM.

Nasscom Strategic Review. 2012. www.nasscom.com, accessed on 16 January 2014.

Ray, P.S. and Sanhati. 2009. *Falta SEZ in West Bengal: A Fact-finding Report on Workers and Environmental Impact*, www.sanhati.com, accessed on 16 January 2014.

Reigstad, Erlend. 2007. *Special Economic Zones: Policy Review and Implications for India*, http://bora.nhh.no/bitstream/2330/1, accessed on 16 January 2014.

Senser, R.A. 2002. 'How Today's Trading System Hurts Women', *Human Right for Workers*, 7 (6), http://www.senser.com, 3 June 2002, accessed on 16 January 2014.

Stiglitz, Joseph. 2003. *Globalization and Its Discontents*, New Delhi: Penguin Books.

UN Office on Drugs and Crime. 2006. *Trafficking in Persons: Global Patterns*, Vienna.

UN Statistical Division. 2000. http://unstats.un.org/unsd/demographic/gender/wistat/, accessed on 16 January 2014.

Varia, Nisha. 2005. 'Maid to Order: Ending Abuses Against Migrant Domestic Workers in Singapore', http://hrw.org/reports/2005/singapore1205/2, accessed on 16 January 2014.

World Bank. 2000. *Global Economic Prospects and the Developing Countries*, Washington, D.C.

World Bank Report. 2012. *Developing Countries to Receive Over $ 400 Billion Remittances in 2012*, http://www.worldbank.org/en/news/press-release/2012/11/20/developing-countries-to-receive-over-400-billion-remittances-2012–world-bank-report, accessed on 16 January 2014.

World Trade Organization Fact Files. www.wto.org, accessed on 16 January 2014.

11

Understanding Gender, Water and Development

A Critical Appraisal

Arvinder A. Ansari

> 'She sends me to fetch water
> Very early in the morning
> Oh! Grandfather it is very difficult for me
> My pot never fills up fully
> The water is so deep
> That my rope hardly reaches it
> The sun rises and also sets
> By the time, I return
> Unable to collect even one pot-full of water'
>
> (A Folk Song of Rajasthan)

Water is the lifeblood of civilizations – a critical component for all life and a necessary foundation for the growth and evolution of society. As cities continue to grow and environmental problems multiply, managing the demands for fresh water becomes a more urgent challenge that affects all aspects of society. A recent poll of 200 leading scientists from 50 countries ranked lack of fresh water as an environmental priority second only to global climate change (UNESCO 2003) millions of people round the globe suffer from the lack of access to clean water, adequate water for food and for other uses. India, despite being a water adequate country, there are still thousands of villages (tanker fed) and many urban settlements either with scarce water supply or without any source of water. If there is no source of water in 2.5 kilometres, then the village becomes no source water village or problem village (NCW 2005: 7).

The four major users of water in India are agriculture, domestic use, and industry and power generation. The water allocation between varying demands for water by different users have tended to ignore the underlying conflicts and have, therefore, aggravated the problems of inequalities and improper distribution. About 84 per cent of the water is estimated to be used in agriculture. Though India is one of the wettest countries in the world in terms of average annual rainfall, yet there are problems of distribution both spatially and temporally with wide variations. Drinking water has always been on top of the priority list of social consumption items, because it is not only essential but also scarce. Keeping this in view in India, the 7th Plan aimed at providing safe drinking water to all sections of the population. The objective of providing drinking water to all sections of people itself was not fulfilled let alone increasing the norm – for instance, only 72 per cent urban people get safe drinking water and the coverage of poor is still lower at 66 per cent. 'More than one billion people are deprived of access to water of sufficient quantity and quality to meet even minimal levels of health, income and freedom from drudgery. Poor women are particularly affected. It is primarily women who bear the daily burden of hauling heavy buckets long distance to meet the domestic water needs of their families ... water has never been a 'free good' for poor women. Meeting the multi-faceted water needs of poor men and women should be a priority in water policy at the international, national, basis and community levels' (Koppen 2002).

I

Why Study Women, Water, and Development

The universal symbolism of water as the primal fluid – purifying, regenerating, creating and destroying – is closely intertwined with women's roles as bearers and nurtures of life, and collectors and managers of household water. In India, water is sacred and rivers are considered goddesses (*Ganga Ma*). Stories abound about the relationship between water and the feminine principle. Connecting the life-bestowing power of water with the sacrifice women make during times of water scarcity. Despite the recognition that access to water is a prerequisite for good health and a sustainable livelihood, and the gender equality and women's empowerment are essential to poverty eradication, no readily agreed upon conceptual framework comprehensively links gender equity, poverty and water deprivation. The World's Women Summit 2010 of the United Nation reports that more women have remained poor, continue to receive minimal pay for their labour, if not totally remain unpaid, and rely mostly on natural resources in their immediate environment for food, sustenance and livelihood, while taking on time-consuming and hazardous household labour for their families, such as water and firewood collection. The report further states that women are assigned the chore of fetching and collecting water, as data in 38 out of 48 countries included in the study show that there are more adult women in households responsible for acquiring the family's drinking water supply than adult men. Girls under the age of 15 are also more likely than boys of the same age to be in charge of water collection (NCW 2005: 5).

'Gavoon Main Paani Aajaye' (Village should get water supply) is still a big slogan and dream of a plethora of villagers, especially women. It has been seen that even women enter politics or contest Panchayati elections to address the water issue. At the global level, it is recognized that women's basic need is having access to an improved, convenient, reliable and safe source of water close to their home. With a focus on this aspect, community water supply programmes (CWSP) have been initiated throughout the developing world, aiming at providing women with improved water supply technologies (Ben and Farhana 2002).

By the mid-1970s it had become clear to scholars, practitioners, and advocates engaged in the field of economic development that development and modernization were affecting men and women in different ways. Rather than acting as a rising tide for all boats, the intended and unintended consequences of modernization were bypassing many women and even adversely affecting others (Tinker 1990). Overall, the process of economic development was radically changing women's roles in the home and in society at large (Boserup 1990). At the same time, it was also seen that especially rural women spent long working days on diverse activities such as farming; agro processing; marketing; handicrafts; animal husbandry; collecting water, fuel, and fodder; and childcare, all of which contributed to the local economy and to development more broadly (Pearson 2000). Women came to be recognized as key participants in efforts to alleviate poverty and achieve social transformation. Water use and water management have been recognized in these literatures as central to sustainable development efforts as well as to women's domestic and economic activities.

II

Outline of the Paper

The primary purpose of this paper is to review marginalization of women's water rights, increasingly under threat of different dimensions of a patriarchal water order by reviewing the diverse and somewhat uneven literature on women's access to and use of freshwater since international agencies have made water for women a cornerstone of their development and humanitarian efforts (Cleaver and Jobes 1996); but much debate surrounds the nature and consequences of their policies. Influential gender-equity advocaters argue that there is a positive synergy between women's interests and the management and conservation of natural resources (Shiva 1989; Merchant 1996). However, critics counter that naively designed 'women-centred' projects can merely add to the responsibilities of already overburdened women, without transforming the power imbalances that constrain their lives at home or in society (Green and

Baden 1995) So are water projects, in which women play a central part, likely to increase their workloads rather than their well-being? Is the participation of women in these projects either necessary or sufficient for project success or for higher levels of access for women? Can more secure access to water spill over into greater economic and decision-making power? These questions can be answered in a tentative manner at best. This is in part because of the general disconnect between scholarship on water policy and scholarship on gender and also in part because the water sector is weak on the kinds of data we would need to get more definite answers.

The women-and-water literature is largely made up of policy documents produced by international agencies. These documents are valuable on their own terms, but also need to be understood in light of the dominant framings of gender, environment, and development that (often implicitly) inform their assumptions, data collection, and conclusions. Specifically, these framings inform the evaluation of interventions aimed at increasing women's access and participation. The field of women and water integrates the broad fields of gender, environment, and development.

III

Gender, Environment and Development

In order to integrate the gendered nature of social, economic, and political processes into development theories, three broad frameworks were proposed over the 1970s and the 1980s. Although the gender and development, has become the mainstream position of most academics as well as donors concerned with development, all three versions can be found in research and policy documents on women and development. The way in which the role of women within development is conceptualized determines, often implicitly rather than explicitly, why access to water is considered important and in which ways access should be sought

In the 1970s, the term 'women in development' (WID) came to stand for the argument that economic development was leaving women out because they lacked the access, or were not allowed access, to the very resources that made development possible. These resources included healthcare, education, voting rights, employment opportunities, credit, and also basic services such as water and sanitation. Women, it was argued, had to be integrated into the development process (Boserup 1970).WID was subsequently critiqued because its focus on access to the tools of development ignored the complexities of gender relations in different societies, and because emphasizing the importance of women's access for development emphasized the instrumental over the intrinsic value of such access. Over the next two decades, the mainstream position on this issue moved from 'women' to 'gender' and from 'in' to 'and', i.e. gender and development. Nevertheless WID-style arguments appear frequently in the water literature, for example, arguing for easier access to water because women can use their freed up time to generate more income for their families.

In the early 1980s, an influential group of feminist scholars attacked the WID approach for assuming that being left out of development was the problem when it was the very process of development that marginalized women. The women, environment, and development (WED) proponents' core premise was that there was a special relationship between women and natural resources because they use them most and know them best. This 'ecofeminist' perspective saw analogies between the marginalization of women and the marginalization of the environment, arguing that the dominant and male forces of development were responsible for both (Plumwood 1992).The WED position had two related strands. The first, argued that women, especially poor women, were victims of a development process that was destroying the environmental resource base. The second, argued that women were the natural defenders of land, trees, and water and could use these efficiently and sustainably if given the chance to do so. WED was, and continues to be, heavily criticized for universalizing women's natures.

Starting from the mid-1980s, gender has been mainstreamed into the literature on resources and on development as well as into the policy documents of international agencies, through the framework of gender and development (GAD). GAD challenges general assumptions about what women need, and emphasizes the importance of understanding gender relations, which are dynamic, in order to understand, for example, how women use water. Gender equality is understood to be contextual rather than rigidly defined. Furthermore, GAD argues that women and men have interests in addition to their gendered interests, such as those associated with ethnic or class identities. Rather than assume that women are naturally sustainable users of resources, GAD asks why women have historically developed these relationships. Rather than assume that access to credit will enable women to irrigate their land, GAD asks what are the common and conflicting interests that men and women have over that land? GAD is explicitly the gender position of the World Bank, the Asian Development Bank, and the United Nations Organization members.

IV

Gender Dimension of Water Use and Management

While the visibility of women in community water management institutions is an important starting point, it does not necessarily lead to their active participation in water decisions. Or challenge the gendered drudgery of water collection. Contextual factors such as culturally defined norms and perception about appropriate female behaviour affects women's participation as much as the facilitating or capacity-building role of a gender-sensitive development organization. So, what does 'understanding gender' mean in the context of water management, and how is it different from the more simplistic notion of 'involving women'?

In 1992, the International Conference on Water and the Environment held in Dublin, Ireland, produced four key principles to guide policies for water and sustainable development. Known as the Dublin Principles, they represent the international consensus, albeit one that was reached after contentious discussions, on best practices in the water sector. They are as follows:

1. Fresh water is a finite and vulnerable resource, essential to sustain life, development and the environment.
2. Water development and management should be based on a participatory approach, involving users, planners, and policy makers at all levels.
3. Women play a central part in the provision, management and safeguarding of water.
4. Water has an economic value in all its competing uses and should be recognized as an economic good.

Agenda 21, the comprehensive action-plan which emerged from United Nations Conference on Environment and Development (Earth Summit, Rio de Janeiro, 1992), re-emphasizes the need for women's participation in water management policy making and implementation calls on governments to promote programme to reduce the workload of women and girls, and facilitate their participation in schemes to adopt alternative technologies, and access clean water and sanitation. But despite principle 20 (political declaration) reiterating the need for the full participation of women in order to achieve sustainable development, there is no analysis of gender differentiated needs, uses and constraints with respect to access and control of water of the complex social process underlying natural resources management (Guijt 1994).

Subsequent international forums, whether on development, Cairo 1994; (Summit for Social Development, Copenhagen, 1995) nor women's issues (Fourth World Conference on Women, Beijing, 1995), have focused on the disadvantaged position of women and children in the context of environmental degradation and their continuing lack of access to clean water and sanitation. As a response to the growing feminization of poverty, multi-stakeholders' action platforms have called upon governments, international agencies and civil society to not only provide technical assistance for women, but also to involve women in policy and research on the structural linking between gender relations, environment and development. In practice, however, most water supply projects have drawn on women's labour without enhancing their technical or managerial skills, and have based their finding on the assumption that involving women will in itself advance the objectives of the project and benefit women (Green et al. 1998: 265).

Such categorizations see women as a distinct but homogenous target group with particular skills and knowledge arising out of their (primary) role as domestic water collectors and users. This conceptualization of women as *privilege water managers* is rooted in the essentialisms of 'eco-feminism' which tends to position women as natural caretakers of the environment because of their childbearing and nurturing roles. While women play an important role in household, water and food security, overlooking the diverse material relationship that women have with water-mediated by class, caste, age and gender – can be problematic. Feminist political ecology (Crow 2001) and feminist environmentalism (Agarwal 1992), on the other hand, share a number of core concerns around a more dynamic conceptualization of women's relationship with the environment. They see relationship as being located in larger social and political structures and cultural practices, and in the 'symbolic' construction of power. Women's 'special' relationship with the environment emerges from the work they do (gender-based division of labour). Their 'incentive' for water conservation is related not just to their dependence on the resources but also to a structure that does not allow them the same access to other economic opportunities as it does to men. For example, in the event of drought and extensive male migration, it is the women who manage the land, care for the old and young dependents, and take part in drought relief work that often involves the construction of water harvesting structures.

A more gender-aware approach to water resource planning and management has largely emerged from the mid-and late-1990s as part of a global agenda not only to *mainstream* gender but also is a technical and political process that requires shifts in organizational culture, goals and strategies, and involves processes of learning and 'unlearning' underlying perception about roles of men and women in water management (Coates 1999). Mainstreaming is more than the mere collection of gender disaggregated data or gender differentiation of project components (technical aspects); it needs to address hard, political questions in water policy and resources allocations.

The most visible of these agendas were the global partnership involving a range of stakeholders, including national and regional governments and civil society organization, in the articulation of *Vision 21: A Shared Vision for Hygiene, Sanitation and Water Supply and a Framework for Action.* Facilitated by the Water Supply and Sanitation Collaborative Council (WSSCC), Vision 21 was shared at the Second World Water Forum at Hague in March 2000. At the core of its ethos is collaborative action by empowered men and women, communities, and water management authorities, to provide safe water and sanitation for the estimated one billion people who are still denied this access. Coupled with improved hygiene practice and food security (through more equitable accesses to irrigation), it is envisaged that this will eventually lead to poverty reduction.

For the first time, a global charter defines gender roles, rights and responsibilities in Water planning, and acknowledgement that past strategies have largely failed because of their simplistic approaches involving women without looking at complex gender inequalities, including male responsibility for water management. Most of the credit for gender mainstreaming goes to the networks and alliance of NGOs, grassroots groups and individual, both men and women, who have been meeting and communicating frequently (often through the internet) to highlight the gender dimension of water use and management. These include the Gender and Water Alliance, coordinated by an international board with a secretariat and a web-site managed by the International Reference Centre for Water and Sanitation at Netherlands,

Since the Vision 21 processes, a plethora of global platforms have reiterated the need to involve women in water and development initiatives with varying degrees of commitment to gender equality. For example, the UN Millennium Development Goals 2000, calls for the promotion of gender equality and women's empowerment, mainly through access to education (Goal 3), while also stressing the need to eradicate poverty and hunger, improve maternal health and ensure environmental, sustainability-goals which impact the lives of poor women worldwide. 'Promote gender equality and empower women' and to ensure environmental sustainability, with a specific target to halve by 2015 the proportion of people without sustainable access to water and sanitation.

The International Conference on Freshwater (Bonn 2001) reiterates the need to involve women and men in participatory water management, and seeks to promote gender equity for access to water and sanitation on the one hand, and access to decision-making and leadership roles on the other. The World Summit on Sustainable Development held in Johannesburg, 2002, builds on other international commitments to promote women's equality with men in decision making, to mainstream gender in all policies and strategies, and to ensure that infrastructure and service for the poor are gender-sensitive. The Plan of Implementation recognizes the inequities in access to resources, information and opportunities that women, particularly from poor, rural and urban women-headed households, encounter.

Access

Access to the minimum quantity of water necessary for domestic use, usually meaning that for drinking,cooking,washing utensils and basic hygiene, can be defined in many ways. The UNICEF/World Health Organization Joint Monitoring Programme (JMP), the main source of national level data on access, defines reasonable access as 20 litres per person per day from an improved source, no more than 1 km distant from the dwelling (WHO/UNICEF 2000). *Improved sources* are household taps, public standpipes, boreholes, protected dug wells, protected springs, and collected rainwater. *Unimproved water sources* according to the JMP are unprotected wells; unprotected springs, rivers, or ponds; vendor-provided water; and bottled water. Current JMP estimates are that 85 per cent of the population has access in Latin America and the Caribbean, 81 per cent in Asia, and 62 per cent in Africa. Globally, approximately 65 per cent of the population without access to safe water lives in Asia and 28 per cent in Africa. Even though these access numbers do not directly translate to those of access for women, they are reasonable proxies because it is almost always women and children who are responsible for the daily provision of domestic water (Table 1).

Table 1

Access to drinking water by region

Region	*Area within region*	*per cent of population with access*
Latin America and Caribbean	Overall	85
	Urban	93
	Rural	62
Asia	Overall	81
	Urban	93
	Rural	75
Africa	Overall	62
	Urban	85
	Rural	47

Source: WHO/UNICEF, Joint Monitoring Programme (JMP) 2000.

The Gender and Water Alliance further claims that gender development indices can be used as indicators of overall water supply development because of the positive correlation between low achievement along the Gender Development Index (GDI) and low water coverage (GWA 2003). For example, putting together data collected by the JMP and the Human Development Report, it can be seen that eight of the ten countries in Africa with less than 50 per cent of overall population served (Burkina Faso, Chad, Democratic Republic of Congo, Eritrea, Ethiopia, Madagascar, Mauritania, and Rwanda) all scored higher than 120 on the GDI out of 146 countries ranked, indicating low level of gender development.

Lack of access to an improved water source, or even difficult or unreliable conditions of access, translates to what the 2006 UN Human Development Report (UNDP 2006) has called 'time poverty' for women and children. NGOs working in rural India report that women in many villages find themselves walking 2 km or more to their drinking water source. Time-allocation studies have found that women in the Konkan region of Maharashtra spent nearly 2 h/day fetching and storing water (Joy and Paranjape 2005) and that, in the dry season, women in the state of Gujarat had to walk 3–4 h/day in order to secure the daily domestic supply (James et al. 2002). The same study from Gujarat also presented evidence that when access to water improved, the women devoted more time to income-generating micro-enterprises. Similarly, an econometric study of time allocation in rural Pakistan found that 15 per cent of an average woman's monthly work effort went to collecting water and that more accessible water infrastructure would allow some substitution of income-generating activities for water collection (Ilahi and Grimard 2000). It has, however, been argued that the implication in many time allocation studies, which is that women would necessarily use time freed up from collecting water in income-generating activities. Freed up time is used for income generation if such opportunities exist and if women are able and willing to use them. Some researchers have made the case that easier access to water is desirable not just for economic reasons but for overall quality of life, regardless of how the extra time is spent (Carr and Sandhu 1988).

Although the disproportionate burden that water provision imposes on women is well documented in the water and development literature, the opportunity cost of girls' time is less studied. The 2006 Human Development Report concludes, mostly on the basis of African case studies and interviews in the course of producing the report, that there is a 'straight trade-off between time spent in school and time spent collecting water' and that this is much less true for boys than it is for girls. This is an area where more research must be conducted. If confirmed by further research, the lack of reliable access to water and sanitation could be a major contributor to continuing gender inequality in education and the opportunities that education can provide.

V

Gender Division of Labour

In examining the different roles of women and men, the gender division of labour provides the underlying principle for differentiating the water-related work that men and women are assigned on the basis of their gender, leaving aside the process of childbearing. Feminist would argue that it is the penetration of (Western) capitalism which has historically and artificially separated productive and reproductive spheres of work leading to the subordination of women, though this does not necessarily mean that gender relation in more traditional or 'primitive' societies are, or were, more egalitarian.

Reproductive Work

Reproductive work, encompasses all the largely unpaid activities which go into the maintenance and reproduction of the households, and are mainly done by women (and girls). Despite attempts to ideologically reinforce it (most notably by religious institutions), the gender division of labour is being challenged by economic policies, technological advances, the media and a range of development catalysts. But to a large extent it is still this division of labour, mediated by class and caste, which determines the water-related roles and responsibilities of women and men.

Water Collection for Household/Domestic Needs

In most poor rural and urban communities, women and girls are responsible for collecting water for drinking needs, including those of livestock, for cooking, cleaning, health and hygiene – work that falls in the reproductive sphere and is therefore, unaccounted and unpaid for, despite the fact that it is acknowledged worldwide as one the most time-consuming and arduous tasks. Not only do women suffer acute physical problems as a result of head-loading heavy pots (up to 20kg in a single journey) but the drudgery of water collection in arid and drought-prone areas constrain their time for productive work. In Banakantha district, SEWA's member used to spend up to four hours a day walking a minimum of 6 km daily to fetch water, where young girls are involved either directly in water collection or indirectly in looking after younger siblings left at home; it has an impact on their access to education and their self-development. Unless educational opportunities are available to young girls, and parents are willing to send them after they have completed their households chores, this situation is not likely to change (Upadhyay 2003).

However, water collection does not affect all women equally – socio-economic status of the households, age and marital status cultural factors such as seclusion and household composition – all affect a women's water-related work. In joint or extended families, the heaviest burden tends to fall on daughters-in-law and, one day, becomes the hounded matron who 'gets her water fetched in her old age'. In better-off households with private ownership of water assets (wells, roof water collection tanks), the time spent on water collection and the drudgery are significantly reduced. Where men do fetch water, it is usually the result of seclusion norms restricting women's mobility in the public domain, or the introduction of water harvesting technologies which have brought water closer to households (e.g. roof-water collection tanks). Apart from the small number of single men who must collect water for their own households, men collect water mainly for business enterprise, for example, for the small restaurants and tea shops they run, or to sell. Moreover, men often have access to transport (a bicycle, wheelbarrow, tractor or animal power) and use lighter utensils such as plastic jerry-cans (Mansuri and Rao 2004). Rarely do men head-load water, it is important to note that in lower caste and Adivasi households, for men to share household tasks is not uncommon – for example, looking after children, or cooking food if women are collecting water or attending meetings. However, given the growing trends towards 'Sanskritization' in Dalit and Adivasi communities, this assumed gender equality is being challenged.

VI

The Care Economy: Health Hygiene and Sanitation

Family health is also viewed as a female responsibility – women bear the main burden of caring for those who are ill; waterborne disease kills 3.4 million people annually, mostly children (GWA 2003) despite the specific problems they face as collectors of water from contaminated water sources. On the other hand, the lack of accessible sanitation coupled with limited privacy, and in many conflict-prone areas, the fear of rape or harassment, affects women and girls the most. Not only has deforestation and the loss of vegetation cover in rural areas forced women and girls to rise earlier and walk farther in search of privacy, but slums and shanty town in urban areas leave them no choice but to squat along railway embankments. In drought-prone areas, women do not have enough water to meet personal hygiene needs, particularly during menstruation and post-natal periods. They may also bathe and change clothes less frequently than men do, as they have to bear the burden of washing them too. The absence of separate sanitation facilities in most schools affects the attendance of girls is higher classes.

Hygiene education to promote the adoption of better sanitation practices (households or community latrines instead of open defecation) and to mitigate the impact of waterborne disease is currently largely directed at women, with little attention being paid to the roles and responsibilities of men in families. But much less attention has been paid to the health risks that women face as water carriers. Women and (usually) girl children fetch water in pots, buckets, or ideally more modern narrow-necked containers, which are carried either on the head or on the hips. A family of five within 1 km of an improved water source would need 100 litres of water a day to meet its minimum needs. The weight of that water is 100 kg (220 pounds) without the container. Plastic is the lightest material for carrying the water; traditional clay pots are much heavier. In these circumstances, women and children may need to walk to the water source two or three times daily, with the first of these trips taking place at around 5 a.m. Globally, more than 50 per cent of poor women suffer malnutrition and iron deficiency (Dasgupta 1993), and thus it should not be surprising that, especially during the dry season in rural India and Africa, 30 per cent or more of a woman's daily energy intake is spent just in fetching water. When several trips are not possible, families manage with 10 litres or less per person per day, even if they live within 1 km of an improved source, and thus have 'access'. Carrying heavy loads over long periods of time causes cumulative damage to the spine, the neck muscles, and the lower back, leading to the early ageing of the vertebral column (Page 1996). Symptoms in the form of constant pain in the neck and the knees may also occur (Joy and Paranjape 2005). The burden of daily carrying is rarely covered in leading public health and epidemiological journals, as it falls outside of the conventional categories of water-borne, water-washed, and water-related ailments.

Productive Work

Productive work includes the production of goods and services for subsistence and income. It is this work which is mainly recognized and valued as work by society, and which is most commonly included in national economic statistics. Both women and men perform productive work, but women's work tend to be undervalued and underpaid.

Water for Production

Both women and men use water for productive purpose, whether it is for substance, rain-fed agriculture, irrigation for cash crops, or for livestock (animal husbandry, dairy farming). In addition, women work in numerous small-scale enterprises which depend on varying types of water – for example, brackish water aquaculture, soft water for dyeing and block printing, and so on. Access to water is vital for sustainable livelihood security, and intervention that changes the control, use of end access to water resources inevitably raise gender equity issues and opportunities (World Bank 2001).

A woman's access to water for productive purpose depends on a number of factors, such as the social status of the household and her economic contribution. In the case of small and marginal farmers or *Adivasi* household where women are involved in agriculture, women need timely access to irrigation as the case study on (AKF/ANANDI 2002) illustrates. When there are competing claims for water, particularly during periods of water scarcity, women's ability to negotiate water rights will depend not only on their position within the households and claims to lend tenure. But also to social networks based on kinship ties, neighbourhood bonds, trust, and reciprocity, as argued later.

Many other economic activities which use water for production – for example, aquaculture development and fisheries, marine (dependent on access to the seas and oceans) or inland – have less documented gender dimension. Although women in fishing communities are involved in a number of tasks such as marketing, making and mending nets, fish processing, and so on, there is little data on the nature and condition of work and even less recognition by the State (e.g. Census figure or budget allocation) on the number of women employed and their gender specific needs and priorities. Data on water use and requirement of the numerous small scale enterprises run and managed by women or women's group are rare.

VII

Gender Priorities for the Development of Water Resources

Women and men have different roles, rights and responsibilities, and exercise different levels of control and power over resources; they often have different needs. The nature and valuation of work in a given context affects the way women and men determine their water priorities and likewise, their capacity to plan and participate in a project, though this is also structured by gender differences in endowments (e.g. technical skills) and asymmetries in right to water resources planner and managers have typically seen women's water need as falling in the domestic sphere, and men's in the productive. However, this false dichotomy is far removed from the practical realities of rural livelihood strategies and overlooks the dynamic inter-linkage between productive and domestic responsibilities for men and women. Gender needs and interest are typically categorized into two areas:

Practical gender needs (PGN): are linked to the daily condition of women's lives, their immediate environment, workload and responsibilities. Access and availability of drinking water is a critical PGN for women, and relatively straightforward to provide, but it is unlikely *on its own* to change gender inequalities underlying who collects water. On the other hand, many drinking water projects designed to reduce the drudgery of women's work can increase their workload. Consequently, women are spending more time collecting water as household water consumption has also increased, partly because the new stand posts and tube wells are located near the roadside. Here women do not have the privacy to bath, so they carry water home for bathing too (Regmi and Fawcett 1999).

Strategic gender interests (SGI): seek to challenge the disadvantaged position of women in society relative to men in terms of labour, power and control over resources. They are not readily identifiable and vary according to particular roles, rights and responsibilities of different groups of women. In relation to water, the most critical SGI is the articulation of water rights. Water rights seek to provide poor, marginalized and landless women with access to water for livelihood needs. In addition, other factors have a bearing on the articulation of SGIs: technological option, capacity-building of women water users to manage community water resources and their collective organization to negotiate access to water resources.

However, practical gender needs and strategic gender interest cannot be neatly separated – practical intervention are not just important points for development agencies, they also have the ability to influence strategic areas of life (power relations and control) if they consciously enable women to challenge embedded gender inequalities. That is, the *redistributive* potential of PGN needs to be considered more actively.

Gender, Water and Participation

In the years leading up to and immediately following the Dublin Principles, water researchers as well as practitioners at the community, national, and international levels have become more gender sensitive than was previously the case. There was broad agreement on the need to include women in water planning and decision making, but perhaps less of a consensus on what such inclusion meant in a concrete sense, in different contexts, and at different scales. During the 1990s, however, researchers and practitioners alike gradually converged on the desirability of local-level or community participation in water management, especially that involving women.

The voices in favour of women participating in decision making over the use and management of local water resources can be (a) WID leaning, arguing that women's involvement is central not only to their empowerment but also to greater efficiency and sustainable development; (b) mixed WID and WED leaning, arguing that women use and know local resources best and, therefore, represent the natural loci of decision making; and (c) GAD leaning, arguing that attention to how both women and men use and negotiate over water resources is important (Jackson 1998), as is the potential for women to increase their self-confidence and status through control over water (Wijk and Sijbesma 1985). The failure of many community-based water resource management projects has been attributed to the exclusion of women at all levels of the project and to the inability of project planners to take their (often hidden) knowledge and priorities into account (Kalbermatten 1991). Numerous reports from Asia and Africa suggest that the inclusion of women as participants and decision makers increased their access to, and control over, local water resources.

Most of these reports are case studies, and some of them may seem more advocacies-oriented rather than analysis driven. Cases for these analyses are purposively rather than randomly chosen; therefore, it is important to recognize potential selection biases in the case study literature on women, water, and participation. Nevertheless, these studies provide insights into the ways in which women-centred policies have affected women's control over water resources, and the consequences of that control.

Early studies from Zimbabwe, Ethiopia and Kenya came to mixed conclusions about the role of women in participation and decision making with respect to the maintenance of village hand pumps … . A study by Cleaver (Cleaver 1991) reported that the national policy of Zimbabwe was to have women take prominent roles in their local Waterpoint Committees but that few women actually were in such leadership positions. Nevertheless, traditional authority figures in many villages were quite effective in managing their local waterpoints. The Ethiopia study, however, argued that the development of successive generations of the Afridev Pump had, with World Bank and UNDP support, evolved to the point where it was simple enough that rural women readily undertook the installation and maintenance of the pumps. Installing and taking care of the pumps gave women much appreciated control over both water and technology (Wood 1992). The Kenya study describes the frequent breakdowns and lack of maintenance of hand pumps in a cluster of Kenyan villages until a FINNIDA grant trained women to repair the pumps. The maintenance improved, as did the self-esteem and authority of the women mechanics. However, the women were expected to install, operate and care for the pumps without compensation, whereas the previously lax and male mechanics had been paid (Hoffman 1992). A more recent study of the Mvula Trust, a South African NGO devoted to community capacity building for sustainable water and sanitation, highlights an example of women-led community participation in KwaZulu Natal. Community standpipes and their diesel pump were maintained and paid for by community members, and all the standpipe 'wardens' were women. The sustained success of this arrangement was partly attributed to the prominence of women, drawn from a long-standing women's community organization, in the Village Water Committee (WSP 2002).

More recently, an ambitious report from UNICEF and the Water and Supply and Sanitation Collaborative Council (WSH 2006) provides case-based evidence on women-centered participation in water and sanitation efforts from around the world. This study is squarely in the WID tradition; its opening paragraph argues for improvements in women's capacity to make water and sanitation decisions and states, 'ultimately, what is good for women is good for the family and the whole community, who share the benefit from all these improvements.' Citing a wide variety of experiences from Pakistan, Nepal, India, South Africa, Kenya, Tanzania, and several other nations, the report finds that 'placing women at the center' of water and sanitation decisions can lead to more households with access to water, more cost-effective service delivery, better placement and maintenance of water infrastructure, better community health and hygiene, and less corruption in financial matters. The report concludes that when women are enabled to make decisions and act on their priorities the overall impacts are always a safer and more sustainable water system and greater empowerment of women.

By contrast, a statistical study of World Bank water projects in rural India is more circumspect about the consequences of women's participation. The analysis of participatory water projects in 45 villages found that women's involvement was often of a token kind; although in the few cases of substantive decision-making power, participation was found to lead to greater self-confidence and status. Overall, though the correlation between community participation and project success was positive and significant, there was no effect of women's participation in project success (Prokopy 2006). Similarly, a comprehensive review of 121 rural water supply projects from around the world found that women were not critical to participation and that it was a myth that involving the community in the project would automatically involve the women (Narayan 1995).

For almost three decades, it has been understood in the water management community that a project plan with no mention of a role for women cannot improve the status quo with respect to empowerment or access to water. However, a review of the case-based literature on women and water in the domestic sphere reveals that it is not clear when and on what terms participation in community projects improves their access to and control over water, or their empowerment overall. Three broad points emerge from a review of gender and participatory drinking water projects. First, although the roles that women play in planning and management can be very important to water projects, effective community participation does not always require women to play a central part. Firmer conclusions from the empirical evidence cannot be reached because there are few projects for which comprehensive gender-disaggregated data are collected. Many well-cited studies of participation in the water sector do not even mention women and instead use the household or the community as the smallest units of analysis (Kleemeier 2000).

Second, participation takes many forms. At low levels of power, although not necessarily of time, it can mean donating labour or perhaps attending meetings without speaking up. At higher levels of power, it can mean active involvement in decisions about water related technologies and priorities as well as the ability to ensure action on these priorities. The original champions of participatory development argued that the priorities, aspirations, and constraints of marginalized groups were to shape the entire process of participation (Chambers 1997). It seems plausible that the level and nature of participation could make a difference both to the success of the water project and to any spillover effects on women's lives. However, many case studies omit the details on what levels of participation were actually achieved and on the mix of project design, social structure and women's preferences that constrained or enabled particular forms of participation.

Third, it may be necessary to read between the lines of a case study to understand whether the central role of women in managing and safeguarding water resources was considered of instrumental or intrinsic value. Embedded in most case studies, often only implicitly, is a WID-, WED or GAD sympathetic worldview. Projects that enable women to supplement their agricultural incomes on account of access to water (Melchior 1992) or that require women to work extra hours to manage their water sources may be successful in WID terms but may contribute less to empowerment (Mayoux 1995). Participatory projects in which women take on water-related responsibilities for which they are not paid have been critiqued for their WED-like assumptions that women are natural protectors of water and thus do not need payment (Cleaver and

Elson 1995). However, the more complex GAD based perspective, which arguably represents the mainstream academic and donor position, is more commonly used as a critique of existing interventions than as the guiding principle for new interventions (Guijt and Shah 1998).

Concluding Thoughts

Gender-sensitive approaches to water resources management are desirable for achieving efficiency, social equity and gender-equality goals. Targets, such as those in the Millennium Development Goals relating to water, are unlikely to be achieved unless gender perspectives are integrated into planning and implementation activities. Instrumental approaches to ensuring more reliable, sustainable and well-managed water supplies are essential to achieving access to water for all, and for ensuring the maintenance of water in the interests of ecological balance and the needs of future generations. However, social and economic targets (such as eliminating poverty, furthering empowerment of marginalized groups, supporting the resilience of the vulnerable and ensuring resources are appropriately managed by those who use them) will only be achieved by a wider focus on social and power relations. Adopting gender-sensitive approaches, therefore, means rethinking water development in a number of ways.

First, it is critical to recognize the need for intersectoral cooperation. People's livelihoods are not divided into subsectors and imposing artificial boundaries on the management of water for different uses is unlikely to yield results. A gender-sensitive approach helps to overcome some of the artificial subsectoral divisions in water as it involves looking at women's and men's lives as a whole and how they are shaped through gender norms and practices.

Second, gender sensitivity necessitates a flexible learning approach to development interventions. Just as natural conditions and the uses of water vary from place to place, so gender expectations and norms differ according to context. Blueprint approaches to project planning and management cannot reflect this, so development agencies need to pay attention to training and capacity-building which allows for a reflective and flexible approach to water resources management at the local level. Gender relations can and do change over time, and, by adopting participatory learning approaches, it is possible for agencies to support and facilitate such changes in progressive ways.

Third, gender relations also impact development institutions, which need to pay attention to the way that such relations impact on the functioning of their own work, as well as on water resources management at the local level. This involves consideration of the type of data collected, the gender balance of staff and the need for gender-awareness training within an organization.

Frameworks for gender analysis of water resources management that encompass issues of social and gender relationships as well as infrastructural provision are required to track both collective and individual actions and recognize both the separateness and interdependencies of women's and men's interests. Such an approach is unlikely to be achieved through the use of checklists alone, and gender analysis cannot be achieved in a one-off event. Rather it requires a rethinking of the way in which development accommodates diversity, complexity and change, while retaining overall goals of an equitable and dignified life for all.

References

Agarwal, B. 1992. 'The Gender and Environment Debate: Lessons from India', *Feminist Studies*, 18 (1): 119–158.

AKF/ANANDI. 2002. *Understanding Women's Experiences in Natural Resources Management*, New Delhi: Aga Khan Foundation.

Ben, C. and S. Farhana. 2002. 'Gender, Class, and Access to Water: Three Cases in a Poor and Crowded Delta', *Society and Natural Resources*, 15 (8): 709–724.

Bonn. 2001. *International Conference on Freshwater: A Report*, http://www.water-2001.de/conferencereport.pdf

Boserup, E. 1970. *Woman's Role in Economic Development*, New York: St. Martin's.

——. 1990. 'Economic Change and the Roles of Women', in I. Tinker (ed.), *Persistent Inequalities: Women and World Development*, New York/Oxford: Oxford University Press, 14–26.

Carr, M. and R. Sandhu. 1988. 'Women, Technology and Rural Productivity: An Analysis of the Impact of Time and Energy Saving Technologies on Women', Occasional Paper 6, UNIFEM, New York.

Chambers, R. 1997. *Who's Reality Counts? Putting the Last First*, London: Intermediate Technology Publications.

Cleaver, F. 1991. 'Maintenance of Rural Water Supplies in Zimbabwe', *Waterlines*, 9: 23–26.

Cleaver, F. and D. Elson. 1995. *Women and Water Resources: Continued Marginalisation and New Policies*, IIED Gatekeeper Series 49, London: Intermediate Technology Publications.

Cleaver, F. and K. Jobes. 1996. 'Donor Policies and Gender in the Water and Sanitation Sector', *Natural Resource Forum*, 20: 111–116.

Coates, S. 1999. 'A Gender and Development Approach to Water, Sanitation and Hygiene Programmes', A Water Aid Briefing Paper, www.wateraid.org

Crow, B. 2001. 'Water, Gender and Material Inequalities in the Global South', Centre for Global, International and Regional Studies, WP#2001–5, University of California, Santa Cruz, http://www@.ucsc.edu/cgirs

Dasgupta, P. 1993. *An Inquiry into Well-Being and Destitution*, Oxford: Clarendon.

Gender Water Alliance (GWA). 2003. *The Gender and Water Annual Report: Gender Perspectives on Policies in the Water Sector*, http://www.genderandwateralliance.org, accessed on December 2004.

Green, C. and S. Baden. 1995. 'Integrated Water Resources Management: A Gender Perspective', *IDS Bulletin*, 26: 92–100.

Green, G., S. Joekes and M. Leach. 1998. 'Questionable Links: Approaches to Gender in Environmental Research and Policy', in C. Jackson and R. Pearson (eds), *Feminist Visions of Development: Gender Analysis and Policy*, London: Routledge.

Guijt, I. 1994. 'Water and Gender on the Agenda: A Review of Water Management and Gender Issues in Agenda 21', proceedings of the workshop on Gender and Water Resources Management: Lessons Learnt and Strategies for the Future, Stockholm: SIDA.

Guijt, I. and M. Kaul Shah. 1998. 'Waking up to Power, Conflict and Process', in I. Guijt and M. Kaul Shah (eds), *The Myth of Community: Gender Issues in Participatory Development*, London: Intermediate Technology Publications, 1–23.

Hoffman, L. 1992. 'Women Handpump Mechanics in Kenya', *Waterlines*, 11: 19–21.

Ilahi, N. and F. Grimard. 2000. 'Public Infrastructure and Private Costs: Water Supply and Time Allocation of Women in Rural Pakistan', *Economic Development and Cultural Change*, 49: 45–76.

Jackson, C. 1998. 'Gender, Irrigation and Environment: Arguing for Agency', in D. Merrey and S. Baviskar (eds), *Gender Analysis and Reform of Irrigation Management: Concepts, Cases and Gaps in Knowledge*, Colombo: International Water Management Institute.

James, A.J., J. Verhagen, C. Wijk, R. Nanavaty, M. Parikh and M. Bhatt. 2002. 'Transforming Time into Money Using Water: A Participatory Study of Economics and Gender in Rural India', *Natural Resource Forum*, 26: 205–217.

Joy, K.J. and S. Paranjape. 2005. 'Women and Water: Relationships, Experiences, Approaches', in B. Ray (ed.), *Women of India: Colonial and Post-Colonial Periods*, New Delhi: Sage, 351–391.

Kalbermatten, J.M. 1991. 'The Water Decade: Personal Reflections', *Waterlines*, 9: 2–4.

Kleemeier, E. 2000. 'The Impact of Participation on Sustainability: An Analysis of the Malawi Rural Piped Scheme Program', *World Development*, 28: 929–944.

Koppen, B. van 2002. *A Gender Performance Indicator for Irrigation: Concepts, Tools, and Applications*, Research Report 59, Colombo, Sri Lanka: International Irrigation Management Institute.

Mansuri, G. and V. Rao. 2004. 'Community-based and -driven Development: A Critical Review', *World Bank Research Observer*, 19: 1–39.

Mayoux, L. 1995. 'Beyond Naivety: Women, Gender Inequality and Participatory Development', *Development and Change*, 26: 235–258.

Melchior, Tellier S. 1992. 'A Lesson in Listening', *Waterlines*, 11: 5–8.

Merchant, C. 1996. *Earthcare: Women and the Environment*, New York: Routledge.

Narayan, D. 1995. *The Contribution of People's Participation: Evidence from 121 Rural Supply Projects*, Washington, D.C.: World Bank.

National Commission for Women (NCW). 2005. *Report on Water and Women*, New Delhi, 5.

Page, Ben. 1996. 'Taking the Strain: The Ergonomics of Water Carrying', *Waterlines*, 14: 29–31.

Pearson, R. 2000. 'Rethinking Gender Matters in Development', in T. Allen and A. Thomas (eds), *Poverty and Development into the 21st Century*, Oxford: Oxford University Press, 383–402.

Plumwood, V. 1992. 'Beyond the Dualistic Assumptions of Women, Men and Nature', *Ecologist*, 22: 8–13.

Prokopy, L.S. 2006. 'Women's Participation in Rural Water Supply Projects in India: Is it Moving Beyond Tokenism and Does it Matter?', *Water Policy*, 6: 103–116.

Regmi, Shibesh Chandra and Ben Fawcett. 1999. 'Integrating Gender Needs into Drinking Water Projects in Nepal', *Gender and Development*, 7 (3): 62–72.

Shiva, V. 1989. *Staying Alive: Women, Technology and Development*, London: Zed Books.

Tinker, I. (ed.) 1990. *Persistent Inequalities: Women and World Development*, Oxford: Oxford University Press.

UNDP. 2006. *Beyond Scarcity: Power, Poverty and the Global Water Crisis*, Human Development Report, New York.

UNESCO. 2003. *Water for People – Water for Life – The United Nations World Water Development Report*, Paris: Berghahn Books, http://www.unesco.org/water/wwap/wwdr

Upadhyay, B. 2003. 'Water, Poverty and Gender: Review of Evidences from Nepal, India and South Africa', *Water Policy*, 5: 503–511.

Water Sanitation Hygiene (WSH). 2006. 'For Her it's the Big Issue: Putting Women at the Centre of Water Supply, Sanitation and Hygiene', UNICEF/Gender Water Alliance (GWA), http://www.genderand water.org/page/5124, accessed on December 2006.

Water Sanitation Programme (WSP). 2002. 'The Mvula Trust in South Africa: An Independent Partner to the Government', Field Note, UNDP/World Bank WSP-Afr.Reg.

WHO/UNICEF. 2000. *Global Water Supply and Sanitation Assessment 2000 Report*, http://www.who.int/, accessed on January 2007.

Wijk, V. and C. Sijbesma. 1985. 'Participation of Women in Water Supply and Sanitation: Roles and Realities', Technical Paper 22, IRC International Water Sanitation Centre, The Hague.

Wood, M. 1992. 'A Handpump for Africa: The Afridev Experience', *Waterlines*, 11: 29–31.

World Bank. 2001. *Engendering Development: A World Bank Policy Research Report*, Washington, D.C.: World Bank and Oxford: Oxford University Press.

12

Development-Induced Displacement

Policy, Information, Awareness and Tribulations in a Globalized World

Manish K. Verma

The era of globalization has visualized a new phase in the history of humankind. Due to incessant scientific advancement and innovations in technology, the scientific world is profoundly taming the natural environment in order to augment development endeavours to turn life comfortable. The era of globalization also converges with the detonation in communication medium and thus turning the world into 'global village', to use the words of Marshal McLuhan. But in analogous to it, another demeaning phenomenon is also shaping up, which is a continuous source of challenge to our development planners. 'The gigantic urge of development activities saw man reaching to such a threshold where their endless craving results in the exclusion and under-development of their fellow brothers' (Verma 2012: 237). Most ironically, at such a moment, when the contemporary global society has entered into 'information age' and known as 'knowledge society' having 'reflex characteristics' (Giddens 1990) in terms of transparency, dissemination of information and awareness. A large number of socially marginalized and economically deprived populace, involuntarily displaced worldwide due to development projects, is forced to live in a communication vacuum having no information at all about their present and future destiny and thus gets exposed to greater 'impoverishment risks' (Cernea 1996). Mathur expresses his despair in a global overview of development projects by saying that, 'Worldwide experience with resettlement has shown that people who are displaced do not easily recover, much less improve, their previous standard of living. Resettlement studies have vividly documented the devastating consequences of failed resettlement projects, which create new pockets of poverty where none existed before. The fact is that for those affected, development has been too often experienced not as an opportunity, but as disruption and impoverishment. Such displacement not only puts affected people to grave impoverishment risks, but also causes a setback to the entire poverty reduction effort. As one observer puts it, 'Truly, development is a very contradictory affair if it reinforces the very poverty that it aims of eliminate' (Mathur 2008: 3–4). The plethora of situation is proving true in the instances of most of the development activities engineered to attain the developmental goals due to illiteracy, ignorance and unawareness on the part of the host communities residing at remote locations in most of the cases.

While acclimatizing with the pace of global challenges, it is well established that no country in the world can survive without development activities. Thus, development is like a life saving device for any country, if it has to survive in the wake of challenges coming from all corners of the globe. It is also true that the goal of developmental activities, by and large, remains void in the absence of barren land. 'Every development project requires large chunk of land, which can be easily available and grabbed in the remotest rural and tribal part of the society, as such the planners generally fall back to these areas. Consequently, overtly the victims of industrial development lose their dwelling and livelihood and covertly their culture, civilization and bonding with the natal place' (Verma 2012: 237). It is not the end of misery for the project affected persons. 'The condition becomes more pathetic when the Project Affected Persons (henceforth PAPs) are not even consulted before, during or after the whole process. Ironically, they remain ignorant and unaware about their future and oscillate between hope and despair. At times, with great amount of expectations, they dream hefty package from the rehabilitation agencies and on un-accomplishment of wishes and desires, they show intense resentment' (Verma 2004: 220–221).

In this background, the main objective of the paper is sociological interpretation of involuntary displacement which awfully disrupts the entire environment of the affected community. Further, attempt is made in the paper to trace the global linkage on the nature and magnitude of development-induced displacement and its aftermath on the affected population between the countries across the globe and Singrauli region of India. By delineating into theoretical acumen, it introspects various policies concerned with the issue and critically evaluates its adequacies. Effort is made to ascertain the level of awareness of the host communities about relocation caused due to development project and tribulations encountered by them due to lack of awareness. In addition to it, the discrepancy between planning and its real execution is also ascertained in the paper. Last endeavour is to comprehend people's perception, reactions and expectations from the entire process. All these investigations have been made in the domain of role played by rehabilitation agencies and its impact on PAPs.

To carry out a systematic study, two National Thermal Power Project (henceforth NTPC) projects – Vindhyachal and Rihand Super Thermal Power, situated in Uttar Pradesh state of India were selected. These projects required vast amount of land for its inception and erection, which displaced thousands of inhabitants. People were forced to resettle at other places in the vicinity. The inductive study includes analysis of affected community living at the new location sites. Around 250 household heads were selected for Interview on the basis of Random Sampling.

For the convenience of analysis, the paper is broadly divided into five parts. The first section provides an overall glimpse of the development-induced displacement scenario across the globe as well as in the Singrauli region of India in the title, 'Development-Induced Displacement: Global Overview and Singrauli'. The second part, 'Policy Planning' discusses in length the origin and evolution of various resettlement and rehabilitation laws in India and World Bank with its highlighted significance. Third section, 'Development-Induced Displacement: Information, Awareness and Tribulations' critically analyses the effectiveness and usefulness of the resettlement and rehabilitation policies discussed in the previous section at the level of information and awareness amongst the land oustees and the tribulations encountered. Fourth segment is devoted to the aspect of 'Development-Induced Displacement: Discrepancy between Planning and Execution' and focuses on various loopholes concerned with the issue. The fifth and last part of the paper is on, 'Development-Induced Displacement: People's Perception, Reaction and Expectations'. It focuses on aspects like how do general masses conceive these policies which becomes the culprit for bringing turmoil in their serene lives. Coupled with this, the kind of reactions and resentment that the locals demonstrate along with their expectations from various rehabilitation agencies are noted in the section.

I

Development-Induced Displacement: Global Overview and Singrauli

Global Overview

At present, no precise data is available on the exact figure of persons affected by development projects worldwide, mostly due to unawareness, negligence and apathetic attitude of the policy planners towards such a vicious problem adversely affecting millions of innocent and marginalized people. Unlike for refugees and internally displaced persons, there are no institutions or publications dedicated to tracking overall development-induced displacement, either on the global or national level. To derive an idea of its magnitude, most scholars, policy-makers and activists rely on the estimates of the World Bank Environment Department which indicates that roughly 10 million people are displaced worldwide every year due to dam construction, urban development and transportation and infrastructure programmes. This number is shockingly high, but still fails to account for large numbers of the displaced. It happens mostly due to the fact that displacement tallies almost always refer only to persons physically ousted from legally acquired land in order to make way for the planned projects. It generally ignores the account of those living in the vicinity of, or, downstream of the projects, whose livelihood and socio-economic milieu might be adversely affected by the project. A detailed enumeration that considers this wider conception of development-induced displacement would be much higher than the estimates of the World Bank Environment Department. Nevertheless, the existing figure available with the policy planners also does not portray any rosy picture. It can well be subsumed in the analysis of various continents across the globe.

Asia and the Pacific

The predicament of involuntary displacement is not new to the world. It dates back to history. It is rightly said that, 'Displacement became a serious issue in the colonial age, intensified after independence, and caused bigger problem in the context of globalization today' (Fernandes 2008: 89). While the plethora of development-induced displacement occurs throughout the world, two countries in particular – China and India – are responsible for a large portion of such displacements just to cater for the developmental requirement of a booming population. According to Fuggle et al. (2000), the National Research Centre for resettlement in China has calculated that over 45 million people were displaced by development projects in that country between 1950 and 2000. Taneja and Thakkar (2000) points out that estimates on displacement in India from dam projects alone range from 21 million to 40 million. The Narmada Sardar Sarovar Dam Project in India, which displaced more than 1, 27,000 people, has perhaps been the most widely researched and discussed project involving forced displacement in the history of mankind. Dreze, Samson and Singh (1997), in their study elucidate a comprehensive picture of displacement and resettlement in the Narmada project.

The Three Gorges Dam Project of China has also been widely discussed and written about as it displaced more than 1.2 million people. The edited book *Yangize! Yangize!* by Quing (1998), is credited for being the first manuscript published from within China for having critical overview of the project affecting such a large number of locals. Cernea's (1993) discussed the displacement of 40,000–50,000 people in Indonesia to make way for a Jabotabek urban development project, which involved the widening and upgrading of roads in Jakarta and nearby cities. He also accounts for the modernization of Shanghai's sewerage system, which displaced 15,000 urban dwellers in the city.

While focusing specifically in the instances of displacement caused by development projects, a worldwide comparative study does not give different picture than India. In a multi-country comparative empirical analysis of the state of displacement in the developed countries like China and France with developing countries such as Bangladesh, Pakistan and Zambia, the results suggest similar pathological experiences. 'The study on China tracks several extended families that were relocated mainly from the Yunyang and Yanzhou districts situated at the head of the Three Georges Dam and resettled to the Shanghai region. The research highlights the issue of forced migration in China and the challenges that lie ahead of the dislodged rural populace. The apprehensive evidences of the study cautioned that the frustration of these settlers, an army of unemployed marginalized group, can be explosive and could unbalance the harmonious social equation. People are 'sacrificed' for the sake of biodiversity and wildlife, and development, is not only unacceptable from the social and human rights perspective, but also from the ecological and economic point of view, because the resultant impoverishment undermines the original intention of poverty reduction and conservation' (Modi 2009: 272–273).

Africa

Cernea (1997) in his paper, African Involuntary Resettlement in a Global Context notes that, while countries like China and India lead the world in the number of persons displaced by development projects, the proportion of population and territory affected by even the largest of projects in these countries is much lower than in some projects in African countries. For example, the Akosombo Dam in Ghana displaced 80,000 people, which is approximately 1 per cent of country's population. The reservoir of the Akosombo Dam flooded 3.5 per cent of Ghana's land, while that of the Narmada Sardar Sarovar Dam covers only 0.01 per cent of India's territory

Colson's (1971) detailed study of the impacts of the Kariba resettlement scheme on the Gwemba Tonga is a classical work, not just from the point of view of the literature on displacement but also from the perspective of anthropology. Kariba Dam project in Zambia displaced approximately 57,000 people. Fahim (1981) offers an in-depth look at the Aswan High Dam Project in Egypt, which displaced close to 1,00,000 people in Egypt and Sudan.

Ghana's Akosombo Dam Project on the Volta River displaced more than 80,000 people, has also been closely examined and discussed by a number of researchers [Chambers (1970), Obusu-Mensah (1996), Hart

(1980)]. Amarteifio, Butcher and Whitham (1966) analysed the displacement and resettlement on Ghanaian village beginning in 1952 in order to make room for the construction of a new port and harbour. The study was completed and published prior to the completion of the Volta River resettlement operations in the hope that its findings could be of use to resettlement planners.

Latin America and the Caribbean

While overall displacement in Latin America and the Caribbean is not as high as in Asia, nevertheless, the region has seen a number of large and controversial resettlement operations. La Rovere and Mendes (2000) provides a detailed account of Brazil's Tucuri Dam Project, Phase I, which was built between 1975 and 1984 and displaced 25,000–35,000 people, despite a pre-project prediction of displacement affecting only 1,750 families in the region. It shows large scale negligence and ignorance on the part of the policy planners towards the land oustees. Guatemala's Chixoy Dam Project is famous for the impunity with which resettlement was carried out. The project involved the resettlement of 2,500 Maya Achi Indians. It began in 1979 and lasted for over a decade and blamed for the massacre of 369 displacees whom local civil patrols and the Guatemalan Armed Forces believed to be 'guerrillas.' Barbaras and Bartolome (1973) enumerate the displacement and resettlement in Mexico's Miguel Aleman Dam Project, which displaced about 20,000–25,000 Mazatec Indians. Guggenheim (1993) narrates the instance of resettlement in the Mexico Hydroelectric Project, which took place in the early 1990s and included two separate dam projects, displaced a total of 3,500 people.

Europe, the United States and Canada

Large scale development-induced displacement and resettlement in industrial countries of Europe and in North America is not much in vogue today. However, history is replete with examples of displacement-inducing projects in these countries, particularly in North America, even if the detailed literature is not available related to it. Scudder's (1996) examined the trend in livelihood displacement and political mobilization amongst the Cree in Canada's James Bay Power Project. The World Commission on Dams case study report by Ortolano et al. (2000) offers a detailed examination of the Grand Coulee Dam Project in the United States – a project that extended over some forty years between 1933 and 1975 and displaced approximately 5,100 to 6,350 people in the region. It also affected adversely the indigenous population residing north of the border of Canada without payment of compensation. Berman (1988) provides a critical discussion of the displacement and resettlement of 300 indigenous families from land protected by treaty to make way for the Garrison dam in the United States in the 1950s.

Thus, it is clear from above discussion that development-induced displacement is the most commonly known phenomena across the globe, which is adversely affecting millions of people annually. Most of those getting affected generally hail from illiterate, deprived and marginalized communities living in remote locations become prone to multifold impoverishment risks (Cernea 1996). Most importantly, large scale unawareness, negligence and ignorance on part of the affected communities towards the issues pertaining to involuntary displacement forms the root cause of the problem. Coupled with it, the apathetic attitude of the policy planners act as catalyst to further aggravate the problem beyond proportion. The pattern of tribulations is similar in cases of underdeveloped, developing and industrialized developed countries. The only difference which is perceived through the analysis across the globe lies in the magnitude of the problem. Whereas, the developing countries like India and China is displacing more people annually, in comparison, the European and American countries are producing far less number.

In this backdrop of the findings of the global overview, attempt will be made to ascertain the linkage between the global and local level by focusing on the Singrauli region, the universe under consideration. Prime focus of attention, here, would be to trace the availability of information and awareness amongst oustees about various Resettlement and Rehabilitation (henceforth R&R) rules and the resultant aftermath at the empirical level.

Singrauli Region: A Retrospect

In the past, Singrauli, known as India's energy capital, was an isolated little known and economically backward rural area with very less population, living in small and scattered settlements. Forest and small fields fence characterized the hilly and undulating landscape, and the locals were mainly engaged in collecting minor products or in subsistence agriculture. After independence the region started getting the attention of the planners basically for its large coal deposits which could be utilized for generation of power for urban-industrial development of the country. In conformity with the national goal, Singrauli region was opened up by rail-road connection.

The first blow to their serene existence was in the form of land acquisition for Rihand Sagar Dam in the beginning of the 1960s. The whole region came under chaos and turmoil and underwent pathetic experiences apropos loss of everything in the stream of the dam. The compensation given was far inadequate in comparison to the loss incurred. Somehow, PAPs settled down, not fully but partially, in the vicinity. They thought, it is once for all and now, no one will embark in future to infringe on their privacy. But the beginning of the 1980s earmarked another doomsday in their life. Now they had to forego the present patch of land to government agencies, on which they were till lately living, for the sake of national goal. Naturally, they were displaced for the second time within the span of 20 years and came across same awesome encountering of distress, agony and anxiety. The agony continued and some of them got displaced for the third time to make way for ongoing development projects. During the past couple of decades many government and private owned development projects came up in the region, viz. NTPC, Coal India Limited, HINDALCO, State Electricity Board and ancillary industries to supplement the bigger industries. In fact, the area lacks proper planning for the execution of various upcoming projects in the region and thus to meet development prerequisites of the nation; the unaware locals are forced to displace more than once in the absence of policy guidelines.

II

Policy Planning

No project can be successfully executed in the absence of watertight policy guidelines. The state of development-induced displacement in the Singrauli region also solicit direction in the form of guidelines of the biggest global donor agency, i.e. the World Bank; and at the national level, the R&R Policy of India, R&R Policy of NTPC, Land Acquisition Act of India for its smooth implementation. An analysis and evaluation of the policy planning incorporated in the power projects of NTPC is done in the following passage.

R&R Policy of World Bank in Retrospect

World Bank (The World Bank 1980; 1990; 1991; 1993; 1994) was the first agency of its kind, to experience the aftermath of development in the form of resettlement and rehabilitation. So, being sensitive towards the growing environmental problems, it was the World Bank that took the first step in formulating an approach to the resettlement question. In 1980, the Bank issued for its staff a statement outlining a policy to guide operations in this sensitive and difficult area with the aim of protecting the interests of the populations displaced by the development projects. This was the first time that any major development agency had instituted a policy to deal specifically with the problems of involuntary resettlement. Since then, the Asian Development Bank, the Inter-American Development Bank and the Organization for Economic Cooperation and Development laid down policies for projects requiring resettlement. As a result, every part of the globe is slowly getting sensitized towards the lives of millions as natural atonement of development.

The World Bank policy on resettlement has undergone several revisions since it first came into existence. Its first revision took place in 1986. A new version of the policy document appeared in 1988. This, too, underwent some revisions lately and an updated operational directive has been in effect since 1990. The Bank policy rests on a collection of basic principles, such as, minimizing displacement, government responsibility towards the relocated, participation of stakeholders, the rights of victims of displacement, protection of host populace interests, a forward-looking definition of objectives of resettlement and so on.

Embedded largely in anthropological research inference and other social science knowledge, the policy on resettlement was developed by a Bank staff group. As per the Bank guidelines, the very nature of involuntary resettlement gives rise to special social and technical problems, which are to a great extent different from, and usually more severe than, those encountered in cases of voluntary resettlement. A feeling of powerlessness and alienation is often engendered in those who are relocated especially when the entire community is uprooted from familiar surroundings. To the extent that pre-existing community structure and social networks disintegrate, and tightly knit kin groups are dispersed to new locations, social cohesion is weakened, and the potential for productive group action is diminished. Thus, the oustees find themselves in a position of helplessness, powerlessness, landlessness, placelessness, alienated and most importantly, joblessness which diminishes the chance of a sustainable livelihood.

Further emphasizing the sociological nature of involuntary resettlement, the guidelines caution against viewing resettlement as a mere environmental issue. Frequently, governments, consulting firms or the media deal with involuntary resettlement under the rubric of 'environmental problems'. But by its very nature, resettlement is a socio-cultural/economic process that happens first to people, rather than to their environment. Therefore, the correct understanding of the sociological nature of involuntary resettlement, with its cultural, economic and psychological ramifications, much exceeds the environmental aspects. It leads to a different course of action on addressing resettlement than if it is regarded only as an environmental problem.

The Bank policy recognizes that in any involuntary resettlement situation, some human suffering is inevitable. As its very first requirement, the policy therefore states that involuntary resettlement must be avoided wherever possible. When it is unavoidable, it must be minimized. Moreover, as Marsden (1996: 6–7) argues:

> Emphasis is on providing resettled people with new and better economic opportunities. The policy states that the families affected by a Bank-financed project should share in benefits from the project and as a result be better off than before. The idea is that since involuntary resettlement dismantles existing production system, all resettlement programs must at the same time be development programs. Emphasis in the policy on preparation of resettlement plans in advance is intended to ensure that

the people affected by projects are helped effectively in their rehabilitation. Specifically, the plans are to provide for compensation in full before departure to the new locations; and assistance in moving to the relocation site and support during the transition period; and see assistance in efforts to improve their living standards.

It is a matter of great concern that the projects designed and appraised in the past had no provision for resettlement and rehabilitation in their guideline. Subsequently, when it was discovered that resettlement would be necessary, the lack of resources for the purpose only complicated the implementation of otherwise sound projects. In order to prevent recurrence of such cases, the Bank guidelines insist that initial planning of any project construction should accompany the budget of R&R and other development activities. As resettlement-planning decisions impinge on the lives of the displaced groups, agencies concerned are enjoined to encourage community participation on planning and implementing resettlement. A related recommendation concerns the involvement of local groups and NGOs. Being closer to the people, they are in a better position to articulate resettlers' needs and defend their rights.

The Bank policy regards as of particular importance the welfare and development of the indigenous groups, ethnic minorities, pastoralists and other marginalized groups, such as women. Even without legal title, these adversely affected groups are to be eligible for full compensation as well as resettlement. Women and dependents even though adults have always been debarred from the resettlement benefits. So is the case with the marginal communities such as tribal and other down trodden citizenry placed at the fringe of the power matrix.

Land Acquisition Act in India

If we go through the history of land acquisition act (MARG 1990; The Land Acquisition Act 1995) in India, it dates back to 1824. The Regulation I of 1824 was the first legislation of its kind. It empowered the government to acquire immovable property at a fair and reasonable price for construction of road, canals or other public purpose. Act VI of the 1857 was the first full enactment, which had application to the whole of British India. It repealed all previous enactment relating to acquisition and its object. Through this act, the principle of Arbitration was introduced for the first time. Collector was empowered to fix the amount of compensation by agreement if possible. If there was no agreement, the dispute had to be referred to arbitration. Later, in the 1870 Act, a detailed procedure for the acquisition of land was provided for the first time. Yet, the provision of this Act did not satisfy the needs of the day and eventually the Land Acquisition Act, 1894 was enacted.

Nearly a hundred year later, in 1984, the Act governing land acquisition in India took another shift and at least on paper, the effort was made to make it more people's friendly rather than government. Prior to 1984, the Land Acquisition Act was not applicable to the states of Jammu and Kashmir, Rajasthan, Kerala and Nagaland, which had their own self-contained Land Acquisition Act. In 1984 the Land Acquisition (Amendment) Act was passed and became applicable to the whole of India except Jammu and Kashmir.

The process of land acquisition starts with the Section 4(1) notification. This is publication of a preliminary notification by the government that land in a particular locality is needed or may be needed for a public purpose or for a company. A notification to this affect shall be published in the official gazette and in two daily newspapers circulating in that locality of which at least one shall be in the regional language. Apart from this, the collector shall paste public notices at convenient places in the said locality.

Then comes Section 5–A by which, within 30 days of the publication of the notification, anyone can file objection to the said acquisition of the land. Every objection shall be made to the collector in writing who, in turn, shall give the objector an opportunity of being heard. Vide Section 9, collector issues public notice and individual notices to persons interested to file their claims for compensation. By Section 11 collector enquires into claims by the litigants. Finally, through Section 11 itself, the collector awards the compensation.

The Government of India drastically amended 1894 Act in 1984. The Amendment Act of 1984 has resulted in the setting down of a time limit for the completion of all formalities between the issue of preliminary notice vide Section 4(1) and the issue of the declaration of intended acquisition vide Section 6(1). This period was three years. Another amendment gave an opportunity to those dissatisfied with the collector's award to apply to him for a re-determination of the compensation payable to them on the basis of an order for higher compensation obtained by any one of them from the reference court by Section 28 A.

These changes introduced by the Amendment Act of 1984 are beneficial to those who seek to minimize the undue delays that characterize acquisition proceedings and compensation payment on a realistic basis. The main point which emerges from the whole process of land acquisition is the immense power that has been conferred to the collector. He is the sole authority and even the judiciary does not have much say in this regard. Once the collector has taken a decision with regard to the land acquisition, it is final and binding on every one.

The Land Acquisition Act does not give the victims any rights to prevent the acquisition of the land. One can only delay acquisition by raising objections. The government is empowered to acquire land for public purpose. But the term 'Public Purpose' is nowhere defined in the act. Objections can be made to the proposed acquisition on the ground that the purpose is not a public one, or that the land is not suited to such public purpose. If someone wishes to argue that the purpose is not a public one, the person will require wider information about the project and the economy as a whole. But, as the time prescribed for raising such objection is only one month, the whole exercise becomes redundant due to inability of the victims to do the needful. Even if a person does, the collector's decision on his objections is final. It is hard to imagine any situation in which the collector goes against government planning and accepts the litigant's objections. Raising objections is, therefore, not a very successful exercise in preventing acquisition as it happened in the case of Narmada, Tehri, Koel Karo, NTPC and other projects of this nature. But one can delay the process. Delay may make the project too expensive to be profitable.

The Civil Court does not have the power to decide whether the purpose behind the acquisition was a public purpose. The Court can only hear the litigant, if the collector does not consider his objections. All that the court can do is to direct the collector to hear the litigant. It cannot decide on merits of the case. The government is also armed with emergency/urgency power, with which it can cut short the entire process of notices and objections. The term 'agency' is nowhere defined in the act. The term 'emergency' is also not defined. If someone's land is acquired in case of an emergency or urgency he cannot raise objections. Nor can the Court question whether there was really an urgency or emergency. 'A study of the Land Acquisition Act and the rest of the existing legislations show that the present legal system favours the state and the industrialists, and ignores the right of the livelihood of the people who depend on the Common Property Resources (CPRs) such as forest etc.' (Fernandes 1996: 3).

R&R Policy in India

Beginning of the 1980s observed sensitization towards the ever growing and all pervasive problems of Resettlement and Rehabilitation. Opinion of the donor agencies as well as national and international organizations concerned with the social cause shifted towards protection of human rights. As a result, there were lots of hue and cry favouring for enactment of a sound National Resettlement and Rehabilitation Policy (GOI 1995; 1996) in India, ingenious enough to solve the problems of the oustees. As such, efforts were constantly paving toward the composition of a national policy.

Till now, there are very few states that have enacted state-wise R&R policies. 'Till 2003 Maharashtra, Madhya Pradesh and Karnataka had rehabilitation laws for irrigation displaced persons. Rajasthan and Orissa had promulgated policies between 1994 and 1998' (Fernandes 2004: 123). Apart from that, some other states issued Government Orders/Resolutions. The Government of Maharashtra developed a progressive policy to deal with people not successfully rehabilitated under previous water projects. There are also company-wise and sector specific policies, such as those developed by the National Thermal Power Corporation NTPC 1993), Coal India Ltd. (CIL 1994) and more recently by some other private and government infrastructure companies.

The attention of the NGO (Non-Government Organization) movement, particularly those concerned with tribal rights, as well as that of some donor agencies have been drawn to the issue surrounding displacement due to increasing interest in matters of sustainability and human rights. These focused around the Sardar Sarovar Project during early 1990s.

There was pressure from World Bank, other major donor agencies as well as national and international organizations concerned with the rights of the displaced. They expressed the need for a national policy. As a consequence, Government of India initiated formulation of a sustainable policy on R&R. Thus, draft of a national resettlement policy has been the subject of widespread consultation amongst state governments and within the NGO community in many parts of the country. In 1989, a National Working Group on Displacement was formed through an NGO initiative which involved government officials also. Draft National Policy on Developmental Resettlement of Project Affected Persons was prepared. Since then, many more drafts have been prepared encompassing an ever-widening group of people – from 'oustees' of 'river valley projects', to 'persons affected by reservoir projects' and now to the 'rehabilitation of persons displaced as a consequence of acquisition of land'.

Ministry of Rural Development prepared the draft of a 'National Policy for Rehabilitation of persons displaced as a consequence of land acquisition'. Responsibility for the progress of government discussions lies with the Ministry of Rural Areas and Employment who has been designated as the nodal ministry for the execution of the policy. This shift in responsibility was perhaps one indication that the responsibility for successful R&R cannot really be given to a particular ministry, but is central to the agendas of many. The draft has been circulated to 15 ministries, including Finance, Water Resources, Social Welfare, Environment and Forests, Power, Coal, Home and Surface Transport, as well as the Planning Commission. This document has been used as a basic text by various NGOs over past years to produce a draft alternative R&R policy after widespread regional discussion. The same draft alternative has also been submitted to the Ministry of Rural Areas and Employment for their consideration.

The circulating government draft emphasized equity, fairness, justice and equality on the distribution of burdens and benefits. The draft recognized the right to settle and reside in any part of the country, the special rights of Tribal, the customary rights to common property resources of those dependent on the forests, and also those classified as 'encroachers'. It recognized that the 'displaced' are actually 'uprooted from the soil'. It emphasized the need to ensure rehabilitation in any displacement process and stressed the need to ensure that the displaced persons actively participate in their own rehabilitation. It was also recognized in the draft that the adverse impacts of involuntary displacement are much wider than those just associated with loss of land. It was formulated on the expectation that future developments will increase rather than decrease the pace of displacement, as more and more land will be required for infrastructure expansion. The revised Land Acquisition Act indicated that R&R is likely to remain a necessary activity, which has to be undertaken prior to developments.

In 1993, in the wake of World Bank withdrawal from Sardar Sarovar due to severe negligence on R&R activities, the ministry of rural development prepared a draft policy in 1993 which was revised many a times after that. After nearly 19 years of continuous debate and discussion, Government of India notified a national rehabilitation policy on 17 February 2004 known as National Policy on Resettlement and Rehabilitation 2003 (GOI 2003).

The main features of NPRR 2003 (GOI 2003) was that it applied to such projects which were displacing more than 500 families en masse in the plains and 250 in the hills or the schedule areas. The policy says that each project-affected families (henceforth PAF), whose house has been acquired, will get a site free of cost. Those families who are below poverty line will be given a one-time grant of ₹ 25,000 for house construction. Land losers will be given one-time grant of ₹ 10,000 per hectare for land development and ₹ 5,000 per family for agricultural production. The best part of the policy was that it made provision for the payment of monthly allowance of 20 days minimum agricultural wages for one year up to 250 days of minimum agricultural wages to the PAFs. Apart from that, a PAF whose entire land is acquired will get one-time financial assistance equivalent to 750 days of minimum agricultural wages for loss of livelihood. Lastly, the policy says that each rural artisan, small trader and self-employed PAF will get financial assistance of ₹ 10,000 for construction of shops or working sheds.

The positive aspects underlying NPRR 2003 was its broad definition of PAF and 'agricultural family'. It made provision of separate rehabilitation agency for the execution of R&R work. Lastly, it restricted R&R benefit to those who have resided in the affected area for 3 years before the notification under section 4(1) of Land Acquisition Act. By this provision, it prevented outsiders from buying land in the affected area and in turn grabbing most of the rehabilitation benefits.

The first central policy on R&R was subsequently replaced by National Rehabilitation and Resettlement Policy (NRRP) by the end of 2007. 'The NRRP's stated objectives include minimizing displacement and promoting non-displacing or least displacing alternatives; ensuring adequate and expeditious rehabilitation with the active participation of the affected families; ensuring protection of the rights of scheduled castes and scheduled tribes; providing better standard of living with sustainable income to affected families; integrating rehabilitation concerns into development planning and implementation processes; and facilitating harmonious relationships between the acquiring body and affected families. While these are useful policy directions, they mean little in practice without specific laws mandating them. Despite acknowledgement of concern for the displaced and democratic process, the policy does not make the right to informed consent explicit. Panchayat or other local bodies resolutions find no mention in ascertaining the views of the affected communities' (Sampat 2013: 43–44).

In another attempt to overcome controversy surrounding payment of compensation and rehabilitation issues in the cities of northern states (Noida and Singur), the Land Acquisition and Rehabilitation and Resettlement Bill 2011 and the Right to Fair Compensation and Transparency in Land Acquisition, Rehabilitation and Resettlement Bill 2012 (RTFCTLARR 2012) was introduced. The LAA Bill 2011 sought to replace the LAA 1894 and bring R&R for the first time within the ambit of a land acquisition law. But the bill clearly conflated public purpose with industrialization, infrastructure development and urbanization.

The third and the latest version of the bill was released in December 2012 after several rounds of deliberations between various ministries. The bill claim that: '… it is a bill to ensure, in consultation with institutions of local self government and Gram Sabhas established under the constitution, a humane, participative, informed consultative and transparent process for land acquisition for industrialization, development of essential infrastructural facilities and urbanization with the least disturbance to the owners of the land and other affected families and provide just and fair compensation to the affected families whose land has been acquired or proposed to be acquired or are affected by such acquisition and make adequate provisions for such affected persons for their rehabilitation and resettlement thereof, and for ensuring that the cumulative outcome of compulsory acquisition should be that affected persons become partners in development leading to an improvement in their post-acquisition social and economic status …' (GOI 2012).

The RTFCTLARR 2012 allows for the acquisition of the land albeit with prior consent of Gram Sabhas or councils. However, no clear procedure as to how consent may be ascertained is mentioned and while Gram Sabhas and urban local bodies are to be consulted, resolutions taken by these bodies find no legitimate mention. The steps to be followed, if there is no consent, are also not mentioned. However, what kind of changes the RTFCTLARR 2012 bill will go through before it eventually passed remain to be seen. In addition, when and how this bill will become law is an open question.

R&R Policy of NTPC

In the beginning of the 80s, when NTPC launched its first project at Vindhyachal, the Policy of R&R was in an infant stage. Socio Economic Survey (henceforth SES) and the formulation of Rehabilitation Action Plan (henceforth RAP) and Remedial Action Plan (henceforth ReAP) was not a familiar concept. In the absence of policy guidelines, the residents were pushed out from their land without any prior SES. First SES of the Vindhyachal Stage-I was conducted nearly 10 years after the inception of the project. The research team conducting SES was not able to trace the Project Affected Families (henceforth PAFs). During the gap of more than 10 years, victims had migrated to distant places in search of livelihood. As a result, the SES report has serious flaws about the whereabouts of the residents. As such, documents lying in the offices of NTPC spell contradictory information about the PAPs.

In the absence of any estimate, land was acquired, which was more than necessary for the project and remained unutilized for years from which PAPs were deriving livelihood. Now, after 15 to 20 years these lands are required for the second stage of the project. This constantly soured the feelings of PAPs.

The small plots (60 feet into 40 feet) allotted to the PAFs in the Resettlement Colonies (henceforth RCs) were too small to accommodate big joint families running into double figure. There was no provision for cattle, which is an essential part of the rural household. Dearth of grazing land, open fields, orchards, open and airy space, ponds, lakes, wells, assembly places etc. suffocating the rural people who are habituated to that kind of living. Sanitation and infrastructure facilities were giving a dismal look in the RCs. The planners forgot while planning that the small patch of land in the RC could no longer solve the problems of those living in purely rural areas. Due to these problems, a large number of families migrated to other places.

At the time of acquisition permanent employment was promised to at least one male member from PAFs. Later, it was realized that large numbers of oustees cannot be employed due to technical nature of the jobs. Thus, NTPC drifted from the earlier promise and started focusing on Income Generation Schemes (henceforth IGSs). They employed renowned consultants for training of IGSs. In the beginning, PAPs showed reluctance towards this. They were not born entrepreneurs and felt scared from indulging in such alien businesses. But due to constant pressure, and also because of lack of source of livelihood, some PAPs dared to venture into these new occupations. But, the whole process failed to take off smoothly in the absence of basic knowledge and business experience. Consequently, daring PAPs who invested money in such business lost capital and were further pauperized.

Land for Land option was the second most viable alternative to permanent employment. Accordingly, NTPC floated with the policy 'Land for Land' option on 'willing buyer, willing seller' basis. Due to breach in demand and supply, price of the land escalated very high. Thus, handful of money received as compensation was not sufficient to buy land for PAPs. In case of forced acquisition, it would have triggered irreversible chain of displacement. Consequently, the option failed to take off.

III

Development-Induced Displacement: Information, Awareness and Tribulations

Due to illiteracy, innocence, unawareness and lack of communication, the PAPs living in the Singrauli region did not realize the advent of catastrophe in their lives in the disguised form of development which eventually led to their displacement from the place of ancestoral inhabitance. Due to unawareness about the devastating effect of R&R, the oustees were not even mentally prepared to face the problem. As such, the oustees had no any other option left with them rather than to be completely dependent on others, especially the project authorities, who were in turn the masters of their destiny.

Awareness about R&R Policy

The illiterate residents found it hard to understand the meaning of section-4 notification. The notice adhered to public places, could not draw their attention. By the time individual notices were served, it was too late. Illiteracy and unawareness about the proceedings of Land Acquisition Act came into their way. Filtered luckiest few, who could manage to interpret and understand the meaning of the notice, were also aware of the complicated laws and proceedings governing land acquisition, lacked money and other necessary resources to go to the court. Mridula Singh et al. (1992: 9) confirm this viewpoint in their observation of Sardar Sarovar and *Tehri*. According to them, 'The process of acquisition remains meaningless to the illiterate masses since they are unaware of any notices that are issued until they receive their individual notices. By then, it is too late to exercise their limited rights under the Amendment Act 1984. Clearly, the opportunities given to the affected people to register their objections against the development process are merely theoretical. In actual fact, they have very little chance to effectively question this process.' Following table supports this viewpoint.

Awareness about Policy

Out of 250 families interviewed in the region, 203 (81.2%) showed ignorance about the policy of R&R and Land Acquisition Act of India.

Awareness	*No. of Persons*	*% age*
No	203	81.2
Yes	47	18.8

Only 47 (18.8%) PAPs, which were semi-literate and conscious enough, were able to acknowledge the meaning of the notice and R&R Act. They were the helpless poor and had everything beyond their control.

Prior Information and Mental Preparedness for Displacement

As discussed above, by the time the residents could understand the meaning of the piece of paper served to them, it was too late and the process of acquisition started well before. In this backdrop, the rich and articulate managed to reap most of the benefits. Mukerji (1997: 22) writes, 'Ironically, the residents of three villages – Harrai, Juwadi and Chandauli, all Brahmin dominated, prospered following their displacement. They located the areas aprior likely to be acquired and bought large portions there, claimed Avdhesh Kumar. It was an excellent investment. The compensation was much higher than the prevailing price of land in those areas.'

Contrarily, the poor, less organized and articulate, lower and middle class people, found themselves in a completely lost predicament and did not know what to do. A resident of the Vindhyachal Resettlement Colony (henceforth RC) opines similar view 'Intelligent people managed to purchase land somewhere in the vicinity with the use of compensation money. Some more intelligent spent it in a more appropriate manner and became richer after displacement. Whereas, those who were foolish, squandered the money and were pauperized to the extent of 'hand to mouth' existence.'

Somehow, the villagers organized themselves and started protesting against forced acquisition. They rallied in mobs to show protest. They hoped that their rally and request would be accepted by the government agencies on sympathetic grounds. But their efforts went in vain. The management, with the help of police forcefully evacuated the land and dwelling. Those not agreeing to vacate easily were beaten and thrown out of their houses.

In Rihand, the whole process of evacuation of Stage-1 residents happened in unfavourable situations. In one instance, it took place during monsoon when the whole area was flooded and it was raining 'cats and dogs'. In this adverse condition, the poor had no place to hide and they spent days and nights in monsoon without roof over their head. Ramnarayan, a displaced from Mithini village and Mahamantri, 'Gramin Kalyan Sangarsh Samiti', Rihand recalls 'NTPC believes in the policy of divide and rule and so the process of relocation took place in multiple stages, not to annoy everyone at the same time. On one such occasion, with the help of 200 police personnel the agricultural land was doused. Till that time, the oustees had not even been paid the compensation for the land. The standing crops were destroyed and the people were forcibly thrown out of the place. Second instance took place next year when the crops were trampled and destroyed again. In this case, 13 inhabitants showing dissent were sent to Mirzapur Jail by the police. Trying to nullify NTPC attempt, a local leader came in front of a dumper and the officials ordered the driver of the dumper to extinguish the leader. Somehow the leader managed to escape and save himself with his leg fractured.'

The tales of Vindhyachal residents are not different. Rewati Raman Pandey, President, 'Sangharsh Samiti', views similarly to his counterpart Ramnarayan. He says, 'In the beginning of the 80s, when the project was in the inception stage, land was evacuated forcefully with the help of CISF, without prior SES. Until then, no compensation was paid and even plots were not allotted in the RCs. Many died due to trauma and concurrent problems. The atrocities continue till date as NTPC is in the process of expansion and needs more land for stage-II and Ash-Dyke. In 1985–86, PAPs were beaten and forcefully thrown out of their land and dwelling. They had to run for their life. Recently, the oustees managed to get attention of the NGOs and international agencies. Consequently, in March 2001, higher officials of NTPC administration were arrested by Police. They were charged with creating nuisance and destruction in the village *Telagwana*, without prior payment of compensation package to the landousetees.' But still after the elapse of more than three decades, the dreaded instances of inhuman activities are practiced in the region by the project authorities and the second and third generations of land oustees are suffering from the same stigma.

Problems After Section-4 Notification

In the words of Singh et al. (1992: 9–10), 'The process of displacement begins long before the land is actually claimed by the authorities. The Amendment Act 1984 prohibits the people and the government from developing their property on the area. Activities related to it come to a grinding halt from the date on which section 4 is issued. This problem is so acute that generations grow up without being able to further their development. No time period is fixed for giving the compensation money after the declaration of the award under section 11 is made. This announcement too can be made after years but the government is under no obligation to extend support for this period. The fallout of such a clause is that life of the people comes to a standstill.'

Similarly, Vindhyachal residents received initial notice in the beginning of the 80s, when NTPC decided and earmarked the region for a power project. The acquisition process started in 1984 and continues till date. Victims know their land shall be acquired in near future but are unknown about the exact date. Those living in close proximity of the project and luckily remained unaffected do not show willingness to invest money for the construction and maintenance of dwelling with the scare of future acquisition.

Problems Faced During Resettlement

The R&R policy says plots should be allotted to PAPs before relocation. They should be consulted for just and proper rehabilitation policy. Contrarily, in the case of Singrauli, due to lack of concrete guidelines, no one was sure about the allotment of plots. Few oustees had no '*patta*' in their name. Several others were cultivating on government lands were since ages. Consequently, some PAPs were forced to evacuate land without allotment of plots in the RCs.

Out of the whole process, most severely affected were those dependent on the village economic structure for livelihood such as carpenter, weaver, barber, priests etc. as they were completely out of business. They had nothing to lose in terms of real estate and thus were denied rehabilitation package. Similar was the case of landless labourers working on other's land.

Monetary Crisis and Unemployment

After the section-4 notification, agricultural activities in the region took a 'U' turn. Since agronomic activities was the prime source of livelihood, consequently other allied activities were also severely affected. As a whole, the region experienced a big loss in the economic base. This was more problematic for those dependent on labour intensive jobs, paid on day-to-day basis. They migrated to other places in search of livelihood. Big landlords and cultivators also had vivid problems. They were unable to cultivate on the notified land because of uncertainty of time of acquisition. Those who dared to think they could had no labourers available, as most of them had migrated to other places. Some, who did not vacate land and cultivated their plots, were forcibly thrown out of the place by the government and their crops were destroyed.

The monetary compensation paid to the oustees was spent on recurrent necessities such as social ceremony, marriage, house construction or maintenance, pomp and show, debt payment etc. Those who didn't spent on these accounts, lost money on other affairs such as gambling, liquor and drugs. The poor residents, having seen such a large amount for the first time, could not able to utilize it appropriately for restoration of their livelihood. After losing everything, PAPs looked towards NTPC for gratification of promise of job and employment. But they were not at all heard

When the PAPs reached other places in search of livelihood, they had to face hostility from the local community because it was creating unemployment problems for the host. Due to imbalance in man-work ratio after sudden influx of many working hands, wage structure experienced sudden decline. In such a grave situation, the victims had to fight for their rights and also from the host community in an effort to firm their foot on the ground.

Ebbing Dependence on Ecology and Environment, Pollution and Health Hazards

Industrial activities completely changed the ecological balance of Singrauli region. Big jungles, trees, forest products, wild animals, livestock, grazing land, flora and fauna, varying climate, ponds, lakes etc. all started decaying. Result was imbalance in ecology coupled with environmental degradation. There is no grazing land in the vicinity. Lack of ponds and lakes are also causing problems. Livestock does not have these amenities. Women walk miles to fetch potable water. Deforestation has directly affected the tribal population as they were dependent on forest products for their livelihood. Destruction of forest was an invisible loss to the tribal people, being highly dependent on forest products, which could never be compensated by NTPC policy.

The dense forest cover has changed into township and ash dyke area, continuously vomiting polluting and hazardous elements into the environment. Ash dyke area, in particular, has become a great problem for the residents. During summer, when the water of the ponds dry up, the dry ash lying in the ponds starts spreading with air and contaminates nearby areas. It is a potential source of various diseases like asthma, sinus, tuberculosis and respiratory problems. There are some more alarming environmental problems. The Rihand Sagar Lake has been found positive for mercury. It can cause skin disease, eye problems and even cancer. People living on the bank of the Rihand reservoir have already developed skin problems. No one is bothered about the pollution generated by projects. All the waste products are falling into the lake. It is observed that fish and other species living in the lake regularly die due to water-borne epidemic.

Role of Rehabilitation Agency

The perpetual episode of R&R requires humane, sociological and anthropological understanding, as it has been engineered by the technocrats, in haphazard, ad-hoc and mechanical manner. Technocrats' first and foremost aim is to construct the project at the earliest so as to maximize profit. The problems mount up ceaselessly due to induction of these apathetic and soulless personnel as frontal officers dealing with the oustees. They do not even intend to hear the sufferer and show a callous approach. This was the constant grievance of the residents. They blamed, the authorities appeared endlessly at the time of relocation and lured them with Santa Clause type compensation packages. But just after relocation, they disappeared from the scene.

When this apathetic approach of the officials reached to the national and international organizations, they raised the issue on various platforms. Under pressure from World Bank, NTPC inducted Sociologists on contract basis. The newly inducted were on the lowest rung of the administrative hierarchy and had no power. They were dependent on engineers for everything. Being fresh postgraduate from the universities they had no formal training for social work. They had to work according to the wishes of their superior. So, instead of abridging the breach between the victims and the management, the skill of the sociologists was used differently. Their main responsibility was to see that land and houses were vacated.

Ms. Madhu Kohli, a social activist working in the region for the rights of the oustees had strong objections with the role of the sociologists. She says, 'The role of Sociologists was to divide the villagers, make them quarrel, so that the management could get the land evacuated from them. At the time of the forced evacuation, they were constantly associated with the police and officers of NTPC. The Sociologists played a very negative role in the region and were disgraceful to their education and skill.' Similar was the observation of Sri Ramnarayan, another activist. He said, 'We no longer want Sociologist. They are more dangerous than any other engineer employed with NTPC. Engineers attack from the front but the Sociologist, on the other hand, skirmish from indoor. They are culprits for the breakdown of various dwellings.'

When contacted one sociologist, he did not show willingness to talk on such a sensitive issue. On request of anonymity, he said, 'We just have to work as per the goal of the management and according to the directives issued by our superior.' When a senior R&R officer of NTPC corporate office was contacted, he said, 'Sociologists are just executing the works assigned to them by NTPC. They are doing the right job. There is no harm attaining the goals of the organization.'

Mistreated by NTPC

Mistreatment	*No. of Persons*	*% age*
No	53	21.2
Yes	Ra197	78.8

In the framework of such a disgraceful attitude of the rehabilitating agencies, efforts were made to ascertain views of the oustees towards NTPC. As per table, it is visible that 197 (78.8%) respondents feel mistreated by NTPC. The victims were not comfortable with the attitude of the management and especially the frontal officials continuously interacting with them. Only 53 (21.2%) families have no ill-feeling for NTPC. They are of upper caste and class, articulate and educated people, who garner the benefits of displacement. They were happy as the process of displacement largely benefited them.

Lack of Participatory Approach

NTPC policy envisioned participation of the stakeholders in the decisions with the objective of minimization of hardships. Coupled with it, PAPs would draw satisfaction for being party to the decision-making tools that determine their own future.

Keeping in mind the overview of friendly relations, several mechanisms were adopted. One such mechanism is known as Village Development Advisory Committee (henceforth VDAC). It comprised of the representatives from District Government, local NGOs, *Gram Pradhan* or *Sarpanch*, representatives of the oustees and officers of NTPC. The idea behind formulation of such a committee was to provide a neutral platform where the problems of PAPs could be discussed and eradicated. But the fanfare, with which the committee was set up, did not function and mostly remained as paper work. The villagers complained that the meetings of VDAC seldom take place and in case of a meeting, the representatives are not informed in advance. So, even in case of a meeting, in the absence of representatives from the oustees, only NTPC agenda is discussed.

The situation with Public Information Centre (henceforth PIC) is quite identical. It was a public library and all-important documents pertaining to land acquisition and; resettlement and rehabilitation were supposed to be kept here so that anyone can consult it. PIC remained closed most of the time. One can see bear, donkeys, goats and cows roaming inside the premises of the PIC. It opens only when outside dignitaries visit the place. So the idea of bridging the gap between the PAPs and NTPC officials remained only on paper.

IV

Development-Induced Displacement: Discrepancy Between Planning and Execution

Huge discrepancies were reported between the entitlement of rehabilitation package and the actual delivery. NTPC claims having given a handsome compensation package and in most of the cases PAPs are better off than before. On the other hand, the oustees narration contradicted this. Following paragraphs will clarify the true reality.

Information about Entitlement

Innumerable promises were made by NTPC officials during relocation. As per the initial R&R policy of NTPC, there were provisions of Land for Land, Permanent Jobs, Contracts and imparting of free training. Main idea was that the oustees could at least regain their lost livelihood, if not able to improve. It was told that the victims would be paid as per market rate for their lost belongings. Even the standing crops would be paid. It was portrayed that the project would generally benefit all and ameliorate the situation of the poor people.

Sri Rewati Raman Pandey, President, *Vishtapith Sangarsh Samiti*, Vindyachal, says, 'Indira Gandhi, the then Prime Minister of India, visited

Singrauli region on 12 November 1982, when the project was announced. She promised permanent job and 5–acre land for each family in lieu of their land and dwelling. So, the oustees were hopeful that as per Prime Minister promise, they will be benefited with the project.'

Actual Delivery of Package

Some PAPs having been lured with the oasis of promises readily vacated their land. Those not willing to surrender were forcibly thrown out of their places. As discussed, the omnipresent and most accessible officials of NTPC became invisible. Even if traced, were found to be unapproachable and seemed to be busy. Very few literate, and those who could potentially create problems in the way of project erection, could afford to get jobs and contracts from NTPC. Rest of others had to struggle for years to legally prove themselves as the real heir of property.

The compensation for agricultural land was far below the market rate and was calculated on the basis of the average of last 10 years' transaction held in the local *Tehsil* office. Jobs and contracts were available to very few lucky persons. Agricultural land was not available in the region, so NTPC waived this option from the list of compensation package. Even if there were some agricultural land for sale, the cost was too high and was beyond the reach of the poor.

Help and Support in Problem

PAPs had deep emotional attachment to the agricultural land and the place of dwelling. They were cultivating the land for food, playing on it for entertainment and worshipping for propitiation of deities. Forced relocation from the lap of mother earth was much painful to the locals. In this adverse condition, efforts were made to ascertain whether they received any help from society?

Help in Problems

Help By	*No. of Persons*	*% age*
None	208	83.2
Neighbour	4	1.6
Friends	5	2
Relatives	19	7.6
NTPC	7	2.8
NGOs	6	2.4
Any Other	6	0.4

As per table, a majority of respondents had negative response. 208 (83.2%) persons narrated that no one gave help and they had to fight all alone with the sufferings. Only 4 (1.6%) received support from their neighbours. 5 (2%) were supported by their friends living somewhere else. 19 (7.6%) families affirmed receiving help from their relatives. Ironically, the agency responsible for the plight of residents did not show much botheration and adopted distant position. As the table indicates, only 7 (2.8%) families received any kind of help from NTPC.

Local NGOs also did not play a positive role. Only 6 (2.4%) respondents received assistance from NGOs. Overall, the help and support provided to them gives a grim picture.

Support of NGOs and Local Party

Two main NGOs showed presence in the Singrauli region. Govindpur Ashram, led by Prem Bhai founded in 1950s. Established some 10 kms to the north of Miurpur on an estate of 240 hectare is based on Gandhian philosophy. The second NGO, Srijan Lokhit Samiti is situated at Navjivan Vihar near VSTPP. Established by Sri Awadhesh Prasad, it is working since the inception of Vindhyachal Project.

These NGOs present in the area credit themselves with extensive list of activities accomplished in order to rehabilitate the oustees. In order to decipher the assistance and sustenance provided to the victims by NGOs, another set of questions was asked.

Assistance by NGOs

Opinion	*No. of Persons*	*% age*
Nothing	231	92.4
Medical	4	1.6
Transportation	7	2.8
Training	3	1.2
Job Counseling	1	0.4
Housing	0	0
Education	3	1.2
Awareness for Job	1	0.4

The result thoroughly discredited the claim made by NGOs. 231 (92.4%) respondents had critical opinion about the work of NGOs. Some alleged, NGOs are the '*spoon*' of NTPC, and others blamed them for being the agents of government. There were still others who voiced that these local bodies are not working for the development of the downtrodden but they are only helping themselves. 4 (1.6%) families authenticated NGOs for their assistance in the area of medical and health. 7 (2.8%) gained with transportation facility. 3 (1.2%) received training in order to regain their lost livelihood. No one availed any housing facility. In areas of education, 3 (1.2%) respondents benefited.

V

Development-Induced Displacement: People's Perception, Reaction and Expectations

NTPC has direct presence in the area due to the power project and the World Bank having funded the project. The residents expressed their annoyance towards the role played by both these agencies. But they are optimistic that one day the attitude of the rehabilitation agencies would be changed and consequently bring about total transformation. So, efforts were made to turn the opinion of the respondents towards NTPC and the World Bank.

Opinion for NTPC Policy

The rehabilitation process took place vide R&R policy of NTPC. Opinion of the PAPs was registered regarding their attitude towards NTPC R&R act.

Opinion	*No. of Persons*	*%age*
No Opinion	1	0.4
Satisfactory, cooperative and helpful	5	2
Hostile and destructive	238	95.2
Neutral	6	2.4

Opinion for NTPC Policy

Positive trend shuns the response as just 5 (2%) persons considered the policy as satisfactory, cooperative and helpful. Majority of PAPs, 238 (95.2%), considered R&R policy of NTPC as hostile and destructive. They felt that the policy was apathetic towards the plight of the victims and favoured the government. 6 (2.4%) families remained neutral. They might not be willing to reflect on such a sensitive issue and endorsed to remain indifferent. Only 1 (0.4%) person reserved his opinion and did not say anything.

Role of World Bank

PAPs know that the monstrous project standing on the graveyard of their home and land is financed by the World Bank. So it was essential to notify the response of the locals towards the role of World Bank.

Opinion	*No. of Persons*	*% age*
Problems, unemployment and distraction	226	90.4
Employment and development	16	6.4
No Change	7	2.8
Any other	1	0.4

Majority of the universe took a negative stride and singled out the response 'problems, unemployment and distraction' – 226 (90.4%) consider the role of World Bank destructive. 16 (6.4%) families showed sympathy towards World Bank and considered their arrival with all round development and employment generation. Only 1 (0.4%) opted for 'any other' response.

Expectation from World Bank and NTPC

In order to decipher the expectations of the litigant from NTPC and The World Bank, queries were made. The result received is tabulated into various codes such as welfare (health, education, training, sanitation etc.), economic (compensation, money, employment, shop, land, contract, IMP Package etc.) and infrastructure (electricity, water, toilet, road, drainage, house, check dam etc.).

Expectation from World Bank & NTPC	*No. of Persons*	*% age*
Welfare	8	3.20
Economic	95	38.00
Infrastructure	3	1.20
Welfare and Economic	75	30.00
Economic and Infrastructure	23	9.20
Welfare and Infrastructure	6	2.40
All the Above	33	13.20
No Expectation	7	2.80

The result shows the victims still have lots of expectation from NTPC and the World Bank, even though considerable time has elapsed since displacement. 8 (3.20%) families opted for only welfare measures. 95 (38%) gave their preference to economic measures for upgradation of their standards of living. Whereas, only 3 (1.2%) preferred to remain with the option of infrastructure facilities. There were many families who aspired for more than one facility. 75 (30%) respondents choose welfare and economic measures. 23 (9.2%) seek economic and infrastructure facilities in the region. 6 (2.4%) remained with welfare and infrastructure activities. There were 33 (13.2%) highly optimistic families who required welfare, economic and infrastructure, all three facilities. Lastly, 7 (2.8%) do not expect anything from these organizations.

Conclusion

The saga of development-induced displacement is a global phenomenon and finds presence at every nook and corner of the world. The only difference lies on its magnitude. Whereas, the developed countries are producing less number of oustees, in comparison, the developing and underdeveloped are accounting for much higher number of toll. The tribulation has taken a giant shape due to manifold reasons such as lack of watertight policy, administrative lacunae, apathetic attitude, unawareness and lack of information. Dwelling at the empirical level, in the case of development-induced displacement in Singrauli region, the findings are not different than what we observed at the global level. In the absence of watertight policy and planning, the episode of R&R undertook in a haphazard and ad hoc manner. Land Acquisition Act of India, being archaic and colonial in nature, is heavily biased against the rights of the masses and favours the government. NTPC R&R Act kept changing as per requirement of time and convenience of the organization. So it also remained ad hoc and situation-centric. Rather than solving the problems of the PAPs, it aggravated it. Most of the victims were caught unaware and unprepared for the problems of R&R. Due to illiteracy, they were not able to decipher the meaning of section-4 notification. Those who understood it could not react properly due to lack of awareness and resources.

Complicated land acquisition act and the resettlement and rehabilitation policy of NTPC destroyed the native completely. They had no house, hearth, agricultural land and sustainable source of livelihood. Tiny amount of compensation was not adequate to recharge them and most of the PAPs lost it into other necessary requirements of the time. Thus, they were left with neither money nor employment. Their dependence on ecology and environment was reduced due to deforestation. It severely affected those landless and tribal people, highly dependent on the forest products. Environmental problems penetrated the health of locals as they fell ill due to various diseases. They had adjustment problems to the new location, which further aggravated when the host opposed their arrival. Infrastructure facility was dismal in the RCs and water and sanitation were the biggest problem. Small plot was not enough to accommodate big joint families.

VDAC was not found acting with full strength and so was the condition of PIC. People felt cheated with the changing parameters of NTPC policy. At this crucial juncture, there were very few helping hands available to them. People felt disgusted with the so-called help of NTPC and found its policy as hostile and destructive. The opinion was same about the World Bank. Lastly, the victims of displacement still have mammoth expectations from NTPC and the World Bank. They believe help and assistance from NTPC and the World Bank could bring back prosperity among the PAPs.

The above findings suggest that while planning a project, the planners should take into consideration even the minute details of it along with the calculation of its profit and loss and accordingly allocate the budget. But the attention on social and environmental aspects is missed on most of the accounts. As a result, the problem of involuntary displacement is taking gigantic shape, day by day, and it is touching the whole world on the same magnitude and scale. The global smack words like information, awareness and communication etc. do not have a significant role to play in the whole epitome as most of the incidences of involuntary displacement are accomplished in the areas inhabited by less educated, unaware and semi-empowered populace.

References

Amarteifio, G.W., D.A. Butcher and D. Whitham. 1966. *Tema Manhean: A Study of Resettlement*, Accra, Ghana: Ghana University Press.

Barbaras, A. and M. Bartolome. 1973. *Hydraulic Development and Ethnocide: The Mazatec and Chinantec People of Oaxaca*, Copenhagen: International Work Group for Indigenous Affairs (IWGIA).

Baxi, Upendra. 2008. 'Development, Displacement, Resettlement: A Human Rights Perspective', in *India Social Development Report 2008*, New Delhi: Council for Social Development, Oxford University Press.

Berman, T. 1988. 'For the Taking: The Garrison Dam and the Tribal Taking Area', *Cultural Survival*, 12 (2).

Cernea, Michael M. 1993. 'Anthropological and Sociological Research for Policy Development on Population Resettlement', in S. Guggenheim and M.M. Cernea (eds), *Anthropological Approaches to Resettlement: Policy, Practice and Theory*, Boulder, CO: Westview Press, 13–38.

——. 1996. 'Eight Main Risks: Impoverishment and Social Justice in Resettlement', paper presented in a workshop organized by the World Bank's Environment Department, NDO, 12–14 March.

——. 1997. 'African Involuntary Population Resettlement in a Global Context', Environment Department Papers, Social Assessment Series No. 045, World Bank, Washington, D.C.

Chambers, R. (ed.) 1970. *The Volta Resettlement Experience*, London: Pall Mall Press.

Colson, E. 1971. *The Social Consequences of Resettlement: The Impact of the Kariba Resettlement upon the Gwembe Tonga*, Manchester, UK: Manchester University Press.

Dreze, J., M. Samson and S. Singh (eds) 1997. *The Dam and the Nation: Displacement and Resettlement in the Narmada Valley*, Delhi: Oxford University Press.

Dubey, Muchkund. 2008. 'Foreword', in *India Social Development Report 2008*, New Delhi: Council for Social Development, Oxford University Press.

Electricite de France International and Group Charbonnages de France. 1990. NTPC Environmental Study of Singrauli Area, Chapter 3, Socio-Economics, EDF and GCF.

Fahim, H.M. 1981. *Dams, People and Development: The Aswan High Dam Case*, Oxford: Pergamon Press.

Fearnside, P.M. 1990. 'Resettlement Plans for China's Three Gorges Dam', in M. Barber and G. Ryder (eds), *Damming the Three Gorges: What Dam Builders Don't Want You to Know*, London: Probe International and Earthscan, 34–58.

Fernandes, Walter. 1996. 'Displacement and Rehabilitation: Principles of the NGO Alternative', paper presented in a workshop organized by the World Bank, NDO, 12–14 March.

——. 2004. 'Rehabilitation as a Right: Where is the Policy?', *Social Action: A Quarterly Journal of Social Trends*, 55 (2), April–June.

——. 2008. 'Sixty Years of Development-induced Displacement in India: Scale, Impacts, and the Search for Alternatives', in *India Social Development Report 2008*, New Delhi: Council for Social Development, Oxford University Press.

Fuggle, R., W.T. Smith, Hydrosult Canada Inc. and Androdev Canada Inc. 2000. 'Experience with Dams in Water and Energy Resources Development in the People's Republic of China', Country Review Paper prepared for the World Commission on Dams, Cape Town, South Africa.

GHK/MRM International et al. 1994. Singrauli Development and Environmental Management Strategy, Displaced People Action Plan, GHK/MRM.

Giddens, Anthony. 1990. *The Consequence of Modernity*, Cambridge: Polity Press.

GOI. 1995. National Policy on Resettlement and Rehabilitation, Ministry of Rural Area and Employment.

——. 1996. National Policy on Resettlement and Rehabilitation, Ministry of Rural Area and Employment.

——. 2003. National Policy on Resettlement and Rehabilitation, Ministry of Rural Area and Employment.

——. 2012. National Policy on Resettlement and Rehabilitation, Ministry of Rural Area and Employment.

Guggenheim, S. 1993. 'Peasants, Planners and Participation: Resettlement in Mexico', in S. Guggenheim and M.M. Cernea (eds), *Anthropological Approaches to Resettlement: Policy, Practice and Theory*, Boulder, CO: Westview Press, 201–228.

Hart, D. 1980. *The Volta River Project: A Case Study in Politics and Technology*, Edinburgh: Edinburgh University Press.

La Rovere, E.L. and F.E. Mendes. 2000. 'Tucuri Hydropower Complex, Brazil', Case study prepared for the World Commission on Dams, Cape Town, South Africa.

Marsden, David. 1996. 'Resettlement and Rehabilitation in India: Some Lessons from Recent Experiences', paper presented in the workshop organized by the World Bank, NDO, 12–14 March.

MARG. 1990. *The Land Acquisition Act and You*, Delhi: Pauls Press and Jain Book Agency.

Mathur, Hari Mohan. 2008. 'Introduction and Overview', in *India Social Development Report 2008*, New Delhi: Council for Social Development, Oxford University Press.

Mathur, Hari Mohan (ed.) 1998. *Development, Displacement and Resettlement: Focus on Asian Experience*, New Delhi: Vikas Publishing House.

Modi, Renu (ed.) 2009. *Beyond Relocation: The Imperative of Sustainable Resettlement*, New Delhi: Sage.

Mukerji, Debashish. 1997. 'Victims of Progress', *The Week*, 23 February: 22.

NTPC. 1993. Resettlement and Rehabilitation Policy, New Delhi.

——. 1994. 'Workshop on Sustainable Development in Singrauli Region', proceedings and paper presented in the workshop at Shaktinagar, Sonebhadra, UP, 11–12 November.

——. 1996. Checklist on Resettlement and Rehabilitation for New Projects, Corporate R&R Group, NTPC Corporate Centre, New Delhi.

Obusu-Mensah, K. 1996. *Ghana's Volta Resettlement Scheme: The Long-term Consequences of Post-colonial State Planning*, London: International Scholars Publications.

Ortolano, L. et al. 2000. 'Grand Coulee Dam and the Columbia Basin Project, USA', Case study report prepared as an input to the World Commission on Dams, Cape Town, South Africa.

Quing, D. 1998. *The River Dragon has Come: The Three Gorges Demand the Fate of China's Yangtze River and its People*, London: M.E. Sharpe.

Sampat, Preeti. 2013. 'Limits to Absolute Power: Eminent Domain and the Right to Land in India', *Economic and Political Weekly*, XLVIII (19): 40–52.

Scudder, T. 1996. 'Development-induced Impoverishment, Resistance and River-basin Development', in C. McDowell (ed.), *Understanding Impoverishment: The Consequences of Development-induced-displacement*, Oxford: Berghahn Books, 49–74.

Singh, Mridula and Others. 1992. *Displacement by Sardar Sarovar and Tehri: A Comparative Study of Two Dams*, New Delhi: Multiple Action Research Group.

Taneja, Bansuri and Himanshu Thakkar. 2000. 'Large Dams and Displacement in India', Submission No. SOC166 to the World Commission on Dams, Cape Town, South Africa.

The Land Acquisition Act and You with Short Notes. 1995. Delhi: Universal Law Publishing Company Pvt. Ltd.

The World Bank. 1980. Social Issues Associated with Involuntary Resettlement in Bank Financed Projects (Operational Manual Statement No. 2.33), Washington, D.C.

——. 1990. Operational Directive 4.30, Involuntary Resettlement, Washington, D.C.

——. 1991. Operational Directive 4.20, Indigenous People, Washington, D.C.

——. 1993. Introduction to Environment and Social Assessment and Requirements and Procedures for World Bank Financed Projects, Environment Department, Washington, D.C.

——. 1994. Resettlement and Development: The Bank Wide Review of Projects Involving Involuntary Resettlement, 1896–1993, Environment Department, Washington, D.C.

Thukral, Enakshi Ganguli. 1995. 'Development, Displacement and Rehabilitation: Is There a Need for a Gender Perspective', *Mainstream*, 11 November: 24–26.

Verma, Manish Kumar. 2004. *Development, Displacement and Resettlement*, Jaipur: Rawat Publications.

——. 2012. 'Policy Planning in Development Projects: Processes and Limitations', in Biswajit Ghosh (ed.), *Interrogating Development: Discourse on Development in India Today*, Jaipur: Rawat Publications, 237–262.

13

Changing Experience of Dalit Youth in Contemporary India

Antara Ray and Ramanuj Ganguly

> The youth are the strength of a strong nation and are capable of substantially influencing polity. In the recent times, the youth have transformed some monarchies into Republics by overthrowing monarchial power. As such, the youth are always at the centre stage of socio-political activities of every country and similar is the position in our nation. To maintain this strength a clear cut youth policy is required to rid the youth of problems relating to education, poverty, nutrition, employment opportunities, self employment, vocational training, health, sports etc. The country at present has no institutional mechanism to harness the potential of our youth and channelise their energy for the betterment of the country. The plight of the youth belonging to Dalit communities such as Scheduled Castes, Scheduled Tribes and Other Backward Classes (OBCs) who have been oppressed for centuries is even worse. Even today the Dalit youth have to face social ostracization though thanks to the reservation policy propounded and given by the messiah of the Dalits Babasaheb B.R. Ambedkar many of them have made some progress but the Dalit youth still require special attention because there is a need to instill a sense of belonging among the Dalit, backward and oppressed youth of the nation by providing them all opportunities for their overall development so that they too contribute to the progress of the country to their full potential. The facilities and opportunities should be provided as a matter of right. (The Dalit, Backward and Oppressed Youth Bill 2010: 4)

This paper attempts to explore sociologically to understand the changing Dalit youth culture in India, tracing their socio-historical factors which facilitated the development of the contemporary youth's identity among Dalits and critically evaluate the ways in which sociology can theorize the relationship between young people, culture and 'old' and 'new' India.

Themes covered by the paper include: youth and Indian society, identity formation, cultural representations of youth style, emerging aspiration criteria, youth culture and new technologies, and positional perception of Dalit youths in global and local youth cultures. The paper will also look beyond youth culture as represented in mainstream discourse in India and would try to undertake a critical examination of trends in post-subcultural youth and neo-tribalism among Dalits. The paper would envisage to develop a critical understanding of the main sociological issues involved in explaining the relationship or break between Dalit youth and the cultural pattern of post-independent India, and to make connections with how the social themes and issues engulfing the Indian society nourish the extensive sociological debates like, the outcome of consumer society, the role of the mass media in the construction of social 'reality', the unstable and shifting nature of caste identity in contemporary social-political settings, and the social relationship are getting redefined in the tug-of-war between the 'local' and the 'global' culture.

The primary background condition for this paper came from the fact that in the socio-historical and technological features that congregated in the post-economic reformed India, when the country moved from welfare economy to get integrated into the market-driven universal consumer culture, the youth were central to this design, and, therefore, the Dalit youths were not left out of this. The Indian youth in general got progressively lulled into the believing that the market advertised stereotypes with a youth tag could be adopted and celebrated with emblematic connotation by them as a means to map a cultural terrain distinctive from the cultural pattern of their parents' generation and from the broader society. Contemporary media representation extol the youth for their energy, enterprise, and ability to engage as a change agent on the one end, and on the other denigrate their gullibility as victims and moral panics, which often lead to erroneous assumption about youth culture and the various forms of social practices that young people in India engage with. Youth groupings among Dalits in particular whose collective identities are still based on structural determinants such as caste, class, and gender in India are getting further divided around new aspirations based on consumer sensibilities of standard of living and taste. Young Dalits, being very similar to any other youth of the country, become victims of cultural, structural and direct violence and become carriers or perpetrators of that violence. There is a strong tendency among politicians, policy-makers, academicians and researchers, though, to see such issues as specific aberration or anomic situation that needs to be resolved. Also, in such cases the youth are underrated as constructive agents with power and potential to bring social change and work as key players in the process of social integration. The problem of bias has also been observed, that Dalit youths face, come from the fact that asymmetric judgment the society has about the process of negotiation with new technology they undertake.

Online communication, mobile telephony or gaming has eroded the public and private space giving a newer meaning to the age-old concepts like social group or community or culture among the youth in general in India. However, the Dalit youth still at times are debarred from using these because of their socio-economic background. This brings to the fore the fact that popular representations and the everyday life experiences of young people among Dalits are encumbered by their caste legacy. Under such legacy, for the Dalit youth, new social pressures, like risk and uncertainty that is confronting youth of contemporary India in general, though, get further magnified. Therefore, for a Dalit youth the popular portrayal of a youth as disaffected, apathetic and apolitical becomes a myth.

> Many of the trends that shaped the transitions of young people in the late 1990s have become more pronounced over recent years ... . Commentators have described this picture in terms of *uneven* and *fragmented transitions*. Uneven, because different groups of young people have very different experiences of the transition to adulthood; fragmented, because the different markers of adulthood are increasingly uncoupled from each other ... . One way of thinking about the changing conditions of young people's lives is to recognise that in becoming adult, young people are pursuing the experience and recognition of *competence*. This is the feeling that you are good at doing or being something. There are rewards gained from the ways in which others react to this such as greater freedom, responsibility, privacy and respect. In the past, competence came as a package, and was founded on economic independence. Increasingly, the process of becoming competent is a piece-meal affair. Different forms of competence may be available in different areas of young people's lives. So for example, a young person may develop a sense of competence by being a valued worker, a parent, a successful student, musician, sportsperson, sexual partner, popular or even feared member of a social scene. Failure to be recognized as competent in one area may encourage young people to invest and seek recognition in other areas. So for example, young people who struggle as students may seek competence and recognition from work, consumption and parenthood.
>
> This perspective makes visible two themes that underlie the contemporary policy agenda on youth. First, a concern with *social exclusion*. We are familiar with the notion that young people's experiences are shaped by economic inequality. However, there is also a cultural dimension to inequality. This is where an initial disadvantage creates the conditions through which it makes sense to young people at the time to make choices and pursue lifestyles that in turn can reduce their life chances and those of their children ... . The second theme that this perspective makes visible is a more nuanced understanding of the role of *timing* in the processes of social inclusion and exclusion, highlighting the potential for *targeting* resources and interventions. Not all young people from disadvantaged backgrounds fall into the stigmatized policy categories ... (Tucker et al. 2004: xiv-xvi)

II

Youths are considered to be the backbone of any country. The change, advancement and innovation rest highly on their shoulders. However, in India, youths as a category are forged with many interest and identity that weaken their collective might. In fact, youth of India reflect just the chasms and fractions that India is mired with as a multi-ethnic and multi-cultural society. Nevertheless, a careful look through is made in various programmes and policies of post-colonial Indian state and/or literature from social sciences, we will observe that similar to the title of the previously cited bill, there is an inherent tendency in the mainstream gaze of the society to homogenize deprivation and marginality among the youth of India. Just like Indian history and politics can never be separated from jati, so is the case of Dalit youth. History of anti-caste movement can be traced as early as thirteenth century starting from various Bhakti traditions throughout India. But anti-caste movement took a prominent shape from early 19th century with Jyotiba Phule. Phule, despite of forming 'Satyashodhak Samaj', couldn't mobilize and organize lower caste people and especially the youth the way Ambedkar did. Here, we can also talk of Gandhi's mobilization of untouchables and inculcating them into his ideology and strategy. However, the mobilizing and organizing of lower caste youth, especially the untouchables took place with Ambedkar. Caste and the exclusive privileges that it bestows were challenged by various efforts to overturn the existing social relations and exploitative structures. Where Gandhi focused on a reformist view towards Hinduism, Ambedkar completely rejected Hinduism and called for 'annihilation' of it and conversion to Buddhism.

The social asymmetry and injustice rooted on caste/jati assigns an ascribed status by virtue of birth, have been engaged with and questioned repeatedly by thinkers and social reformers throughout the history of India. 'Dalit' is the widely used post-colonial term for the former untouchables and Avarna or Pancham of so called 'Hindu religion' belonging to the Indian society. The term 'Dalit' does not have a uniform definition. Various people, academicians, and theoreticians have understood and defined it in different ways. The non-Dalit writers and intellectuals mostly have found its origin in Sanskrit texts locating it in the term 'Dal' which means broken, crushed, scattered, downtrodden etc. In the pre-colonial period the most prominent figure who raised the issue of untouchability was M.K. Gandhi (1869–1948). Though Gandhi called untouchability an evil but he did not go against Varnashrama Dharma. He believed untouchability as a subversion of true Hinduism. But first time in history of Indian caste system, untouchables or Ati-shudras or Pancham were named as 'Harijans', or children of God, which marked a relative change in status and a new discourse emerged with the terminology.

Although there has been an increasing interest in youth of Dalit descent, most of the studies are paying attention principally on the strong points in their character in comparison to the youths of other higher caste groups, especially their high cumulative attainment in production process in primary and secondary sector of Indian economy. The emphasis on mechanisms underlying successes or failure of Dalit youths in the various domains has contributed to the maintenance of the popular myth of the success of Indian democracy. However, recent studies, in the first decade of 21st century India, indicate that caste-based discrimination continues to remain a salient feature in the experiences of Dalit Indian youth, including those who have converted to other religion like Christianity or Sikkhism, and that such discrimination has negative consequences for their psycho-social well-being. Researches that present thick descriptive insight into Dalit youths' perceptions of discrimination, most have focused on high school and college students, and relatively little is understood about the incidence, nature, and correlation of caste-bigotry among youth that have been done. Moreover, few studies have examined the extent to which affirmative Dalit identity works as a protective shield and if such identity beliefs provide any form of security to the Dalit youths in contemporary India.

This paper attempts to examine the rate of recurrence and variables of perceived caste discrimination from peers among Indian youths, and the extent to which, aspects of caste identity may serve as a protective in-group identity. Engaging in this kind of argument about the Dalit youths here is critical, because it is in such a phase of life for individual issues of caste identity and discrimination becomes particularly relevant as they begin to come to terms more profoundly with questions of who they are as Dalits. The Dalit youths even in urban areas of Indian states like Uttar Pradesh, Bihar, Madhya Pradesh, Rajasthan, Gujarat, Maharashtra and Andhra Pradesh report alarmingly high contact caste-based discriminatory practices from their peers, which exposes the fact that protective in-group identity of Dalit youths often moderate the harmful effects of such exposure for their mental well-being. On the basis of the above it can be argued that Dalit identity today does shield Dalit youths' mental health in the face of caste-based discrimination. Media hyped stereotypes have made the society believe that caste-based discriminations faced by Dalits are only of physical nature laced with violence; though it is our finding that Dalit youths' daily experiences from peers some form of rudeness, disrespect and caste bigotry, and discrimination in the form of exclusion and mockery. Salience of caste-based discrimination among Dalit youths harms their sense of self and creates poor self-esteem, resulting in distorted group and self-identity issues among them. The distance to schools/colleges is also considered a huge barrier for Dalit children, and a significant part of the explanation for the low enrolment rate and the high dropout rate. Due to the unwillingness of higher caste groups to live side by side with Dalits, Dalit families often live in remote areas, away from the main villages and schools. This residential pattern has two major implications. First, the location of schools within the main villages, and hence within higher caste areas, makes it difficult for Dalit children to gain access to schools, due to caste tensions. Second, the great physical distance to schools often result in Dalit children dropping out, as the distance is simply too far to walk on an everyday basis (UNICEF 2006: A). Similarly, migratory labour is another factor that adds to the high dropout rates. Many Dalits are landless and are forced into migrant labour, as this is often the only way to ensure the economical survival of their families. The continuous migration in search for labour implies a frequent disruption of the Dalit children's education and makes them incapable of keeping up with the academic advancement of other children (HRW 2007). The caste bias manifests itself in the way teachers ignore Dalit students and unjustly fail them in exams, in social exclusion and physical abuse, and in the unwillingness of the university administration to assist and support Dalits. As a grave consequence of this harassment, a disproportionate number of Dalit students have committed suicide (The Death of Merit 2011: A). Indeed, in India alone, 18 Dalit students have committed suicide in one of the country's premier institutions between 2008–2011, and this number only represents the official cases. Counting all the Dalit students whose families did not protest against the incessant discrimination that eventually led to suicide, the number is likely to be much higher (The Death of Merit 2011: B). According to the Briefing Note (2010) by Navsarjan Trust (www.navsarjan.org), more than 200 million people in India are vulnerable to discrimination, exploitation and violence simply because of the caste into which they were born. The caste system relegates Dalits, formerly known as 'untouchables', to a lifetime of segregation and abuse. Caste-based divisions dominate in housing, marriage, employment, education, and general social interaction – divisions that are reinforced through economic boycotts and physical violence. Dalits are forced to perform tasks deemed too 'polluting' or degrading for non-Dalits. The reports provide irrefutable evidence that public servants and community members in India, i.e. state as well as non-state actors, violate a number of human rights protected by domestic laws and international human rights treaties. The various rights that are violated for Dalits are: The right against 'untouchability'; deny Dalit children access to an equal education; deny Dalit children their equal right to health by forcing them into hazardous work that includes cleaning human excrement and disposing of dead animals; child labour and manual scavenging-work that is likely to be hazardous or to interfere with the child's education, or to be harmful to the child's health or physical, mental, spiritual, moral or social development; teachers and community members force Dalit children into unpaid labour, primarily cleaning schools, homes, and toilets, in what constitutes a modern form of slavery.

However, those Dalit youths who have a high self-perceived regard about their caste background are found to be confident and are often engaged in raising the consciousness and standard of living of people from Dalit caste group. Thus, they, by affirming their Dalit identity negate the outcomes that have been associated with discrimination based on caste origin. However, this does not mean that such engagements have dented the overall caste-based discriminatory scenario among the youth of contemporary India. Nor has it changed the mental make-up of the educated urban youth of the country because their perception of others' opinion about their own is invariably negative and full of suspicion. Those who readily accept their subjugated position in the society among the Dalits, and are still engulfed with prejudice, are a sizable section of the Dalit youths who are socially and mentally better off because they do not hurt the status quo, thus, instead of hostility enjoy the security of ordered inequality.

The National Campaign on Dalit Human Rights and International Dalit Solidarity Network in their 13th Session of the Universal Periodic Review Report of the Human Rights Council – India: 'Caste Based Discrimination in India – Key Recommendations and Questions' found the following to be key areas of human rights violations for Dalits in India: Impunity and non-implementation of laws for the protection of Dalits; Exclusion in access to basic services and extreme poverty; Diversion of economic benefits allocated for Dalits; Lack of political participation; trafficking and forced prostitution of Dalit women; Manual Scavenging and Bonded Labour; systematic exclusion of Dalits during disaster management; Infrastructure inaccessibility and investment related displacement is most acute in Dalit populated area; and Continued Discrimination in Education.

In a small survey keeping the above in mind, it was observed that Dalit youths perceived their role not as an age-group but a stage in their life where they represent a life-characteristics and optimistic personality stance, thus, have certain option and opportunity to assert their autonomous outlook. It was seen that the Dalit youths consider that conditions are ripe to correct the skewed life-chances of their community; however their optimism was tempered by their reading of continued dominance of caste hierarchy and emergence of newer forms of caste-alliances; therefore, there was a sense of meaninglessness among them about their educational attainments and employability. Similar to youth of any place and culture, for the Dalit youths, education, employment, travel, friends, music and sensitivity about current social issues were found to be the spot of highest concern and attention, but what was unique was their very high concern about environmental sustainability and distortion of local culture due to 'cultural imperialism' due to spread of mass media. Another cause that was perceived by them with apprehension was the 'homogenizing-tendency' by the market in terms of availability of goods and services, thus, signifying their concern about the fact that unique cultural trends, tastes and skills are getting destroyed, making many from their community unsuitable for the job-market and forcing them to accept unsustainable wage. Though they are upbeat about the enabling and life-transforming role of technology but are very conscious about the 'digital divide', and their limitation in matching the performance of those who have access to technology. With regard to politics and civil society movement, a sense of disenchantment and disregard seems to have crept in among the Dalit youths as they feel – the Dalit question compounded with reservation issue – has queered the field for them, and personal agenda fulfilling ambition of politicians and civil society workers who do not have much to offer them.

Nussbaum (2000) broadening the capability approach of Amartya Sen indicates that for expansion of human capabilities to achieve the human potentials, various other lived criteria of individual has to be taken into consideration like their life-span, nutrition and health, safe housing and shelter, choice of mate selection and time of procreation, informed decision making, right to education, security against discrimination and threat of violence, and ability to make choices about ones politics, environment and relationships. Taking these as parameters to check the condition of Dalit youth, it was found that they perceive under most of the factors they remain both deprived and excluded, as well as strive much more than other young people because of their caste background. In the long run, such passive condition of exclusion restricts their space for participation in the society. It was also observed that there are numerous and multifarious connection between basic rights and health of the vulnerable groups like the Dalits. Violations or lack of attention to basic rights can have serious health consequences for certain groups like Dalits who generally experience cruelty, dishonour and bigotry, unsafe customary practices etc. The approach with which health procedures and schemes are designed are favourably disposed to higher caste groups for which it fails to protect the rights of accessibility to services, information and awareness; dignity and privacy; and cultural sensitivity of Dalits. The health concern of a Dalit youth depends on the circumstances such as adequate earnings, livelihood, access to safe drinking water etc. However, opportunity to avail all the above aspects are seriously affected by their belonging to a caste group that still has to endure inequity. Empowered Dalit youth today are aware of social milieu related abuses, so they can recognize and confront their caste groups' face, understand the reason for which they suffer group-specific health hazards that also affect their health awareness.

Poverty is a significant feature that adds to social exclusion of the Dalits. Though poverty has an effect on all poor, it has distinct impact on the Dalits. Poverty, illiteracy and age-old caste based domination make them more vulnerable to environmental and health problems as they have scarce ability to take advantage of the facilities of essential services that in the last two decades in India have progressively become commercialized and valued in terms of money. Heterogeneity of multicultural India and the accommodation process practised due to the political compulsions has given rise to newer forms of intense tension, conflict and violence against the vulnerable Dalits. The apathetic state machinery is antagonistic and intimidating; therefore, when their support is sought by such groups they become victims of red-tape and policy inertia, bribery, benefaction and clientalism. The insecurity and inequality that marginalize the youth among the Dalit poors in India also deprive them of constitutional rights and social security entitlements.

It will be misleading if one fails to see the Dalit youth in relational terms because their identity is constructed in relation to other groups, social and economic processes and institutionalization. Freire's (1996) argument seems very pertinent for such youths today as he suggested to recognize the dehumanization process they face that has to be struggled against without taking any 'generosity' from the oppressors who act humane to perpetuate their oppression.

The mobilization of youth around the issue of caste and untouchability differed from region to region in pre-colonial India, despite of Ambedkar's influence and charismatic leadership. Examples can be drawn from various states like West Bengal where a small section of youth got involved in Namashudra's movement which was later on, quite unsuccessful because of lack of leadership and motivations for youths. In case of UP, pre-independence era saw the mobilization of Chamar youth against the caste system. Though they supported Ambedkar but their standpoint and actions were different from Mahars of Maharashtra who were under Ambedkar's direct leadership. If we look at the lower caste youth of other two states prominent in anti-caste movement that is Karnataka and Andhra Pradesh, we will again find different political ideologies mobilizing the youth. Where Karnataka was mainly pro-Congress and supported Harijan Sabha, Andhra and the Telengana regions were mainly under left ideology. There was no prominent Dalit antagonization in terms of movement but there was Adi-Andhra movement in coastal Andhra and Adi-Hindus movement in Hyderabad region where youth, not particularly belonging to Dalit, participated but youth of other exploited class and caste participated against the exploitative character of caste. Hence, if we look at the scenario of Dalit politicization and participation of youth in pre-independence period, then we will find that other than massive participation of youth in Dalit movement at Maharashtra there was no significant unified political action carried forward by the Dalit youth in India. Anti-caste agitations and politicizations, though present in various pockets throughout India, but that was dependent on the localized problem of that area and was guided in a specific directions by the dominant political ideology out there. Like, for example, since Andhra was dominated by Left politics mainly, hence they ignored many caste realities; rather concentrated on the issues related with land which was more economical in nature (Omvedt 2010). Therefore, no exclusivist

Dalit politicization took place. There, youth were guided to take part in issues that are being taken up by the leftist parties. The upper caste and non-Dalit youths of colonial period were basically involved in the nationalist movement against the colonial rule and there was no visible involvement of them in the anti-caste and Dalit politics. Moreover, it can be seen that the Dalit youth of India in the pre-independence era were not under one ideology or political affiliation but under many and were scattered in terms of their political action against caste structure and relations producing out of it; since then there existed ideologies of Ambedkar, Gandhi, Nehru, Leftist ideologies and other regional specific motivations. Dalit movement in particular and anti-caste movement in general, of pre-colonial India, was value-oriented or anti-systemic movements (Omvedt 2010).

The term 'Dalit' and 'Harijan' both have two different political and social connotations. In the post independent era the term 'Harijan' lost its achieved popularity among the youth belonging to lower caste and untouchables, whereas the term 'Dalit' gained its power and influenced them. It is argued that the term 'Dalit' is not merely a rejection of the very idea of pollution or impurity or 'Untouchability', it reveals a sense of a unified class, of a movement toward inequality (Michael 2007). 'Dalit' as a subject emerged as political which does not put forward one unified identity; since the concept is having many interpretations and definitions, no one homogeneous interpretations can be found. Youth of contemporary India is getting absorbed but scattered within multiple ideologies, strategies, associations and political parties associated with this term 'Dalit'. Dalit now has no one discourse but numerous divisions which in the process are channelizing the youth into caste politics in diverse ways. Dalit grouping has emerged as a unique kind of political subject taking youth in its folds and enabling constitutive contradictions in Indian political modernity. Political liberalism, democracy, growth of human freedom, neo-religions, media and diverse theoretical discourses have taken away the binary and paved ways to various contradictions, dispersions and disaggregations rather than cohesion on the issue, demands and rights among the youths of India.

Ambedkar, with his death in 1956, left behind a vast political legacy which nationalized the Dalit identity that further got deep-seated with Dalit Panther movement. Greatest mobilization and agitation of Dalit youth which established the concept 'Dalit' and 'Dalit Discourse' took place in Maharashtra during 1970s with Dalit Panther movement. The members of Dalit Panther movement were young men and women mostly between the ages of 20–30 years. The militant Dalit Panthers Party founded by two writers Namdeo Dhasal and Raja Dhale in April 1972, antagonized and organized the Dalit youth of Maharashtra within its folds (Zelliot 2010). But this radical movement was unable to mobilize the

Dalit youth from other parts of India into their fold, the way it happened among the Dalit youth of Maharashtra, since there does not exist one homogenized identity, ideology and political affiliations. This can be drawn from the instances that Dalit Panther Party was formed only in Maharashtra and in no other state of India. A radical movement carried forward by Dalit Panther Party did not influence the other parties based on Dalit ideology existing elsewhere like UP, Andhra Pradesh, Karnataka, Tamil Nadu, Kerala etc. Rather the Dalit Panther organization got scattered within two years (founded in 1972 and split in1974) and remained important only in Maharashtra among the few youths (Zelliot 2010).

If we analyse the contemporary period we will find that there is no one ideology or one set of motivations that are guiding the youth of India to get involved in the Dalit Politics. The politics which surrounds the identity of 'Dalit' where youth are taking a leading role in terms of actions is not one but many. Post-independent India, rather than extracting out from the situation of caste-exploitation has crystallized new kind of discriminations in terms of creating 'secular-Brahmanism', 'elite-Dalits' and as a result fragmented youth political involvement as too many choices and directions are being laid down before them.

Dalit youths of contemporary India are placed across different regions, economic standards, educational levels, social groups and sartorial choices. Youth reside in multiple worlds and Dalit youths are no exception in this matter. These numerous worlds socialize youth in different ways. The processes of globalization and modernity have created a huge impact in these worlds. This brings out multiple responses, activities and inclinations. S.M. Michael (2007: 35) writes:

> Dr Ambedkar's followers formed the Republican Party of India immediately after his death, mainly for representing the interests of the Scheduled Castes and other weaker sections. But the leadership crisis in the Republican Party and the growing attacks on the Dalits made the Dalit youth reject their leadership and adopt a militant method. These Dalits, especially the educated Dalits in Maharashtra, came forward and took up the task of bringing all the Scheduled Castes onto one platform and mobilizing them in the struggle for their rights and justice. The caste stigma, which remained even after its abolition, was painful to these sensitive youths … . They struggled at various levels in villages, cities, educational institutions and working organizations by means of a number of political and social organizations like the Dalit Panther, mass movements, Dalit Liberation Army, Youth Republican, Dalit Sangharsh Samiti and through Dalit theatre, Dalit art, Dalit literature and among Christians, Dalit theology etc. These organizations, dominated mostly by youth wings of the Dalit force, worked as pressure groups for educating and mobilizing the Dalits and demonstrating in order to get their problems resolved. Dalit consciousness was also expressed in the formation of the Bhim Sena, Dalit Sena, Dalit Sahitya Movement, the emergence of Dalit Rangbhoomi (Dalit Theatre) BAMCEF

(The All India Backward SC, ST, OBC and Minority Communities Employees Federation), the Bahujan Samaj Party (BSP) and the Bharatiya Republican Party (BRP).

With the growth of democratic institutions and so many political parties, the Dalits have begun to assume importance in national politics of contemporary India. The leaders of these political associations by taking advantage of this situation are mobilizing youth in their favour. Hence, the youths involved in these associations are only acting as pressure groups. The political initiatives taken up by Dalit leaders have marked a beginning of a new era of involvement of youth in democratic politics. Rajni Kothari maintains that Andhra Pradesh and Bihar are examples where one can see that there is a rapid succession of various caste groups into factional networks of politics which provided the best channels of mobility (Kothari 2009). Youth of the contemporary India who are always inclined towards upward mobility and more so true for Dalit youths who have a history of being a depressed and exploited class are getting easily antagonized under political folds.

Religion plays a major role in forming diverse ideologies and sources of motivations for the mobilization of youth in the contemporary Indian situation. Just like every other social institution, religion has also gone through immense change and we call this new age religion to be 'neo-religious institutions'. The use of religion in politics can be seen from India's pre-independence era but the formation of neo-religious institutions, the mixing of religion and politics has been consolidated with the interdependence on each other. The party organizations and voting habits are significantly getting affected by religious divisions and affiliations. One important reason for having multiple political parties and ideologies is due to this numerous neo-religious associations. One of the key political parties Bharatiya Janta Party (BJP) is an example of a dominant political party of India which is closely linked to Hindu-nationalist agenda that is getting strong support from neo-religious association like Rashtriya Sayamsevak Sangh (RSS). India's political-religious divides, in turn cross-cut caste loyalties. Youth of India being active, politically inclined and ambitious are getting absorbed by these various parties and neo-religious institutions. Institutions like RSS are saffronizing Hindutva and motivating a large section of youth to participate in their activisms. These institutions are not only attracting youth from upper caste but also from Dalit caste. Hinduistic ideology, in a relative subtle manner, is swaying the mind and attitude of lower caste youth into its domain and thus forming an ideology completely contradicting the ideology of Dalit associations and the youths involved there. But in case of Dalit politics too, we see acute political motives to gain state power. If we look at the case of UP's, BSP (Bahujan Samaj Party), first for the sake of votes mobilized the Dalit people and especially the Dalit youth of UP where the

Dalit population is very high. As a result, BSP came into power and Mayawati became the first Dalit woman to be a chief minister. But later on the ideology of the BSP changed and they tried to incorporate the Brahmins and other upper castes youth into their party ideology for votes. Gopal Guru (2004) has very rightly argued:

> In the post-Ambedkar period, however, dalit perception of political power supplanted the need to capture state power. Instead of attacking the state, both from within and outside the parliamentary structures, they have made the state and other institutions a sole repository of politics, reflected in the routine, everyday petty politics and the manipulation of the state structure for personal aggrandizement. The strategy of bargaining, adopted by the dalits in the post-Ambedkar period, howsoever, ineffective it may have been, has definitely contributed to stable, coherent and manageable politics of the dominant forces who are always on the edge of the unstable equilibrium in this country … . Dalits after Ambedkar have not fully realized the nexus between the state and the dominant castes. They also have shut their eyes to the fact that the political power of the dominant class is based on an unstable equilibrium of compromise of which the dalits themselves are a part, such as the BSP-BJP compromise in UP.

Caste loyalties are increasingly being taken up by the religious divisions and as a result many youth from the poorest caste are most likely to convert out of Hinduism. One such religion is Christianity where we can find a large number to people formerly belonging to the class 'untouchable'. But the rise of Hindu religious power in many parties has resulted in lack of inclusion of the Dalit converts to Christianity and Muslims into the constitutional safeguards. A Dalit who is a convert to Christianity or Islam will lose affirmative action rights. Dalit youths are getting fragmented with respect to the issues whether to remain within the Hindu fold or to get converted into Buddhism (but this conversion is more or less restricted to the Dalits of Maharashtra) or to some other religion. Dalit leadership has failed to give one unified direction to them. As a result, there are multiple political parties and associations in the country attracting and catering to the Dalit youth. If we look at the states of Kerala and Karnataka, we will find Christian-Dalit youths whose demands are different from those of Dalits under Hinduism. Dalit converts into Christianity does not have the privilege of protective legislation. So their main motive is to get included as a category under protective legislation as Scheduled Castes of India. But they are not yet included and hardly any significant actions can be seen either from Dalit Christians or from the Dalit-Hindu activists working under Dalit ideology. Same is the case of Dalit converts to Islam. Dalit ideology and leadership are unable to take their demands and basic need into account and Dalit youths, are further getting segregated between Hindu, Muslim and Christianity. The religio-ideological and associational position of Dalit youth of

India is creating multiple ideologies and thereby, Dalit youths are picking up political mobilization in order to achieve their personal aims as well as broader associational aim. But no one party or political ideology is all encompassing diverse issues and as a result, multiple political associations are mushrooming. Hence, Dalit youth belonging to Christianity or Muslim has separate path of actions than those Dalit youth who are within Hinduism. As a result not only within the various political parties but Dalit youths are getting segregated within various religions and therefore, no one unified action or ideology and reality.

The other important factor which is acting as a catalyst for the further consolidation of youth in Dalit politics is media. Media, rather than focusing on the identity of a youth as simply 'youth', is viewing and constructing their identity with their *jati* affiliations. The political reality is continuously being constructed, deconstructed and reconstructed by media, and is being analysed in terms of emergent power relations invariably centring on *jati*-quotient. Indian media fail to serve the public in the matter of caste tensions. Indian media does neither have a religious beat nor caste beat, per se, and thus lack reporters with expertise. Guru (2004) argues:

> If we focus on the social background criterion of representation, it was difficult to come across any Dalits representing their own case in the mainstream press electronic media. It was clear from reading the press that it was the non-Dalits who spoke on the issue of caste ... there was a serious mismatch between the discourse in the print media and Dalit mobilization, without giving us an authentic sense of the people's perceptions ... . Scholars wrote on the issue without bothering to take any feedback from those who are the victims. Since they have language competence they have the privilege of being invited by the media, darlings' of the press are able to write on anything and everything. Their interventions, however, raised more dust at the polemical level than elevating our understanding through the sociological expansion of the debate.

The media often fail with poor reporting and mislead and mislabel events and thus create constructed reality. It will be a utopian view if it is considered that media only deal with reality of today's youth and Dalit politics since in fact, reality is far too complex to be comprehended by one single way. The recognition of Dalits and non-Dalit are forcefully enacted by this new age media. Rather than portraying youth as only youth, media is now attaching them with their jati affiliation and giving a new turn to the issue ignoring the reality. Dalit politics as conducted in the mainstream media (now interchangeable with PR, post paid-news) is all symbolism and no substance. This might explain the lack of outrage at the government's opposition to the inclusion of caste in the United Nation's 2009 draft principles and guidelines for the effective elimination of discrimination based on work and descent. The comprehensive framework – the first of its kind – could have lead to the establishment of an international monitoring mechanism on caste discrimination. Some might question the need for global guidelines, believing that we live in a peaceful post-caste world with the occasional cacophony of vote-bank politics. But a three-day documentary film festival in Chennai called 'Imaging Dalit Reality: Politics of Visual Representation' challenged such notions.

The aspirations of youth of contemporary India are completely different from those of traditional India and it is more so true in case of Dalit youth. This change is due to the difference in vision perceived by the untouchables of the past and today. Visions set the agenda for both thought and action; philosophical, political, or social theories are built on them. Believers in one vision will view themselves in a very different moral role from the followers of another. The ramifications of such conflicting visions extend into economic, judicial, military, philosophical and political spheres (Michael 2007). Increased intellectual activism has marked Dalit-Bahujan cultural life at both national and the regional levels in recent times (Guru and Geetha 1997). Under the backdrop of various theoretical and academic discourses, the mobilization and antagonization of youth for/against the Dalit politics is taking place.

There is no one ideological expression and framework for the youth of India for supporting or rejecting the Dalit politics. Here, not only exists an egalitarian framework but also strong pro-religious forces. There is a conflict between political motives, ideologies and motivations which are further getting consolidated by neo-religious institutions, media and numerous theoretical discourses and thereby, putting forward options to be picked up by the youths of India as they desire. In one hand there is high militancy of Dalit youth in Bihar, Andhra Pradesh and Maharashtra, on the other there is lack of involvement of youth in Dalit politics in West Bengal and Punjab, and highly politically mobilized youth of Uttar Pradesh. The antagonisms between youths belonging to various associations are creating contradiction of Indian politics now. The recognition and politicization of public space and its all encompassing capacity to involve the youth have been possible due to the state structure, multiple political associations, neo-religious institutions, media and theoretical discourses. These multiple areas have been created due to the tension between 'Dalit' as a subject formation and individual rights which is enveloping youth into multiple realities of Dalit politics. This politics of identity and recognition of youth is giving rise to politics of difference among the youths of India on the issue of Dalit politics. This is, in turn, perpetuating inequality and new age discriminations among the youths. Political, media's and theoretical discourse's intervention increasingly in the name of human rights is further crystallizing the issue of exclusion among the youths.

References

Amirthanathan, Amala Jeyarayan. 2009. *Marginalization to Citizenship: Marginalization and Strategies of Responses of Urban Poor Youth in Salem, India*, The Hague, Netherlands: International Institute of Social Studies.

Ansell, N. 2008. *Children, Youth and Development*, New York: Routledge.

Arun, C. John. 2007. *Constructing Dalit Identity*, Jaipur: Rawat Publications.

Atkinson, A.B. 1998. 'Social Exclusion, Poverty and Unemployment', in A.B.A.J. Hills (ed.), *Exclusion, Employment and Opportunity*, London: London School of Economics.

Bhagavan, Manu and Anne Feldhaus. 2009. *Claiming Power from Below: Dalits and the Subaltern Question in India*, New Delhi: Oxford University Press.

Case Stories of Dalit Children in India – Victims of Caste Discrimination: A Survey by Navasarjan Trust. 2001. Published by the Robert F. Kennedy Centre for Justice and Human Rights in the Report 'Understanding Untouchability'.

Cheesman, N. 2002. 'A Hereditary Curse: Inequality and Indian Education', *Human Rights Solidarity*, 12 (5): 17–20.

Dahiwale, S.M. 2006. *Understanding Indian Society*, Jaipur: Rawat Publications.

Death of Merit, 2011 a and b (Part I, II), Documentary on Caste Based Discrimination in Indian Higher Education by Manish Kumar of IIT, Roorkee, Available on Youtube.

Durham, J.C.A.D. 2007. 'Introduction: Age, Regeneration, and the Intimate Politics of Globalization', in J.C.A.D. Durham (ed.), *Generations and Globalization: Youth, Age, and Family in the New World Economy*, Bloomington and Indianapolis: Indiana University Press, 1–18.

Fisher, C.B., S.A. Wallace and R.E. Fenton. 2000. 'Discrimination Distress during Adolescence', *Journal of Youth and Adolescence*, 29 (6): 679–695.

Freire, P. 1996. *Pedagogy of the Oppressed*, London: Penguin.

Garratt, D., J. Roche and S. Tucker (eds) 1997. *Changing Experiences of Youth*, London: Sage.

Girija, K.P. 2009. 'On Suicides, Caste and Higher Education', *Insight–Young Voices*, February–March, http://thedeathofmeritinindia.wordpress.com/2011/04/26/84/

Guru, Gopal and V. Geetha. 1997. *Dalit Intellectual Activism: Recent Trends*, Mumbai: VAK Publications.

Guru, Gopal. 2004. 'Politics of Representation', in Sukhadeo Thorat and Umakant (eds), *Caste, Race and Discrimination: Discourses in International Context*, Jaipur and New Delhi: Rawat Publications.

Hardgrave Jr., Robert L. 1970. 'Political and Primordial Solidarity: The Nadars of Tamilnad', in Rajni Kothari (ed.), *Caste in Indian Politics*, New Delhi: Orient Blackswan Private Limited.

Herrera, L. 2006. 'What's New about Youth?', *Development and Change*, 37 (6): 1425–1434.

Human Rights Watch. 2007. *Hidden Apartheid*, Shadow report for the UN Committee on the Elimination of Racial Discrimination (CERD) for the review of India.

Ilaiah, Kancha. 2009. *Why I am Not a Hindu!* Calcutta: Samya Publications.

Irwin, S. 1995. *Rites of Passage: Social Change and the Transition from Youth to Adulthood*, London: UCL Press.

Judge, Paramjit S. and Gurpreet Bal. *Mapping Dalits*, Jaipur: Rawat Publications.

Kothari, Rajni. 2009. *The Writings of Rajni Kothari*, Hyderabad and Delhi: Orient Blackswan.

Mannheim, K. 1952. 'The Problem of Generations', in K. Mannheim (ed.), *Essays in the Sociology of Knowledge*, New York: Oxford University Press, 286–322.

Michael, S.M. 2007. *Dalits in Modern India*, New Delhi: Sage.

Nussbaum, M. 2000. *Women and Human Development: A Study in Human Capabilities*, Cambridge: Cambridge University Press.

Omvedt, Gail. 2006. *Dalit Visions*, New Delhi: Orient Longman Private Limited.

——. 2010. *Dalits and the Democratic Revolution*, New Delhi: Sage.

Riele, K.T. 2006. 'Youth "at Risk": Further Marginalizing the Marginalized?', *Journal of Educational Policy*, 21 (2): 129–145.

Rosenberg, M. 1965. *Society and the Adolescent Self-image*, Princeton, NJ: Princeton University Press.

Shah, Ghanshyam, Harsh Mander, Sukhadeo Thorat, Satish Deshpande and Amita Baviskar. 2006. *Untouchability in Rural India*, New Delhi: Sage.

Srinivas, M.N. 1952. *Religion and Society among the Coorgs of South India*, Oxford: Oxford University Press.

The Dalit, Backward and Oppressed Youth (Development and Welfare) Bill. 2010. http://164.100.24.219/billstexts/rsbilltexts/asintroduced/dalit.pdf, accessed on 10 October 2010.

Tucker, Stanley, Jeremy Roche, Ronny Flynn and Rachel Thomson. 2004. *Youth in Society: Contemporary Theory, Policy and Practice*, London: Sage.

UNICEF 1946–2006: Sixty Years of Children. A State of World's Children, Special Report, Unicef.

Upadhya, C. 2008. 'Employment, Exclusion and "Merit" in the Indian IT Industry', in S.M.D.K.S. Babu (ed.), *India's Development: Social and Economic Disparities*, New Delhi: Manohar Publishers.

www.countercurrents.org/dasani110610.htm, accessed on 13 October 2011.

Zelliot, Eeanor. 2010. *From Untouchable to Dalit: Essays on the Ambedkar Movement*, New Delhi: Manohar Publishers.

Part IV

Economic Dimensions

14

Industrial Policy Reforms Since 1991

The Contexts of Globalization and Domestic Politics in India

Siddhartha Mukerji

The Indian economy has evolved over a period of time to assume a global character in contemporary times. It has undergone several twists and turns due to the volatile political environment internally and the rapidly changing global economic and financial environment. Starting with the adoption of a socialist pattern of close-door economy in the 1950s, the Indian state thereafter has tried to create a business-friendly environment for private companies to make inroads into the global market and compete with the economic giants in Asia like China, Japan and South Korea. Indian companies today have made their mark in the fields of software technology, pharmaceuticals and textiles.

The liberalization drive that was initiated by Rajiv Gandhi in the mid-1980s created favourable conditions for future integration with the global market. Globalization of industrial sectors was not in the immediate agenda of policy makers. But a gradual push to export-oriented sectors was certainly an objective. As globalization is naturally seen as a logical culmination of liberalization, the chain of events that unfolded since the mid-1980s automatically led to the adoption of structural adjustment programme that thoroughly globalized the Indian economy in general and industry in particular.

This chapter attempts to analyse and understand the structural changes that were brought about by the government since 1991, with the underlying objective of opening up the Indian market to outside players. These outside players consisted of the big multinational companies located in the Western countries and some of the newly developed countries of East Asia. Therefore, the corporate sector in India today includes both domestic companies and the multinational giants.

The new economic policies, first addressed the need to remove the structural constraints that inhibited private investment and then facilitated global integration of economy by creating a durable environment for FDI (foreign direct investment), FII (foreign institutional investment) and international trade. The economic transformation and the associated industrial reforms resulted from a convergence between two kinds of environmental influences, namely the 'global economic' and 'internal political' environments. However, the latter was not as consistent as it appeared to be for facilitating the industrial reforms. Before I begin to account for these changes and the kind of politics that drove them, it is essential to capture the theoretical advocacies of the globalization-model for industrial policies.

I

The Globalization Theory of Industrial Policy

The globalization argument advanced by Baldev Raj Nayar, Robert Wades and Pallan and Abbott holds the assumption that, it was the impact of the ever-increasing influence of economic globalization, i.e. a tendency of the market towards relentless expansion to bring the entire globe under its reign, that led some of the countries in the Third World to rethink their industrial and trade strategy. The movement on the path of globalization involved reduction in financial costs and large increment in benefits flowing from international economic transactions. The benefits of globalization propelled the East Asian states to integrate their economy with the international market. On the other hand, countries like India and Brazil, following protectionist policies in the 1960s and 1970s remained unsuccessful in harnessing the benefits.

Robert Wades (1999: 346) also observes that the East Asian states like Taiwan, South Korea and Japan made inroads into western markets when the growth in world output was high. The international market in the late 1960s and the early 1970s was in an expansionary phase.[1] While the East Asian states made full utilization of the incentives provided by the international market by adopting export-oriented industrial strategies, countries like India and Brazil failed to do so by following protectionism and high levels of regulation on the private sector.

The growth in world output plummeted from 41 per cent in 1970–79 to 2.6 per cent in 1980–87. Protection in the markets of the developed nation accelerated in the early 1980s (Wade 1999: 347).[2] India began opening up its economy to the outside world in the early 80s when the conditions were not ripe to harness the benefits of globalization. However, even in the 80s external liberalization was not in the immediate agenda. The first step was to increase pace of domestic liberalization followed by opening the market to external competition (Kohli 2006: 1257–1258).[3]

Ronen and Abbott examines suggests seven competitive strategies adopted by the states in both the developed and developing world to reap the benefits of globalization (Ronen and Abbott 1999: 6).[4] The important fact that these scholars observe is that the state acquires a centre stage in devising strategies for global integration. According to them this new incarnation of state in the global world represents an 'institutionalization of globalization in the state system'. Pallan and Abbott state that, the system of states has accommodated itself to the process of globalization and provided the institutional infrastructure upon which globalization and markets function. The state and the global world are mutually reinforcing. The states initiated industrial transformation by harnessing the opportunities provided by the global market and that globalization itself induced a transformation in the nature of state in the process. On the other side, the global market has restructured itself by the inclusion of the new developmental nations in the Third World within its rubric.

National distinctiveness and uniqueness has been emphasized as a consequence of socio-economic settlements between social agencies and institutions within the national boundaries. In other words, the relation between capital and labour, state and firms, capital and capital etc. is said to have influenced the contours of industrial policy making significantly. However, policy making in the states have also been affected by changes in the global environment. Some nations have had a positive response to such changes while others have been left behind in the global race. This fact is clearly reflected in the export-oriented policies of the East Asian states in the early 1960s and the regulatory, inward-looking strategies of India and Brazil during the same period. In the latter states, especially in India the realization came as late as the early 1980s when the political dynamics of global integration found its initial start-up. This transnational dimension of industrial policy making acquired further momentum under the leadership of Rajiv Gandhi. Business in India, since the mid-80s, attained a novel dimension with greater engagement between state, local firms and foreign enterprises.

The year 1991 is marked by a significant breakthrough in the economic history of industrial reforms in India when the nature and structure economic policy framework underwent tremendous transformation. It brought with it a new set of policies that marked a complete departure from the past practices. The policy framework that existed for nearly four decades since independence was characterized by import-substitution industrialization; high levels of control over private sector; restrictive technology policies; monopoly of public sector especially over strategic industries; restrictions on foreign equity participation; high degree of bureaucratization and stringent taxation measures for the private corporate sector. The underlying idea behind rethinking past strategies was the urgency to promote rapid industrial growth and modernize the Indian industry by seeking the active role of the local entrepreneurs and the foreign firms.

The needs of the time required extensive liberalization of the industrial and financial policies to promote rapid economic growth. There has been a broad political consensus at all levels regarding the elimination of unnecessary state regulation of private and foreign investment. Although there have been criticisms regarding certain reforms' measures from both the left and right parties, the fact of the matter is that the new economic package received unanimous approval of the parliament, clearly indicating a wider consensus on economic liberalization in general and industrial liberalization in particular. As Ashutosh Varshney points out that, 'if the Indian parliament had not passed the budget embodying the economic reforms, the economic reforms would have not become a reality, not because of faulty economic logic but due to institutional constraints of a parliamentary system' (Varshney 1999: 245).[5] However, it must be noted that there has been an inconsistency in the implementation of economic reforms, and the pace has generally been sluggish due to various political constraints.

The changes advanced in the 1990s have been in continuity with the liberal measures adopted by the Rajiv Gandhi government since the mid-80s. Although, the 1991 economic reforms brought about an overhaul of past policies, its roots can be traced back to the 1980s when momentous advances were made in the direction of industrial liberalization. Elaborating on this account Montek Ahluwalia points out that, 'rethinking on economic policy had begun in the mid-80s when the failure of past strategies such as import substitution, public sector dominance, and pervasive control over the private sector had become clearly evident' (Ahluwalia 1999: 26).[6]

There is a deluge of literature that accredits the advancement of economic reforms to the 1991 Balance of Payments (BOP) crisis that hit the economy adversely in the year 1990. While the BOP crisis undoubtedly triggered the readjustment of economic policies on 1991, it can only be held as a necessary and not a sufficient condition for economic restructuring. Therefore, apart from adopting measures to restore macroeconomic stabilization, there was an urgent need to reorient the industrial and trade policies in the direction of liberalization, debureaucratization and deregulation.

The transformation of economic ideas was also conditioned by the success of East Asian and South-East Asian countries, which were able to achieve high growth rates and rapid industrialization as a result of their export-oriented strategies, and provision of suitable environment for private corporate investment. It was widely recognized that the compulsions of a globalized world necessitated greater autonomy of the private sector and foreign companies, and greater integration with the international market. The underlying objective of the new economic policy was to promote technological advancement, industrial modernization and managerial efficiency.

The larger question that confronts us is that, to what extent was the attainment of policy changes possible within the context of high degree of political uncertainty and disparate demands of various interest groups. It is noteworthy that a minority government succeeded in pursuing radical changes, while there was immense political repercussion to the partial liberalization advanced by the previous government. Despite a general endorsement of the economic reform package, the divergent ideological stances of various political parties and those within the ruling party have checked the rapidity of liberalization subsequently.

II

The BOP Crisis of 1990–91

The later half of 1990–91 faced the initial blow of the BOP crisis with the sharpening of oil prices as a result of the Gulf war. A drastic regression in the foreign exchange reserves to about $ 1 billion, and the loss of confidence on the part of external financial institutions to cut off foreign lending, further aggravated the situation. As per the estimates made by the Economic survey, the foreign exchange reserves declined from a level of ₹ 5,480 crores in August 1990 to ₹ 1,160 on January 1991 (Ahluwalia 1999: 4).[7] Such a drastic fall in the foreign exchange reserves culminated in an acute BOP crisis.

The rapid loss of foreign exchange compelled the government to compress the imports. It was felt that curbing imports would serve as a first hand measure to manage the crisis. As a result the government embarked upon a series of measures to control the inflow of imports. These included: imposition of a cash margin of 50 per cent excluding the capital goods; restrictions on capital goods imports; putting an extra surcharge of 25 per cent on the prices of petroleum products except domestic gas and raising the auxiliary custom duties (Ahluwalia 1999: 8).[8] This step taken on the part of the government seemed unviable as it led to further predicaments on the industrial front. The resultant fall in imports, especially those that served as key inputs to certain key industries led to a sharp decline in industrial production.

The cumulative effect of external shock and the deteriorating fiscal situation resulted in high inflation and an ever-increasing public and current account deficit. In IMD Little's and Vijay Joshi's view the roots of the crisis can be found in the deteriorating current account position since the early 1980s when the oil prices almost doubled. The situation got worsened due to complete stagnation of exports between 1982 and 1985. The current account deficit kept increasing throughout 1980s, until it reached a peak level in the early 1990s. There was no substantial effort on the part of the government in the direction of current account adjustment (Ahluwalia 1999: 14).[9] Little and Joshi further elaborates on this account and state that the external oil shock was only an additional liability on the vulnerable economy that had been following unsustainable macroeconomic policies since early 1980s (Ahluwalia 1999: 14).[10] The tremendous pressure exerted by the combination of fiscal imbalance and the external oil shock, directs our attention towards two important factors that resulted in the macroeconomic crisis. These are adverse international circumstances and governmental mismanagement of public finances.

As the short-term credits had already begun to exhaust, the BOP position exacerbated. The termination of commercial loans by the lending agencies in 1990 added to the predicament. This was followed by a massive outflow of NRI deposits in October 1990.[11]

In order to restore fiscal stability, the government resorted to a stand-by arrangement with the IMF under which the IMF was to provide a sum of $2.3 billion over a two-year period. A comprehensive programme of structural adjustment in areas of industrialization, trade and finance accompanied this. While the immediate aim was to reduce inflation and the fiscal deficit, and improve the BOP position, the structural adjustment programme was attempted to fulfil the long-term vision of promoting rapid industrial growth by creating better prospects for private and foreign investment. The structural reforms involved greater opening of the industrial, trade and financial sectors to the international market, minimization of governmental controls over private and foreign firms, and gradual expansion of private sector enterprises over sectors that were earlier reserved for the public sector. The underlying idea was to induce greater managerial and operational efficiency within the PSEs and seek greater participation of private enterprises in the existing and new industrial ventures.

Although the year 1991 marks the beginning of radical economic restructuring towards greater industrial liberalization, the reforms have been implemented gradually to eschew unnecessary political repercussions and economic uncertainties. The next section analyses the nature of industrial reforms that have been implemented gradually over a period of time. The areas of significant industrial policy reforms include: industrial licensing, foreign investment, taxation, public-private participation in industrialization and trade policies.

III

Industrial Licensing

Industrial licensing is a regulatory instrument employed by the government to resist the private and public sector enterprises from investing in a new unit and capacity expansion in existing units beyond a certain limits. Following loopholes could be located in the system of industrial licensing:

1. High rigidities
2. Delays inherent in the system of bureaucratic control
3. Corrupt practices associated with the procurement of industrial licenses

The New Industrial Policy announced on 24 July 1991, brought far-reaching changes in industrial investment and production. The first step taken in this direction was the abolishment of industrial licensing for all except 18 industries (Ministry of Industry 1991).[12] However, the industrial policy was still considered to be restrictive as there seemed to be no clear justification for the continuation of licensing requirements for industries such as sugar, edible oils and consumer durables such as motor cars, refrigerators etc. The Eighth Five-Year Plan pledged to shorten the list and suggested that industrial licensing must be restricted to only defense explosives and hazardous chemicals (Planning Commission 1992).[13] To realize this objective, later, in a series of steps the list was brought down to alcoholic beverages, cigars and cigarettes, electronics, aerospace, defense products, hazardous chemicals and pharmaceuticals. The MRTP Act was amended to abolish the special permission required by the large industrial houses for investment (Ministry of Industry 1991).[14] MRTP was an additional onus imposed on the big businesses to check monopolization in the domestic market. The MRTP provision indicated a strong prejudice on the part of previous regimes against private corporate investment and proved to be detrimental for the germination of private entrepreneurship. In the wake of providing greater space for small business associations and ensure public sector dominance in strategic industrial sectors, the government stifled the initiatives of big businesses. The abolishment of this stringent provision in 1991, undoubtedly provided handsome opportunities for promising entrepreneurs amongst the big ones to invest in new industrial projects, especially in upcoming sectors such as IT, telecommunications and electronic consumer durables.

Further, in order to create better opportunities for private sector enterprises in the new industrial projects, especially those in the core and basic industrial sectors, 9 out of 17 industries earlier reserved for the public sector was opened to private sector. Also, very recently the government decided to de-reserve 51 items for small-scale industries in order to procure economies of scale in such areas through the entry of large private enterprises (*Economic Survey* 2002–03).[15]

The abolishment of industrial licensing was the most significant step in the direction of providing healthier prospects for economies of scale and optimal growth that had otherwise remained hindered over the years due to anti-monopoly legislation. The underlying objective was to deregulate the industry in order to promote an efficient and a competitive industrial economy. In this context paragraph 21 of the Industrial Policy Statement observes that,

'In order to achieve the objectives of the strategy for the industrial sector for the 1990s and beyond, it is necessary to make a number of changes in the system of industrial approvals. Major policy initiatives and procedural reforms are called for in order to actively encourage and assist Indian entrepreneurs to exploit and meet the emerging domestic and global opportunities and challenges' (Ministry of Industry 1991).[16]

There has been a wider consensus in political circles on the abolishment of industrial licensing. It was widely recognized that high levels of bureaucratization, high instances of corruption, power aggrandizement of bureaucrats, political leadership and the technocrats characterized the system of industrial licensing. The stringent requirements of industrial licensing only resulted in inefficiency and erection of a high cost industrial structure. Unfortunately, the changes advanced in 1980s in the sphere of industrial licensing were ad hoc in nature. It was only in 1991 that a complete elimination of state control through industrial licensing and MRTP provisions was found to be desirable.

IV

Foreign Investment

The vitality of foreign investment, especially that of FDI lies in the innumerable opportunities that it provides to the industry in areas of technological upgradation, access to global managerial skills and practices; optimization of the usage of human and natural resources, and enhancement of international competitiveness promoting high efficiency (*Economic Survey* 2003–04: 142).[17] The importance and significance of FDI was well recognized by the government in 1991, since when it became an integral part of the reform package. Underscoring the importance of FDI, the Industrial Policy Statement of 1991 notes that,

'Foreign investment would bring attendant advantages of technology transfer, marketing expertise, introduction of modern managerial techniques and new possibilities for promotion of exports. This is particularly necessary in the changing global scenario of industrial and economic cooperation marked by mobility of capital. The government will therefore welcome foreign investment which is in the interest of the country's industrial development' (Ministry of Industry 1991).[18]

The FDI policy has received prior attention of the policy makers and has contributed significantly in the economic growth of the country since early 1990s.

The second mode of foreign investment is called portfolio investment that concerns investment in the capital market. It involves buying and selling of shares in the stock market. Portfolio investment has been a prominent source of foreign capital inflow in the economy. In the early 1990s, the government only paid a lip service to the enhancement of portfolio investment. It gained momentum only since the mid-1990s when the government came up with a host of liberal measures for the foreign institutional investors.

A radical restructuring of FDI policy was initiated in 1991, when the limits of foreign equity holding was raised from the earlier level of 40 per cent to 51 per cent in a wide range of priority industries. The establishment of FIPB provided the requisite institutional framework for negotiations with large international firms (*Economic Survey* 1991–92: 13).[19] The new initiatives taken by the government with respect to FDI fetched encouraging results. Between August 1991 and December 1992, 2,154 foreign collaborations including 894 cases of foreign equity participation was approved by the government. The aggregate value of foreign equity investment figured to as high as ₹ 42.9 billion in 1992.

Acknowledging the significance of FDI in overall industrial growth, the provisions were further liberalized in the mid-1990s. On December 1996 the equity limit of FDI was raised to 74 per cent for nine categories of industries. In the mining sector three industries were made eligible to receive automatic approval by the RBI for 50 per cent equity participation. Apart from this, 13 additional industries were brought within the purview of 51 per cent equity participation (*Economic Survey* 1996–97: 110).[20] The Union budget of 1999–2000 further extended the 74 per cent FDI provision under the automatic approval route to bulk drugs and pharmaceuticals.

The liberal policies adopted by the government resulted in a dramatic jump in FDI inflow since 1991. The total approval of FDI that amounted to a total of $200 million in 1991 shot up to $14.6 billion in 1997. However, in 1998 the figures plummeted to a level of $2.3 billion on account of the South-East Asian crisis (Kumar 2000: 807).[21]

Regionally, the FDI inflow from the United States and European Union has been significantly high. Recently, South Korea, Singapore and Japan have become major contributors of FDI. India's strong commitment towards its 'Look East policy' has necessitated greater engagement with East and South-East Asian nations in the areas of trade, investment and security concerns. The vitality of FDI lies not only in improving efficiency and intensifying domestic production, but promoting India as an important economic player in the globalized world. Some of the major foreign investors in India include General Motors, Coca Cola, Samsung, Du Pont, AT&T, Shell, Hewlett Packard, Texas Instrument and LG International. This large list of participants clearly indicates that India has emerged as a favourite destination for FDI. Looking at the ongoing trend on the FDI front, one can ascertain larger volumes of FDI inflow in the near future.

The second area of foreign investment that received notable attention, especially since the mid-1990s is portfolio investment. Such investments made by the FIIs have proved to be an important source of foreign capital. Recognizing the significant role of portfolio investment in economic growth, the government introduced certain liberal measures since mid-1990s. In 1996, the FIIs were permitted to make equity investments in certain companies and the investment limit of 5 per cent of the total equity of a single company by an individual FII was raised to 10 per cent.

In order to regulate the inflow and outflow of portfolio investment, SEBI was established. The two important objectives of SEBI can be stated as follows:

1. regulatory reform of capital markets
2. market modernization (Kumar 2000: 805)[22]

As per the guidelines announced in 1992 for the FIIs in the Indian capital market, the investors were granted the permission to 'invest in all types of securities traded in primary and secondary markets with full repatriation of benefits without restrictions on either volume or trading' (Kumar 2000: 805).[23]

There have been some other remarkable institutional innovations to promote and regulate the activities of foreign investors. The most significant amongst these has been the establishment of Foreign Investment Promotion Council in the year 1996. The main task of this body was to prepare project reports in certain selective areas of prior importance and accordingly facilitate the inflow of foreign investment in such areas (*Economic Survey* 1996–97: 111).[24] Additionally, the Union budget of 1999–2000 laid the provisions for the establishment of FIIA to provide a single point interface between the foreign investor and the government (*Economic Survey* 1999–2000: 116).[25] The underlying idea behind such moves was to provide an institutional platform for greater interaction and engagement between government and the foreign investor, and facilitate free flow of information on both sides.

There was much criticism in both left and right political circles regarding indiscriminate opening up of Indian industry to FDI and FIIs. It has been argued that the government has succumbed to the pressures of foreign companies, which has resulted in loss of state autonomy. An increased level of portfolio investment over the years has been particularly problematic on certain counts. Pointing out the unsuitability of portfolio investment for the economy, Jayati Ghosh says that,

> 'An increasing reliance on portfolio investment is not a healthy sign because the portfolio investor does not have a real long-term stake in the enterprise or the investment concerned and resort to capital markets when returns seem to be higher. Such investments are deemed to move out of the country if the returns are not up to the expectations. They can move out of the country in case of adverse political developments, or even at anticipation of instability, or in response to adverse circumstances in the international markets' (Ghosh 1997: 112).[26]

There has been the other side of the argument that points towards the benefits accrued from portfolio investment. Its significance has been located in its contribution to foreign capital inflows in the domestic economy. In Ahluwalia's view, despite growing concerns about volatility and reliability of portfolio investment, India's experience has been most encouraging on this front (Ahluwalia 1999: 56).[27] The rising figures of portfolio investment since mid-90s are a clear indication of this fact. This argument can be further validated by the fact that although the portfolio capital fluctuated, it did not turn negative in 1997–98, when the South-East Asian crisis hit the world economy.

V

Taxation

The tax reforms in India since early 90s were based on the recommendations of the Raja Chelliah committee appointed in 1991. The recommendations were based on two important objectives, which are as follows:

1. to make the tax structure more flexible
2. improve the administration of tax so as to generate higher revenues for growth

Although there have been previous attempts to reform the tax structure, most of such moves were ad hoc in nature. There has not been a serious attempt to make the tax administration more efficient and workable. Consequently, there have been innumerable cases of tax evasion in the past, particularly since the early 70s. As IMD Little and Vijay Joshi observe that, a dismal performance has been most prominent in the area of indirect taxation due to increasing inefficiency, evasion and corruption (Ahluwalia 1999: 81).[28] The fundamental idea behind introducing new taxation reforms was to create better milieu for private investment and production.

To implement the taxation reforms recommended by the Chelliah committee the government embarked upon a new set of policies for taxation. In the arena of direct taxation, personal income tax has shown a sharp decline from 56 per cent at the beginning of the reform period to 30 per cent in 1997–98 (Ahluwalia 1999: 56).[29] During the same period, corporate tax was brought down from a level of 53 per cent to as low as 35 per cent. The Union budget of 1997–98 brought a greater reduction of these taxes and abolished the tax dividends in the hand of shareholders for pushing up industrial investment and consumer demands (*Economic Survey* 1997–98: 98).[30]

In the area of indirect taxation, especially in case of excise duties, the performance has been exceedingly dismal. In 1991–92 and 1992–93 the excise duties mostly on final consumption goods were raised to a reasonable limit. Later, the government introduced MODVAT under which credit was given for excise duties paid on inputs against excise duties due on output. In the budget year 1999–2000, the government attempted to rationalize the excise duty structure and restore 100 per cent MODVAT credit. In addition to this, custom duties on steel were removed; 5 per cent custom duty was imposed on a majority of zero duty inputs and countervailing duties were introduced for capital goods sector.

In 2003–04 a special 'Task Force on Indirect Taxation' was installed by the Ministry of Finance to suggest reforms in areas of custom tariffs and exemptions, custom procedures and trade facilitation, export promotion and improving automation in tax administration (*Economic Survey*

2002–03: 113).[31] Very recently, the Union Budget of 2003–04 introduced a series of measures to relax and rationalize indirect taxation. The custom duties were brought down to an abysmally low level of 25 per cent. A three-tier duty structure of 8 per cent, 16 per cent and 24 per cent was also introduced in case of excise duties. The overall progress of the IT sector necessitated a reduction in custom duty for project imports in IT sector (*Economic Survey* 2003–04: 22).[32]

With the advancement of taxation reforms since 1991, there has been a marked improvement in the taxation system. The new measures have instilled immense confidence amongst the local entrepreneurs and the exporters to locate new areas for investment and production. The case of IT sector involving wider participation of local firms is exemplary in this regard. The only area that has not received adequate attention, despite government pronouncements, is the reformation of tax administration. There is an impending need to strengthen the tax administration to generate higher revenues for growth and equity.

VI

Public Sector Disinvestment

The sale of public equity as an effective means of ensuring fiscal discipline, and improve performance of PSEs was announced in the Industrial Policy Statement of 1991. Paragraph 31 of the Industrial policy statement observes that, 'serious problems are observed in the insufficient growth in productivity, poor project management, over-manning, lack of continuous technological upgradation, and inadequate attention to R&D and human resource development' (Ministry of Industry 1991).[33]

In 1991–92 the government undertook the task of partial disinvestment of a part of public sector equity through public finance institutions. The government justified this step on the ground that it would ensure greater accountability on the part of managers and create a new culture of efficiency and productivity in the PSEs (*Economic Survey* 1991–92: 14).[34] Despite some initial steps taken in the direction of disinvestment, an outright sale of PSEs was postponed to eschew political repercussion. Later, as the process of disinvestment picked up momentum, it invited the wrath of various political parties and the social groups, particularly the labour. Disinvestment became a major contentious issue by the late 90s when it was pursued more rigorously.

Disinvestment has been necessary from the point of view of providing greater autonomy of management, and ensuring strict accountability. Elaborating on this account, Bimal Jalan argues that, 'the combination of partial sale of equity and outright sale of loss making enterprises can generate substantial resources' (Jalan 1993: 36).[35] The iron-fist control of the government over public sector enterprises in the last four decades had made them redundant and unworkable. By the early 1980s, a large number of PSEs had turned into sick units.

Realizing the prior necessity of public sector reforms, the NDA government undertook the task of 'real' disinvestment as an essential component of second-generation reforms. The Union budget of 1998–99 announced disinvestment in a number of PSEs such as GAIL, VSNL, CONCOR and IOC. It was also decided that the government would dilute its share to 26 per cent in most of the cases, retaining its control only on certain strategic sectors (*Economic Survey* 1998–99: 97).[36] The Disinvestment Commission, which was set up in 1996 to identify PSEs for equity participation as well as working out the necessary modalities, recommended strategic sale of Air India Ltd., and deferment of disinvestment in case of Central Electronics Limited. In 2001–02, the government went further ahead in bringing down government equity in all non-strategic PSEs to 26 per cent. In January 2002, the Disinvestment Commission made a series of recommendations in its 13th report with respect to four major PSEs, namely, NLC, RHES AND Project and Equipment Corporation Ltd (*Economic Survey* 2001–02: 164–170).[37]

By the turn of the 20th century, the disinvestment policy of the government had become politically contentious. The disinvestment of BALCO, HPCL and BPCL was challenged in the Supreme Court. While in the case of BALCO, the Supreme Court gave a verdict in favour of disinvestment; in the other two cases it reversed its judgement and restrained the government in proceeding with disinvestment.

The disinvestment policy has been politically the most sensitive area of economic reforms. It not only created discontentment amongst various political parties but also the individual members of the ruling party and its alliances. Debroy and Mukherjee points out that, 'the disinvestment programme took a new and controversial turn when Arun Jaitley was replaced by Arun Shourie as the Union Minister for Disinvestment in August 2000' (Guha 2004: 37).[38] However, it is noteworthy that the government went ahead with its policy of full-fledged disinvestment despite intermittent political predicaments.

VII

Trade Policy

The main objective of the new trade policy adopted by the government since 1991 was opening up of the Indian industry to international market, and providing strong impetus to export-oriented sectors, especially those that exhibited a certain degree of comparative advantage. To restore the BOP position, the imports including those that served as essential inputs to export-oriented industries were restricted in 1991. As the situation improved, the import policy was liberalized. In 1991 itself, certain steps were taken to reform the policies. The most prominent amongst these is the introduction of 'Eximscraps' in place of import entitlements (*Economic Survey* 1991–92: 12).[39]

While the exports increased sharply between 1991 and 1996, it faced stagnation in 1998 on account of the East Asian crisis and the economic sanctions imposed on India for conducting nuclear tests in 1998. To revive the export sector, the government embarked upon series of export-promotion measures since 1998. This included: exemption of special additional duty of 4 per cent on exports; provision of zero duty for export promotion capital goods extended to specific bio-tech and small-scale engineering industries; and provisions to set up private software technology parks (*Economic Survey* 1999–2000: 86–87).[40] Later, the zero duty 'export promotion capital goods scheme' was further extended to chemicals and textiles. In order to boost software exports, exemptions were granted in case of custom duty on import of capital goods in the software sector. In the Union budget of 2001–02, further incentives were extended to IT and telecom sectors, and an integrated package for the modernization of the textile sector was introduced. To promote exports of gems and jewellery, custom duties on 'cut' and 'polished' gemstones were reduced. The 10 per cent surcharge on custom duties was removed with the objective of pruning the custom duties from a level of 38.5 per cent to 35 per cent on these items.

One significant change in the area of trade has been a marked improvement in trade relations with the countries in South, South-East and East Asia. This has been a part of the larger agenda of India's Look East policy. Due to adverse circumstances created by the East Asian crisis in 1997–98, there seemed to be a certain degree of stalemate in trade with these countries during this period. However, the momentum of trade picked up thereafter, on account of serious efforts taken by the government to renew trade relations. The trade relations were enhanced both at bilateral and multilateral levels. The former includes the conclusion of free trade agreement with Sri Lanka and Thailand.

The government also took series of initiatives to emerge as a major trading partner within SAARC. The most significant step in this direction was a unilateral elimination of quantitative restrictions on imports of around 300 items imported from SAARC countries. This has been a historic move on the part of India and other SAARC countries in the direction of South Asian Preferential Trade Arrangement (SAPTA). Additionally, the custom duties charged on the imports from SAARC were relaxed. Despite all such moves, much needs to be done to increase the volume of trade. India's negative list of items of trade with SAARC still remains lengthy. There is a pressing need to promote greater integration of the South Asian economies through South Asian Free Trade Area (SAFTA).

The overall political support for an export-oriented strategy of industrialization and liberalization of trade policies has provided better prospects for integration with the international market. It has been subsequently realized that India needs to explore potential export areas, and derive the benefits of global trade by promoting such products in the international market. Theoretically, the comparative advantage of the product determines its level of competitiveness in the international market. For instance, the comparative advantage of Indian software in the international market has been widely recognized. This can be indicated by a sharp rise in software exports between 1993–94 to 1998–99 by 2.5 per cent of GDP as compared to 1 per cent in the late 1980s (Kumar 2000: 809).[41]

The other objective for promoting potential exports must be to direct FDI in export sectors. Till very late, FDI remained confined to production for the domestic market. Recently, the vitality of export-oriented FDI was widely recognized and accordingly steps were taken to remove all impediments for efficiency seeking export-oriented FDI. In this context, lessons must be drawn from Chinese experience where the MNCs account for 40 per cent of manufactured exports.

There still remains a large hiatus between the manufacturing capabilities of domestic entrepreneurs and those of its foreign counterpart. To catch up with the international standards, the Indian firms need to strengthen their manufacturing capabilities. One major reason for such inefficiencies is a certain amount of negligence on the part of Indian firms in developing in-house R&D, and excessive dependence on foreign technology transfer for industrial production. Although, foreign technology is essential for building domestic technological capabilities, it must be complimented by equivalent efforts to develop in-house R&D, and improve upon existing technology. Also it must be noted that, with indiscriminate opening of the domestic market to MNCs, product differentiation is taking precedence over building complimentary assets. Such a strategy has resulted in dismal R&D efforts on the part of domestic firms. Secondly, it has adversely affected the competence and competitiveness of the domestic vis-à-vis their foreign counterparts (Basant 2000: 821).[42] To overcome such shortcomings, both the public and private sector enterprises must invest extensively in developing indigenous technology through in-house R&D.

VIII

The Political Context

The advancement of industrial reforms converged with the genesis of coalition politics in India since 1991. The beginning of 1990s is characterized by high levels of political uncertainty with the collapse of two subsequent governments within a span of two years followed by the installation of a minority government under the leadership of P.V. Narsimha Rao. The Congress-ruled minority government brought about a radical restructuring of economic policies with the advancement of a comprehensive package of economic reforms. The most interesting fact about this development is that in the midst of high political uncertainty and a severe economic crisis a minority government could pursue its radical policies. Although, there were marked ideological differences amongst various political parties on the new direction of economic policies, the fact of the matter is that the economic reforms received unanimous approval of the parliament.

Although there has been a broad consensus over the basic framework of economic reforms, certain policy measures have aroused wide discontentment amongst political parties and certain social groups. While policy changes with respect to industrial licensing, taxation and export promotion received a unanimous approval of all concerned political parties, those related with disinvestment and foreign investment created much political havoc especially since the mid-1990s. These policy measures have been outrightly rejected by both the Left and Right parties. While the Left parties interpreted such moves as succumbing to the pressures of the international financial institutions particularly the IMF and the multinational companies, the Right organizations such as the 'Swadeshi Jagran Manch' saw it as a clear repudiation of the idea of 'swadeshi' and 'self-reliance'.

Until the mid-1990s the disivestment issue did not come up forthright as the government had paid only a lip service to it till then. Considering its political sensitivity, the policy of disinvestment was pursued partially till the mid-90s. The outright sale of public sector enterprises was not in the immediate agenda of the government as any radical step with regard to privatization would have been politically detrimental for continuation of the ruling party. This would have, not only, created discontentment amongst the political parties, but also certain social groups, particularly the labour. According to Ashutosh Varshney, the hurdles for privatization were certainly political in India. To substantiate his argument on this account, he says that, 'since there has been overstaffing in most of the PSEs, a policy of privatization would entail massive lay-off and this would create resentment amongst the labour in the organized sector' (Varshney 1999: 254).[43] He further elaborates on this and says that, 'in a society where millions are unemployed, the economic idea of firing existing employees on grounds of rationality is hard to present politically' (Varshney 1999: 255).[44]

Once the issue of disinvestment picked up momentum since the late-1990s, it created much hue and cry in the political circles. The disinvestment policy was pursued full-fledgedly after Arun Shourie replaced Arun Jaitley as the Union Minister of Disinvestment in August 2000. His grand designs of disinvestment and privatization was not simply opposed by the other parties, but his own colleagues within the cabinet as well as RSS, the parental organization of the BJP. The BJP leaders holding other ministries showed reservation over the sale of PSEs that fell within their Ministry. It was seen as a loss of autonomy on the part of these ministries. They were most unwilling to loose their strategic control over such enterprises. In this regard, Paranjoy Guha Thakurta cites the controversies surrounding the disinvestment of BPCL, HPCL and NALCO. While the

Union Minister of Petroleum and Natural Gas, Ram Naik showed tremendous anxiety over the disinvestment of BPCL and HPCL, the Coal and Mines Minister Uma Bharti felt unhappy with the manner in which NALCO was sought to be privatized (Guha 2004: 38).[45] These two cases clearly reflect the power and ideological conflict that existed amongst respective ministries on the issue of disinvestment. These issues became politically so much controversial that they were ultimately taken to the Supreme Court for the final verdict. In its historic judgement, the Supreme Court restrained the government from proceeding with disinvestment of the two PSEs. A similar observation is made in the case of privatization of Videsh Sanchar Nigam Limited (VSNL). The management of this enterprise was handed over to the Tata Group of companies. The Union Minister for Communication and IT opposed this move on the ground that the new private owners did not fulfil their commitments of the deal and transferred large portions of money in other companies.

The policy of disinvestment and privatization has been contentious for two other important reasons. First, the money earned out of selling the equities was neither used to dissolve the ever-increasing public debt, nor did it go into investment in social sectors such as health, education and rural development. It was used for undesirable purposes such as reducing budget deficits. Second, instead of tackling the problem of sick units effectively, the government decided to sell the most viable and profitable PSEs referred to as 'navratnas' and the 'miniratnas'.

The economic history of disinvestment suggests that once the policy of privatization was initiated there was no possibility of reversing the strategy. The policy of disinvestment has certainly been beneficial from the point of view of managerial efficiency and technological advancement of the Indian industry. In this context, Bimal Jalan has substantiated the need to pursue privatization more rigorously. He highlights the vitality of private ownership, in that, it establishes a market of efficient managers and thereby improves the quality of management. Stating the organizational efficiency of private sector over public sector he says that,

> 'Political interference is unavoidable in public corporations and is much a cause of inefficiency ... . On the contrary, the board of private corporations where ownership is widely dispersed are more accountable that tightly controlled public enterprises' (Jalan 1993: 31).[46]

The historical experiences with tightly state-controlled PSEs suggest that in such enterprises, political consideration become more important than business or commercial considerations.

While there has been reasonable acceptability of FDI liberalization in political circles, the indiscriminate opening up of the capital markets is seriously contended. It has been pointed out that at present no political party including the CPI(M) has overly opposed the entry of multinationals. This is further validated by the fact that the state of West Bengal, which has been under CPI (M) rule for quite sometime has recently shown a strong acumen towards facilitating FDI in the priority sectors. On the contrary, the extreme Right organizations such as RSS and SJM have been highly critical of liberalization of foreign investment. However, such organizations have not overtly condemned such a policy assuming it as a vital step towards the growth of domestic firms.

In the recent past, there has been a major controversy over unchecked opening up of the power sector to the international giant, Enron. The central government in connivance with the Maharashtra state government finalized a deal with Enron to supply power in Maharashtra. Instead of using conventional sources of power such as coal or hydelpower, a highly unviable and expensive substance called 'Naphtha' was used for the purpose. Resultantly, the Enron-led power project in Maharashtra involved exorbitant capital costs that seemed nearly impossible for the power distributors to bear. With passage of time, the Maharashtra State Electricity Board realized the impossibility of meeting the financial obligations to Enron, and was almost caught up in the web of bankruptcy. Finally, the state government had to freeze all its payments to Enron on the ground that the company had failed to meet some of its obligations. This backtrack on the part of the state government invited much criticism from the advocates of FDI. It was found to jeopardize all prospects of FDI in future. While there was an immediate repercussion to the step taken by the government in the Enron case, the fraudulent schemes of Enron was soon realized after its international collapse (Chandrashekhar and Ghosh 2002: 135).[47]

The conclusion of such shady deals, as that of Enron, between the government and foreign investors only reflects the ineffectiveness and unaccountability on the part of central and state governments in implementing the economic reforms. Recognizing the contribution of foreign investment to industrial growth, there is a greater need to make the negotiation process more transparent and corruption-free. Any hasty decision, as in the case of Enron would only lead to loss of faith in government's policies and high levels of popular discontentment.

FDI and disinvestment would not in itself have generated much opposition had it not been besmirched with scams like Enron and VSNL. While most issues concerning FDI, FII and disinvestment are generally confined to 'elite politics', any exposure of scams and controversies brings these policy issues in the arena of mass politics in a negative light, hence creating suspicion and stalling the process of genuine reforms.

The liberal prescriptions accorded to the foreign investor must not go to the extent of subserving the interests of the local entrepreneurs. The interest of the latter must be given utmost priority even if it requires a certain degree of protection to the local firms. The most viable strategy would be to follow a balanced approach towards accommodating the interest of both the contenders. The underlying objective behind assigning such responsibilities must be to promote rapid industrial growth, propel the export sector, and provide viable public infrastructure. A proper realization of these objectives would certainly eliminate the political hindrances to the implementation of industrial reforms.

Conclusion

To begin with, the industrial reforms in India received a strong impetus from both global economic dynamics and encouraging political environment since the mid-1980s. Though the macroeconomic crisis may have triggered its adoption out of economic compulsions and the pressures of international finance institutions, the economic success of East and South-East Asian countries resulting from global integration were nevertheless equally motivating. At the same time, the structural changes initiated in 1991 and thereafter were not short-sighted but embodied a futuristic vision of providing the best possible environment for private and foreign investment. The institutional constraints on investment and capacity utilization for private firms were, therefore, completely revamped. All these measures were found to be necessary for promoting technological advancement in the Indian industry, facilitating competition in the domestic market, invigorating the export sector, and improving managerial efficiency.

The macroeconomic crisis that hit the Indian economy in 1990 served as the immediate cause for advancing fundamental changes in the economic policy framework. The restoration of fiscal balance had to be followed by a comprehensive structural adjustment programme that would eliminate the existing bias against the expansion of private and foreign companies. While most policy changes received unanimous approval of all political parties, those concerning FDI, portfolio investment and disinvestment had to be postponed or pursued gradually to eschew unnecessary political repercussion. The political sensitivity of such policies can be determined by the extent of political havoc that they created by the turn of the century.

Despite intermittent political hindrances to the implementation of industrial reforms, the importance of its continuance from economic point of view was thoroughly realized. This is clearly indicated by the fact that all governments since 1991 have accelerated the pace of industrial liberalization. There has been no indication of a reversal in the strategy. Specific incentives extended to priority sectors such as textiles, software, gems and jewelry, iron-ore and aluminum industries has led to an improvement in the export potential of such sectors.

The overall health of the industrial economy has drastically improved since 1991. There has been a rapid increment of FDI inflow in most sectors and more recently in the export-oriented sectors. India has emerged as the most favoured destination for FDI and portfolio investment. In the recent years the IT sector in India has shown tremendous international competitiveness. Such positive indications of industrial growth are accredited to the liberal measures adopted by the government to provide a healthy germination of private and foreign firms, and enhance the quality of management in all enterprises.

Notes

1. Robert Wades. 1990. *Governing the Market: Economic Theory and the Role of Government in East-Asian Industrialization*, New Jersey: Princeton University Press, 346.
2. Ibid., 347.
3. Atul Kohli. 2006. 'Politics of Economic Growth in India: 1980–2005', *Economic and Political Weekly*, 41 (13), 1 April: 1257–1258.
4. Palam, Ronen and Jasm Abbott. 1999. *State Strategies in the Global Political Economy*, London: Printer. Competitive strategy has been defined as a set of policies that are explicitly aimed at improving the climate of business, national or multinational companies, and hence at enhancing competitive advantage of such countries in the global economy.
5. Ashutosh Varshney. 1999. 'Mass Politics or Elite Politics? India's Economic Reforms in Comparative Perspective', in Jeffery Sachs et al. (eds), *India in the Era of Economic Reforms*, New Delhi: Oxford University Press, 245.
6. Montek Ahluwalia. 1999. 'India's Economic Reforms: An Appraisal', in Jeffery Sachs et al. (eds), *India in the Era of Economic Reforms*, New Delhi: Oxford University Press, 26.
7. Ibid., 4.
8. Ibid., 8.
9. Ibid., 14.
10. Ibid.
11. In April–June 1991 the net outflow of NRI deposits amounted to a total of $952 million.
12. Ministry of Industry. 1991. *Statement on Industrial Policy*, Government of India, 24 July, para 23.
13. Planning Commission. 1992. *Eighth Five-Year Plan*, Government of India, 87.
14. Ministry of Industry. 1991. *Statement on Industrial Policy*, Government of India, 24 July, see government decisions on MRTP: section E (i).
15. *Economic Survey*, 2002–03.
16. Ministry of Industry. 1991. *Statement on Industrial Policy*, Government of India, 24 July, para 21.
17. *Economic Survey*, 2003–04, 142.
18. Ministry of Industry. 1991. *Statement on Industrial Policy*, Government of India, 24 July, para 24.

19. *Economic Survey*, 1991–92, 13.
20. *Economic Survey*, 1996–97, 110.
21. Nagesh Kumar. 2000. 'Economic Reforms and their Macroeconomic Impact', *Economic and Political Weekly*, XXXV (10), 4 March: 807.
22. Ibid., 805.
23. Ibid.
24. *Economic Survey*, 1996–97, 111.
25. *Economic Survey*, 1999–2000, 116.
26. Jayati Ghosh. 1997. 'India's Structural Adjustment: An Assessment in Comparative Asian Context', *Economic and Political Weekly*, XXXII (20–21), 17–24 May: 112.
27. Ahluwalia, op. cit., 56.
28. Ibid., 81.
29. Ibid., 56.
30. *Economic Survey*, 1997–98, 98.
31. *Economic Survey*, 2002–03, 113.
32. *Economic Survey*, 2003–04, 22.
33. Ministry of Industry. 1991. *Statement on Industrial Policy*, Government of India, 24 July, para 31.
34. *Economic Survey*, 1991–92, 14.
35. Bimal Jalan. 1993. *India's Economic Policy: Preparing for the 21st Century*, New Delhi: Penguin Books Ltd., 36.
36. *Economic Survey*, 1998–99, 97.
37. *Economic Survey*, 2001–02, 164–170.
38. Paranjoy Guha. 2004. 'Ideological Contradictions in an Era of Coalitions: Economic Policy Confusion in the Vajpayee Government', in Rahul Mukherjee and Bibek Debroy (eds), *India: The Political Economy of Reforms*, New Delhi: Rajiv Gandhi Institute for Contemporary Studies, 37.
39. *Economic Survey*, 1991–92, 12.
40. *Economic Survey*, 1999–2000, 86–87.
41. Kumar, op. cit., 809.
42. Rakesh Basant. 2000. 'Corporate Response to Economic Reforms', *Economic and Political Weekly*, XXXV (10), 4 March: 821.
43. Varshney, op. cit., 254.
44. Ibid., 255.
45. Guha, op. cit., 38.
46. Jalan, op. cit., 31.
47. C.P. Chandrashekhar and Jayati Ghosh. 2002. *The Market that Failed: Neoliberal Reforms in India*, New Delhi: Left Word, January, 135.

15

New Capitalism and Violence

The Case of Special Economic Zones in India

Manisha Tripathy Pandey

> Capital is dead labour, which, vampire-like, lives only by sucking living labour, and lives the more, the more labour it sucks.
>
> —Karl Marx, *Capital*

> The forces of a capitalist society, if left unchecked, tend to make the rich richer and the poor poorer.
>
> —Jawaharlal Nehru, *New York Times*, 7 September 1958

The pace of technological transformations, the mobility of capital, temporary employment and the privatization of the welfare state leading to demographic changes has created a new global economy and a risk society. Contemporary globalization and technological change have given rise to a transformation of work, characterized by insecurity and impermanence within the employment marketplace. This 'new economy' – elastic, flexible, mobile and networked – has meant rapid decline of lifetime employment and a work and career developed within a single organization. This emergence of new economy and the flexible regime of the new capitalism, propelled by the stock market crash of 2008, have led to the radical restructuring of the labour market.

Risks, networks, liquidity and flexibility are the key trends of contemporary social processes in the new capitalism. Richard Sennett (1998) uses the term 'flexible capitalism' to describe the modern economy. It defines the goal of most modern companies which is to be able to continuously change to suit the market. Companies no longer provide job descriptions or long-term contracts, but rather an opportunity to participate in a competitive market. The flexibility of modern capitalism is a form of oppression, where the transnational activities of multinational corporations restructure investment in the West away from manufacture to finance, and industrial production is exported to labour-intensive Third

World countries. The victims of 'flexible capitalism' face new burdens and dominations and are treated as dispensable and disposable. They are challenged by the inconsistencies in the workplace and thus lose a sense of identity and security.

The disorienting effects of new capitalism lead to the 'corrosion of character' (Sennett 1998). Sennett spells out deeper emotional consequences of such organizational changes. People fear being displaced, sidelined, or underused (Sennett 2006). This new capitalism is the latest phase of late capitalism or multinational capitalism. We have moved from a world of rigid, hierarchical organizations, in which self-discipline shaped the durability of the self, to a brave new economy of corporate re-engineering, innovation and risk, in which the fragmented or dislocated nature of self-experience moves to the fore. Sennett (1998) writes:

> In place of organizations as pyramids, management wants now to think of organizations as networks ... the network is constantly redefining its structure.

In his work *Late Capitalism*, the Marxist economist Ernest Mandel (1973) argues for three periods in the development of the capitalist mode of production. The first was *freely competitive capitalism*, which occurred from 1700 to 1850 and was characterized largely by the growth of industrial capital in domestic markets. Secondly, there was the phase of *monopoly capitalism*, which lasted until approximately 1940, and was characterized by the imperialistic development of international markets as well as the exploitation of colonial territories. Finally, there is the period of *late capitalism* emerging out of the Second World War, whose dominant features are the multinational corporation, globalized markets and labour, mass consumption, and the space of liquid multinational flows of capital.

In late-capitalism, the rise of digital-based electronic technology and computer-driven production replace unskilled and semi-skilled labour, both in production and management, thus leading to violence. There is inherent violence of globalization and capitalism. Engels argued that capital has a logic of its own that justifies the use of whatever means available to achieve economic prosperity. And violence, many a times, is seen and taken as the legitimate mechanism of transformation and social change, essential to create a new and developed order.

Hubs, corridors and zones are the new spaces of neo-liberalism and represent the neoliberal language of development. This language carries with itself a certain degree of aggression and violence. The marking of a zone on a continuous landscape becomes a surgical operation where a spot is highlighted and the rest is made blackout. This paper is identifying violence in global economic processes and explores how it operates at the level of lived experience, as in the case of Special Economic Zones (SEZs) in India. The 'end of history' saw the emergence of post-socialist capitalism. Market has become the leading lens through which development is viewed. Unregulated markets treat human labour as a commodity to be bought and sold, vulnerable to exploitation. The so-called modern welfare state is supposed to be inclusive in its approach towards the poor and the marginalized. But the creation of SEZs as a product of neo-liberal policies of the Indian state has led to structural violence (Pandey 2010).[1] People are excluded because they lack the means to participate in the market. This kind of development basically promotes 'accumulation through dispossession' (Harvey 2005), i.e. accumulation of material wealth through inequality and deprivation.

Ethical assessment of the means and ends of development is important. The paper deals with the means and ends of development with respect to the postmodern capitalism or new capitalism. India, through its five-year plans, has followed development as a philosophy and ideology of the state. But with neo-liberalism, there has been a shift in the means of development from the state to the market in many vital areas. This paradigm shift from growth with social justice and empowerment to neo-liberal development has led to systemic and structural violence. This has been made possible through Public-Private Partnerships (PPP) serving the interests of crony capitalists. The paper explores how the practice of crony capitalism in free market economy in India, as in the case of SEZs, increases the 'crisis of modernity' leading to exploitation and social inequality. The proposed universal values of development like life-sustenance, equity, security, identity, capability expansion and 'well-being' are definitely undermined. The new capitalism and the market promote exclusion and isolation and the state is not being accessible, responsive and accountable. This kind of development is certainly not democratic and participatory and thus violent.

SEZ policy is inherently unfair and biased; its emphasis is on industrialization, science and technology, ruthless exploitation of natural resources and letting loose of market forces, completely ignoring and eroding the cultural and social sensibilities, identities and distinctiveness of its subjects. Thus, establishment of SEZ is another way of alienating the poor from the developmental process. Instead of being people-centred development where people are empowered to make their own choices, emphasizing the relevance of local knowledge and values as guidelines and tools, development through creation of SEZs is definitely anti-poor and anti-nature. This structural violence leads to an overt and relational kind of violence with the affected people agitating and taking law into their hands. The SEZs are promoting the expropriation of community resources for the profit of domestic and multinational corporations and the financial market. Violence occurs when a number of SEZ projects which have no 'public purpose' usurp land from property owners, with the help of the Land Acquisition Act, 1894 at below the market value of these properties. Thus, new capitalism and new practice of imperialism in the form of SEZ is very oppressive and violent.

I

Understanding Violence

Slavoj Zizek (2008) illustrates the contradictions of contemporary capitalism. Violence, Zizek argues, takes three forms – subjective violence (acts of assault, murder, terror and war); objective violence (the 'symbolic' violence embodied in language and its forms, racism, hate-speech, discrimination); and systemic violence (the catastrophic consequences of the functioning of our economic and political systems) – and often one form of violence blunts our ability to see the others. This systemic violence is called structural violence. Zizek expounds on this in his book, *First as Tragedy, Then as Farce* (2009). He writes:

> Herein resides the fundamental systemic violence of capitalism, much more uncanny than any direct pre-capitalist socio-ideological violence: this violence is no longer attributable to concrete individuals and their 'evil' intentions, but is purely 'objective', systemic, anonymous.

The recent financial and economic crisis has demonstrated the fragility of the free market system that its defenders believed had triumphed in the cold war. Zizek (2011) discusses this situation. The global capitalist system is approaching an apocalyptic zero-point. Its 'four riders of the apocalypse' are comprised by the ecological crisis, the consequences of the biogenetic revolution, imbalances within the system itself (problems with intellectual property; forthcoming struggles over raw materials, food and water), and the explosive growth of social divisions and exclusions. This is what is called as the crisis of modernity and this crisis get illustrated by the consequences of setting up of SEZs in India.

The term 'violence' is explained in general terms as 'needs-deprivation' or 'avoidable insults to basic human needs' (Galtung 1969). Galtung defines violence as 'avoidable impairment of fundamental human needs or, to put it in more general terms, the impairment of human life, which lowers the actual degree to which someone is able to meet their needs below that which would otherwise be possible'. He constructs a typology of violence composed of three categories: personal, structural, and cultural. He asserts that structural violence, as opposed to personal or direct violence, is indirect in that there may not be any person who directly harms another person in the structure. The violence is built into the structure and shows up as unequal power and consequently as unequal life chances. The power to decide over the distribution of resources is unevenly distributed, which is the pivotal causal factor of the structural inequalities. Because structural causes are responsible for constrained agency, it leads to behavioural violence. It is the effect of structures on individual agency that results in this gap between potential and actual fulfilment of rights.

II

SEZs as New Means of Systemic Violence and Imperialism in India

Systemic violence is inherent in a situation whose intentional or unintentional consequence is injury and destruction. It becomes a form of institutionalized social injustice when development projects like SEZs bring about displacement and loss of livelihood of the weaker sections of the society. The resources are extracted by the economic elites of the society and those who are 'resource poor' have no involvement in this process, taking away the little they had, and in the process reducing them to 'objects' of development.

The doctrine of 'eminent domain' is invoked by the state and for the 'greater common good', private and community land is acquired from the people compulsorily to facilitate the SEZs, which invariably necessitate involuntary resettlement of people. Land-grabbing policies are tools of exploitation and anti-people, thus involving violence. Even as SEZ creates wealth for a few, it leaves the others with an irreparable sense of loss and impoverishment and violence wrought in their wake. And when these people make an attempt at resistance, they inevitably come into conflict with the law. Already marginalized, they are further faced with the state power. There are many dimensions of this systemic violence: oppression, humiliation and discrimination along the lines of class and ethnicity; and regular harassment, violence and torture by arms of the State.

Zones or geographically delimited enclaves within a country are products of the process of globalization. First, such modern zone was set up at Shannon in Ireland in 1956. These zones are distinguished from the rest of the land in terms of their specific administrative authority, benefits enjoyed by industries located in them and availability of better business facilities. India was one of the first in Asia to recognize the effectiveness of the Export Processing Zone (EPZ) model in promoting exports, with Asia's first EPZ set up in Kandla in 1965. With a view to overcome the shortcomings experienced on account of the multiplicity of controls and clearances; absence of world-class infrastructure, and an unstable fiscal regime and to attract larger foreign investments in India, the Special Economic Zone Policy was announced in April 2000.

Five years later, SEZ Act, 2005 was also introduced and in 2006 SEZ Rules were formulated. According to the Act, a Special Economic Zone is an especially demarcated area of land, owned and operated by a private company, which is deemed to be foreign territory for the purpose of trade, duties and tariffs. SEZs enjoy exemptions from customs duties, income tax, sales tax and service tax. They are projected as duty free area for the purpose of trade, operations, duty and tariffs. SEZ units are self-contained and integrated having their own infrastructure and support services.

SEZ can be sector specific or multi-product enclaves. It helps in the development of infrastructure of the area around the SEZ, provides employment to people, and makes the exports more viable. The government argues that this will help the country's products to become more competitive vis-à-vis providing all round development of region. Within SEZs, a unit may be set-up for the manufacture of goods and other activities including processing, assembling, trading, repairing, reconditioning, making of gold/silver, platinum jewelry etc.

The main features of Indian SEZs are as follows: modern harbour, airport, financial institutions, a vibrant industrial city with abundant supply of skilled and semi-skilled manpower, well connected with network of public transport, local railways and cabs; pollution free environment with proper drainage and sewage system; in-house customs clearance facilities; and uninterrupted power supply. Goods and services coming into SEZs from the domestic tariff area or DTA (area outside of the SEZs where the laws of the country will be applicable) are treated as exports from India and goods and services rendered from the SEZ to the DTA are treated as imports into India. There are simplified procedures for development, operation, and maintenance of the Special Economic Zones and for setting up units and conducting business. There is single window clearance for setting up of a SEZ.

The main *promise and rationale* of SEZ Act, 2005 of India are described in the guidelines provided in Section 5 which is as follows: (a) generation of additional economic activity; (b) promotion of exports of goods and services; (c) promotion of investment from domestic and foreign sources; (d) creation of employment opportunities; (e) development of infrastructure facilities; and (e) maintenance of sovereignty and integrity of India, the security of the State and friendly relations with foreign States.

III

Crony Capitalism and Violence

SEZs are industry units aimed at exports. This has resulted in the displacement of poor farm families and villagers, brutal land acquisition, and gross human rights violations by the state at the behest of private capital. No doubt, the SEZs have resulted in immense social and political injustices. SEZs are basically a manifestation of the state bowing down to private and monopoly capital. The SEZ scheme was and continues to be largely driven by big businesses and monopoly capital, with the state merely becoming an implementing agent. This is a manifestation of the shift from an import-substitution economic model to an export-promotion one. SEZs have been implemented using a variety of institutional structures across the world ranging from fully public (government operator, government developer, government regulator) to 'fully' private (private operator, private developer, public regulator). In many cases, public sector operators and developers act as quasi-government agencies in that they have a pseudo-corporate institutional structure and have budgetary autonomy. SEZs are often developed under a public-private partnership arrangement, in which the public sector provides some level of support (provision of off-site infrastructure, equity investment, soft loans, bond issues etc.) to enable a private sector developer to obtain a reasonable rate of return on the project. But when some economic elites receive preferential treatment and privileges (favouritism in the distribution of permits, special tax breaks, grants from government etc.) for accruing wealth by taking the support of the state, violating the principles of the market economy, violence happens. This is crony capitalism. It refers to a capitalist economy in which success in business depends on close relationships between people in business and government officials. The 2G and coal blocks allocations in India are recent examples. The policy of setting up of SEZs has favoured corporate interests. The practice of crony capitalism in free market economy in India, as in the case of SEZs, increases the 'crisis of modernity' leading to exploitation and social inequality.

IV

Economic Vulnerability and Social Deprivation

SEZs have rendered many people homeless and pauperized. SEZs involve prosperity of certain groups and sufferings and sacrifices of the rest of the population. The former reaps benefits and the latter bears the cost. These economic enclaves get labourers – daily wage, temporary and permanent – from the neighbouring villages and towns. This leads to the formation of 'ghettos' with distinct 'culture of poverty'. There is an imperialism of a new kind where the cultural values of the more powerful class is gradually imposed on the weaker class leading to 'particularization' of dominant values and colonization of the culture of the poor by the rich. This leads to more deprivation, violence and therefore resistance.

The new technology often displaces labour and goes against the interest of labour. The unskilled labourers are always at a loss. Workers in SEZs are subjected to gross violations of their labour rights. This includes job insecurity, prevention from unionization, extensive casualization and contractualization of labour, lack of benefits, unjust wages, and overtime without pay. There are many examples of rampant labour abuse in SEZs that provide a picture of what deregulation and labour flexibility in 'flexible capitalism' amount to. The Cochin SEZ, which according to trade unions has relatively better working conditions than other SEZs, has workers often paid below minimum wages, rampant contract/casual work, 10–12-hour work days with no overtime pay, frugal meals, no proper health practices and complete suppression of workers' right to collectively bargain. The SEZs in Bangalore, Delhi and Noida have had cases of workers getting below minimum wages, working 6 or even 7 days a week, and getting laid off with no notice.

The infiltration of the 'outsiders' in the land of the natives, keeping the people distanced from the decisions which affect their lives, relegating them and dividing them from within, are the acts of violence by the developers of SEZs. This structural violence facilitates behavioural violence as a response to the inequalities and exploitation that underpin structural violence. Nandigram and Singur incidences suggest that structural violence contains intrinsic capacity to provoke agential violence from the deprived and displaced groups.

Moreover, there is an ongoing debate on the violation of democratic principles in these SEZs. Are these SEZs becoming new corporate City-States? Would the SEZs turn ultimately into sovereign city-states – treasure islands of prosperity in a sea of poverty and misery – unaccountable to the vast majority of citizens in the neighbourhood, is a pertinent question (Shrivastava 2008).

V

Human and the Environmental Costs

Displacement constitutes systemic violence as it does not only amount to physical relocation but also reduces the quality of life of the communities into subhuman conditions. The local economies are dismantled with the creation and expansion of industrial capitalism. The attack on the local, takes the form of primitive accumulation and the gains to the capitalist, the form of rent appropriation. Primitive accumulation involves the dispossession of the commons – land, water, forest, knowledge, technology etc. – that are shared at the community level (Patnaik 2008; Basu 2008). Thus, the process of establishing SEZs in India is essentially a classic unfolding of the process of 'accumulation by dispossession' where the violence of capitalism is subsumed in the rhetoric of development (Banerjee-Guha 2008). A major part of the growth envisaged in the SEZs is through real estate and infrastructure development. This kind of corporatized development helps the consumer culture to grow, creating more and more deprivation and unrest among the marginalized.

The growth of employment in the entire organized sector since the inception of the economic reforms in 1991 has been negligible for the poor and the unskilled. With more automation, organized services also require limited supplies of labour. SEZs attract modern industry and services in order to succeed. To that extent they are unlikely to generate too many jobs. Moreover, the few jobs that will be generated will be for highly skilled labour, usually not available in the countryside. Furthermore, if the experience of existing SEZs in places like Noida is to be evaluated, the working conditions – poor wages, non-existent benefits, long working hours, occupational hazards, and discrimination and so on – under which people are employed inevitably violate human rights, apart from keeping the benefits of growth away from the poor. A study of the Nokia SEZ (Dutta 2009) reveals the enormous costs that the state government has incurred to attract Nokia to invest in the state and the few benefits such investments have created.

In Jamshedpur, Tata Steel increased annual production from 1 to 5 million tonnes between 1991 and 2005 but nearly halved its workforce from 85,000 to 44,000. India's growth has been capital-intensive, with very minimal employment growth and increases in wages and benefits. Over 8 per cent annual economic growth has resulted in just 1 per cent growth in regular employment.

Development of SEZs has meant displacement and pauperization of the tribals. In the name of development, tribals are displaced from their traditional habitats and livelihoods with little or no rehabilitation, and are rendered destitute and bewildered by the development process. They are pushed into a vortex of increasing assetlessness, unemployment, debt bondage and hunger due to loss of access to traditional sources of livelihood, viz., land, forests, rivers, pastures, cattle etc.

Also, any loss of access to traditional sources of livelihood marginalizes women in the labour market. It is only when land and other sources are replaced that women, at least partially, regain their economic status. Women not only suffer in terms of health and nutrition, they also lose the capacity to provide a secure future for their children. By resorting to seasonal migration they have unwittingly denied their children access to school, health care, child welfare, and other welfare services. Further, considering widow, deserted and unmarried adult daughters as dependents, ceases their right to claim separate or independent rehabilitation package with no land allotment in their names.

Ironically, the displaced lot does not include the landless, the non-asset class, the seasonal farmers, the fishermen, the boatmen, those engaged in work of art, culture and handicraft, and small shopkeepers in the affected village. They are ignored from benefitting from the rehabilitation package. Besides the profound vulnerability of the landless agricultural workers, there are people like the destitutes, beggars, the uncared for, aged, women victims of violence and abandonment, the disabled, leprosy patients, the mentally ill and children deprived of adult care, who are anyway condemned to the margins of society and are likely to be the first to fall by the wayside (Hemadri et al. 1999).

Displacement forces people to leave their usual habitation to which they are used to. In doing so, they leave behind, besides their personal belongings, the common utility assets, the social bondage and security. Not only the landowners are evacuated but the landless agricultural labourers, carpenters, blacksmith, cobbler, barber, tailor and so on, those dependent on these lands and the village society lose their livelihood too. The common properties like grazing land, ponds, wells, sacred grooves, worship places, playgrounds, and fuel sources are lost in the village; on the other hand, in urban context, people lose their place of work like hawking/vending, self-employment etc. The worst form of social impoverishment is caused due to loss of networks which are built up over generations; people who are used to traditional way of living are exposed to entirely new and alien living conditions and environment that is hostile to them. The variations and inconsistencies in the workplace lead to a loss of a sense of identity and security.

People start to confront and fight within the family for compensation money; the callous settlement of different ethnic and caste groups develop inter-community hostility. Many are rendered jobless and even if jobs are offered, they find it beneath their dignity of once being landowners to work as labourers. The affected people have no political voice, and as they are resettled in new fragmented conditions where there is lack of uniformity; thus the living conditions at different places of resettlement are different and the common interest vanishes leaving the marginalized ones further to the periphery.

Special Economic Zones have produced another class of vulnerable group, the urban poor or the new proletariat. The rural population, who are displaced or move out in search of livelihood (forced migration), end up living at the periphery of the urban society. A landowner in the village becomes an urban labourer, thus, creating a new class of people in the urban social hierarchy, viz., the slum dwellers. The person who had already lost his usual life and livelihood, thereby, also loses his marginal life.

Another allegation on the SEZs has been that they are accentuating regional and sectoral disparity and imbalances. Most of the approved SEZs are in the high-income states. States like Bihar, Assam, Arunachal Pradesh and Manipur do not have SEZs. The SEZs connected with Information Technology have proliferated post-2005 and provide bulk of employment. These SEZs are generally close to already existing urban centres. SEZs have not helped spread industrial or service activities to the remote areas or rural hinterlands.

There are controversies regarding snatching of arable or cultivated land or buying land at much discounted rates from farmers or poor people and then selling it to corporates at exorbitant prices. The practice of crony capitalism is prevalent here. The total amount of land to be acquired across India is 150,000 hectares. This land – predominantly agricultural and typically multicropped – is capable of producing close to 1 million tonnes of foodgrains. If SEZs are seen to be successful in future and more cultivated land is acquired, they might endanger the food security of India.

The environmental cost of setting up of SEZs would also be no less. SEZ would submerge huge areas of forest cover causing irreversible loss to varieties of flora and fauna besides the land area. The pollution (air, water, soil and noise) caused by the industries accentuate the miseries of the present as well as the generations to come. Added to this pollution by industries and urban centres, is the garbage and toxic waste that is generated over time. At the new rehabilitation site, the main environmental impact is the destruction or degradation of natural ecosystems. This is due to clearance for cultivation and houses, roads etc.

Despite being on the seismic map of the world and even after facing numerous fatal tremors, SEZs are not planned and constructed in accordance with the safety norms with respect to earthquakes. Many of the SEZs are planned at the coastal region which faces cyclones every now and then. No preventive measure is seen at any level of governance, be it policy choice, planning, implementing or enforcing. The philosophy seems to be redressing the after-effects of such disasters. Much of the loss of life and property is caused not only because of the intensity of the disaster but due to absence of efficient post-disaster rescue, resettlement and rehabilitation (immediate, short-run and long-run) operations, escalating the losses (of both life and property) manyfold.

VI

Adequate Compensation

The government promises 'humane' displacement followed by relief and rehabilitation. All that exists today in terms of law and policy is the Land Acquisition Act, 1894 that lays down that the government is not bound to provide a displaced person anything but cash compensation. The Rehabilitation and Resettlement (R&R) Policy of 2007 clearly mentions the principle of three 'minima'. These are: (i) minimize displacement; (ii) minimize acquisition of land, and (iii) minimize agricultural land for non-agricultural use. Under the new policy, a set of terms and conditions have been given if there is a project which involves physical displacement of 400 or more families en masse in the plains to 200 or more families in tribal or hilly areas. There are several components of this policy. It introduces the concept of Social Impact Assessment (SIA) along with the current norm of Environmental Impact Assessment. The SIA would also involve public hearings on displacement-related issues, loss of livelihood, compensation, effects on family. But the question remains that when rehabilitation of people displaced by dams like Hirakud, built way back in 1950s is not yet complete, what will happen to those displaced by SEZs? Moreover, there is always a discrepancy between policy and practice or formulation and implementation of policies.

VII

Resistances and Protests

There have been differential responses to the implementation of SEZs across states in India. The political landscape of India presents small and large people's movements to fight back the appropriation of their natural resources, livelihood and survival by their own governments and large national and international corporations. The villagers in India, whether they are from Nandigram in West Bengal, Raigad in Maharashtra or Santhankulam in Tamil Nadu, express their strong feelings of discontentment against the establishment of SEZs. But the resistance against SEZ in India became massive when political parties and social activists also joined the farmers. The dissent, uproar, and opposition from the farmers have led to behavioural violence of the farmers against the state.

These resistance movements have been across the regions in India. In the southern region, Muthanga forest land struggle, Waynad, Kerala by Adivasi Gothra Maha Sabha and farmers' rally against Reliance SEZ in Navi Mumbai. In western and central India, farmers protest against SEZ in Raigad, against land acquisition by Reliance in Greater Mumbai. In the eastern region, there is a struggle in Singur and Nandigram (West Bengal) against SEZs and displacement and Adivasi struggle in Jadugoda against uranium mining and displacement. In the north-east region, there have been people's movements in Doyang and Tongani, Assam against forcibile eviction from forests. In the northern region, farmers protest against Reliance SEZ in Jhajjar, Haryana and farmers struggle against land acquisition for Trident SEZ in Barnala, Punjab.

The Nandigram violence is a very famous incidence related to SEZ controversy. Nandigram is a rural area in Purba Medinipur district of the Indian state of West Bengal. In 2007, the West Bengal government decided to allow Indonesian-based Salim Group to set up a 'chemical hub' at Nandigram under the SEZ policy. Farmers of that village were against it. So, on the order of the Left Front government on 14 March 2007, more than 3,000 heavily armed police stormed the Nandigram area. The main objective was to remove the protestors in order to expropriate 10,000 acres of land for SEZ. During this incidence, police shot dead at least 14 villagers and wounded 244 persons including children and women (Bandyopadhyay 2007).

The farmers of Jamnagar have resisted by moving the court. In November 2006, farmers from the Jamnagar District in Gujarat moved the High Court of Gujarat and later to the Supreme Court in order to challenge the setting-up of a 10,000-acre (approx. 4,000-ha) SEZ by Reliance Infrastructure. They claimed that the acquisition of large tracts of agricultural land in the villages of the district not only violated the Land Acquisition Act of 1894, but was also in breach of the public interest. This led the government to 'consider' putting a ceiling on the maximum land area that can be acquired for multi-product zones and decide to 'go slow' in approving SEZs.

Mundra SEZ is another enclave situated on the northern shore of the Gulf of Kutch with the Adani Group as its developers. This SEZ will cause displacement and destroy livelihoods for fishermen. There are other serious concerns such as the destruction of mangroves affecting the marine ecology of the Gulf of Kutch, according to environmentalists.

VIII

Re-engineering Capitalism

In this era of contemporary globalization, capital has achieved a new-found global mobility leading to centralization of command and decentralization of production. Any development project is brutal, ruthless and violent if it does not take people as active participants. Large pieces of land are acquired by business tycoons to develop business or industrial city on such land. The lands are acquired by the state at lower cost with promises of giving jobs and other benefits, and then sold to these companies at huge profits. The companies then forcefully evict the landowners and most of them become labourers in the new industry. The establishment of SEZs on agricultural land leads to the conversion of a peasant into a proletariat. The new proletariat finds it difficult to confront and cope with machine technology and participate in industrial life where the skills of handicraft or agricultural labour get undermined. This creates a situation of 'corrosion of character' and loss of identity and job insecurity. This is the violence of new capitalism.

Further, there are many apprehensions against the establishments of SEZs. Given the concessions on import duties, there are likely to be foreign exchange losses (rather than gains). This results in loss of public revenue and creation of serious economic imbalances. There may be only a few generation of new activities and thus employment. Large-scale land acquisition would lead to not only displacement of farmers and others, but also have serious implications for food security. There could be misuse of acquired land by the developers for real estate. The establishment of SEZs involves many legal violations:[2] it violates the letter and spirit of the Indian Constitution; it infringes the fundamental rights of the citizen guaranteed in Part III of the Constitution; relaxation or inapplicability of many labour laws; no environmental clearance needed; violates Panchayat Raj Act, 1996 for local self-government; violates laws granting rights and control to Adivasi communities over their land; and violates many international conventions on human rights. In the SEZs, the sacred law and the Constitution of the country are overlooked (Ali 2007). Doing away with the regulatory hurdles leads to the making of SEZs as exploitation zones. Instead of being for public good, SEZs are producing goods for few at the cost of the rest of the population.

When people protest, the response of the State has been to crack down on the peaceful protestors, levy false charges against the leaders and arrest them and often also resort to police firing and violence to terrorize the people. We only need to remember Singur, Nandigram, Kalinganagar and countless other instances where peaceful and democratic forms of protest were crushed by the state with ruthless force. It is, thus, the action of the State that blocks off all forms of democratic protest and forces the poor and dispossessed to take up arms to defend their rights. Instead of addressing the source of the conflict and the genuine grievances of the marginalized people, the Indian state opts for the extremely myopic option of launching a military offensive.

The contemporary development process has its beneficiaries as well victims. But the victims, sufferers and the threatened outnumber the beneficiaries. Economic growth does not necessarily ensure equitable distribution of 'well-being'. This is the predicament of development. Development has to be unbiased and humane in its consequences. The socio-cultural costs of developmental projects like SEZs can be minimized, if not eliminated. The neo-liberal policies of the state marginalize the poor and the underprivileged. When a centre of prosperity develops in the vicinity of poverty, the widening disparity creates more and more tension, cultural conflict and violence. This definitely increases pain and frustration among the displaced lot with a manifestation of all the problems of core-periphery relationship and unequal exchange.

Capitalism as an economic system is inherently violent and has been perpetuating violence in all its stages. With the US becoming a victim to the global capitalist order and trying to wriggle out of the crisis of 2008, there is a need for change from the perilous present to a more promising future and a process of reorganization of capitalism is imminent. Efforts have to be made to re-engineer capitalism or restructure economies so that they remain effective in delivering economic prosperity and sustainability. The new model of capitalist economic system should be inclusive, equitable and sustainable; and should minimize the violence of the market and the state by creating a new balance between the two.

Notes

1. SEZs and structural violence has been discussed in my earlier paper, Pandey (2010).
2. Adapted from 'SEZs and Land Acquisition: Factsheet for an Unconstitutional Economic Policy' prepared by Citizens' Research Collective, New Delhi, March 2007, available at http://www.sacw.net/nation/sezland_eng.pdf, accessed on 23 April 2010.

References

Ali, R. 2007. 'Economic Zones of India and the "China Model": What is Going to Happen?', http://sanhati.com/articles/231, accessed on 23 April 2010.

Bandyopadhyay, D. 2007. 'Buddha's Killing Fields of Nandigram', *Mainstream*, XLV (50), 1 December.

Banerjee-Guha, Swapna. 2008. 'Space Relations of Capital and Significance of New Economic Enclaves: SEZs in India', *Economic and Political Weekly*, 43 (47), 22 November: 51–60.

Basu, P.K. 2008. *Globalization an Anti-Text: A Local View*, New Delhi: Aakar Books.

Cernea, Michael M. 1999. 'Development's Painful Social Costs: Introductory Study', in S. Parasuraman (ed.), *The Development Dilemma: Displacement in India*, London: Macmillan Press.

——. 2000. 'Risks, Safeguards and Reconstruction: A Model for Population Displacement and Resettlement', *Economic and Political Weekly*, 7 October.

Dutta, M. 2009. 'Nokia SEZ: The Public Price of Success', *Economic and Political Weekly*, 44 (40), 3–9 October: 23–25.

Galtung, Johan. 1969. 'Violence, Peace, and Peace Research', *Journal of Peace Research*, 6 (3): 167–191.

——. 1990. 'Cultural Violence', *Journal of Peace Research*, 27 (3): 291–305.

Government of India. 2009. http://sezindia.nic.in/html

Harvey, David. 2005. *The New Imperialism*, Oxford: Oxford University Press.

Hemadri, R. et al. 1999. 'Dams, Displacement, Policy and Law in India', *Business News*, www.Indiabiznews.com/biznews, accessed on 23 April 2010.

Infochange. 2010. www.infochangeindia.org/200703246254/other/features/faqs-about-sezs.html, accessed on 23 April 2010 and http://en.wikipedia.org, accessed on 23 April 2010.

Mandel, Ernest. 2008 (1973). *Late Capitalism*, London: Verso Classics.

Mander, Harsh. 2003. 'A People Savaged and Drowned', *Frontline*, 20 (8), 12–25 April.

Mohanty, Biswaranjan. 2005. 'Displacement and Rehabilitation of Tribals', *Economic and Political Weekly*, 26 March, 1318–1320.

Pandey, Manisha Tripathy. 2010. 'Ventilating Predicament of Development: New Economic Enclaves and Structural Violence in India', in Ravi Kumar (ed.), *The Heart of the Matter: Identity, Development and Violence*, New Delhi: Aakar Books.

Patnaik, Utsa. 2008. 'Imperialism, Resources and Food Security with Reference to the Indian Experience', *Human Geography*, 1 (1): 40–53.

Sennett, Richard. 1998. *The Corrosion of Character: The Personal Consequences of Work in the New Capitalism*, London: W.W. Norton.

——. 2006. *The Culture of New Capitalism*, New Haven: Yale University Press.

Shrivastava, Aseem. 2008. 'Towards Corporate City-States?', *Seminar*, 582.

Zizek, Slovaj. 2008. *Violence: Six Sideways Reflections (Big Ideas/Small Books)*, New York: Picador.

——. 2009. *First as Tragedy, Then as Farce*, London: Verso.

——. 2011. *Living in the End Times*, London: Verso.

16

Does Excess Manpower Continue to be a Liability for India?

Some Introspections in the Context of Globalization

Biswajit Ghosh

Social scientists and demographers have for long debated the relation among population, environment and development. In recent times, the old debate has yet again become popular due to new but opposite evidences like expansion of labour market owing to economic globalization on the one hand, and ecological crisis because of global warming on the other. Though the problem of 'over' or 'under' population hunts many countries of the world today, the societies of the Third World are particularly faced with the problem of 'population bomb' (Ehrlich 1968). In order to attain the cherished goal of sustainable development, these societies need to improve the quality of life and material security of the billion poorest people as well as to realize an ecological wisdom of intergenerational equity in the use of natural resources (Mehta 1997). Although human resource is a significant factor in development, any sustainable society also requires an optimum population to realize the goals of equity and efficiency. As our earth has a carrying capacity,[1] it cannot sustain any unlimited number of people for maximum period of time. There is no doubt about the fact that any additional number of human population puts demands for more houses, schools, hospitals, roads, jobs etc. and when a society fails to meet these increasing demands, the socio-cultural, economic and political environment degrades. Contrarily, the 'success' of a society to meet the additional demands of its population puts constraint on environment. The overpopulation question has, therefore, occupied a pivotal place in any discourse on development in the countries like India. But, in recent years, globalization has, at least theoretically, enhanced the possibility of labour migration, mobility and involvement of 'excess' (in the form of unemployed or underemployed persons) labour in the expanding areas of our economic activities. It is argued that our population itself is our strength as huge population offers a bigger pool of human resource and hence a bigger consumer market too (Roche 2012). The trend towards shifting of manufacturing and information technology-based industries from the advance industrialized countries to those where labour is cheap appears to be our 'advantage' now. Looking into this aspect of utilization of 'human resource' and the existence of huge 'youth power' in the country, some have argued that India's overpopulation is a prerequisite to development and not a liability in the changed context. Yet, as noted by the Asian Development Bank, 'The demographic dividend is not, however, an automatic consequence of demographic changes. Rather, it depends on the ability of the economy to productively use its additional workers' (ADB 2011). This essay seeks to introspect into the population-environment-development debate with particular reference to India in the context of globalization.

I

Population Growth in Developed and Developing Countries

Let me begin the discussion by looking into the contrasting features of population growth in the developing and developed countries of the world. By the middle of 2012, growth of world population was 7.058 billion (bn), 84 per cent of which (5.93 bn) are in the developing and poor nations of Asia, Africa and Latin America (PRB 2012). In other words, world population growth is far from systematic as nearly 37 per cent (2.6 bn) of world population today is in two developing countries, viz. China (1.35 bn) and India (1.25bn) only. Data stated in Table 1 reveal the tragic aspect of population growth: poorer the country, the higher the rate of population growth. As a result, despite decline in fertility rate in most areas of the world, world population has increased by 71 million during 2011–12 (PRB 2012). The more developed countries of the West today have low death (10 per 1000 population) and low birth rate (11 per 1000 population). One important reason for such a 'demographic transition' is the high standard of life of their population as revealed in their HDI rank and Gross National Income in Purchasing Power Parity (GNI PPP) per capita. But, due to steadily declining population, these societies are also faced with strains of fading and aging populations.[2] Changes in the composition of population during the last century also have left adverse effect on the environment because different population subgroups behave differently and their requirements also differ. This aspect of the population-development relation once again reminds us about the fact that

Table 1

Population and Development

Countries (by Population Size)	Population Mid-2012 (Million)	Per cent of Population Ages		Populati on Per sq Km	Rate of Natural Increase	Infant Mortality Rate	Total Fertility Rate	Projected Population (Millions)		GNI PPP Capita (2010 US$)	Per cent Population below $US2/Day 2009	HDI Rank 2012
		<15	65+					2025	2050			
China	1,350.4	16	9	141	0.5	17	1.5	1,402.1	1,310.7	7,640	36	101
India	1,259.7	31	5	383	1.5	47	2.5	1,458.2	1,691.1	3,400	76	136
USA	313.9	20	13	33	0.5	6.0	1.9	351.4	422.6	47,310	–	3
Indonesia	241.0	27	6	127	1.3	29	2.3	273.2	309.4	4,200	51	121
Brazil	194.3	24	7	23	1.0	20	1.9	210.1	213.4	11,000	10	85
Pakistan	180.4	35	4	227	2.1	68	3.6	229.6	314.3	2,790	61	146
Bangladesh	152.9	31	5	1,062	1.6	43	2.3	183.2	226.3	1,810	81	146
Nigeria	170.1	44	3	184	2.6	77	5.6	234.4	402.4	2,240	84	153
Russia	143.2	15	13	8	–0.1	7.5	1.6	140.8	127.8	19,240	1	55
Japan	127.6	13	24	338	–0.2	2.3	1.4	119.8	95.5	34,610	0	10
Mexico	116.1	29	6	59	1.5	15	2.3	131.0	143.9	14,400	9	61
Philippines	96.2	35	4	321	1.9	22	3.2	117.8	154.5	3,980	45	114
Vietnam	89.0	24	7	268	1.0	16	2.0	101.6	110.2	3,070	38	127
Ethiopia	87.0	41	3	79	2.4	59	4.8	115.0	166.5	1,040	78	173
Egypt	82.3	32	4	82	2.0	24	2.9	102.0	135.6	6,060	19	112
Germany	81.8	13	21	229	–0.2	3.4	1.4	79.2	71.5	38,100	0	5

Source: World Population Data Sheet 2012 and Human Development Report 2012. Data related to ' per cent of Population below $US2/Day 2009' is taken from World Population Data Sheet 2011.

population is a prerequisite to development only to the extent that a society requires more manpower to carry out its developmental activities.

It has generally been argued that population explosion is responsible for many of the ills faced by countries like India. This is mainly because these developing countries have not been able to match their rate of development with rising population (Ghosh 2005). Consequently, research has quite credibly established the linkages among population-poverty-pollution (called three Ps) in the Third World. The poor, says World Development Report (1992), are both victims and agents of environmental damage. They, with few assets, have to depend on environmental resources available in their surrounding. As they care more about extracting common property resources (CPRs) on daily basis than about conserving them for future, they end up exploiting the natural habitat excessively. On the other hand, environmental degradation affects the health of a poor by exposing him to the damages caused by polluted water, air or unsanitary living conditions. Poverty is, therefore, identified as a 'disease' that sickens the society and promotes larger families. Many poor parents produce children not because they are ignorant but because they need them to supplement family income. The problem of child labour thus emanates from economic hardship. As opposed to this, the Western countries of the North face the problem of environmental pollution caused basically by over-industrialization and consumerism. This again suggests that 'development' alone cannot guarantee a pollution free sustainable society for the South. In the context of such contrasting realities, let me now look into the theoretical arguments about the population, environment and development debate.

II

Population, Environment and Development: A Fresh Look into Old Debate

Since the anonymous publication of Thomas Robert Malthus' *An Essay on Population* in 1798, in which a concern for human population outstripping the means of support was first raised, we have come across several varieties of approaches on population control. The Cornucopians in particular believed that human ingenuity would be able to find new resources to match rising populations. In the Marxian alternative to Malthusian pessimism, for instance, the role of science and technology in a Socialist economy in increasing food production and inducing advance birth control measures were highlighted. These optimistic propositions of the Marxists have proved to be true so far as the Western Capitalist and European Socialist economies are concerned. Almost all of them have by now succeeded in containing their population by raising their means of subsistence through agricultural modernization and industrial alternatives. Even some developing countries like India could raise their food production level many more times through Green Revolution. To the Marxists, therefore, one should rather concentrate on solving the problems of inequality and exploitation under capitalism as proper development would do away with the problem of overpopulation. Without denying the efficacy of the Marxists' 'equity perspective' of development, it is important to note that the entire interconnection between economy and ecology is too complex to grasp. Marx made several errors in his counterblast to Malthus. First of all, population explosion is no more a problem of the 'capitalist world'. The international political context of the population debate has shifted today. Second, unlimited economic growth and technology-based model of development cannot always be eco-friendly. There are limits to what we can expect science and technology to do. Technology depletes natural resources, contributes to environmental problems and help in perpetuating social inequality. In several parts of the world, per capita income and food availability did not rise much even after the Green Revolution. Over and above, the recent natural devastation in Japan and consequent nuclear disaster proves the limits of new technology to work for the larger benefits of mankind and society at large. Third, affluence alone could not contain population in several countries of the Middle East. Going by the experience of 'developments' in the twentieth century, it rather appears that rapid population growth increases conflict, encourages corruption and violence, brings pressure for rapid solutions and reduces our ability to solve problems peacefully and quickly. Contrary to the geo-economic reality of Soviet Union, the Chinese Communists have, therefore, adopted a rigid yet pragmatic approach since 1979 to contain population that recognizes the reality of demographic compulsion as opposed to the Marxian optimism. Hence, they carried out nearly 330 million abortions since 1971 (*The Times of India* 2013a). Chinese Communist leaders today claim their one-child policy during the last 40 years has prevented overpopulation and boosted economic development. Even though the Malthusian fears about food production are premature, all fears about population growth at all times must not simply be erroneous. Indeed, as Sen (2000: 209) argues, there are many other issues related to population growth and overcrowding.

The Demographic Transition Theory of recent origin has also attempted to relate population growth with the levels of socio-economic development in the long run through three stages. This theory particularly looks into the experience of the Western countries to argue that population growth stabilized in the long run in the path of development. But the timings of change in different Western societies do not always match with the projections of this theory. Thus, countries like France experienced fertility decline at relatively low levels of development, while countries like England begun fertility decline at a much higher level of development (Coale 1986). Similarly, American fertility has remained steady around replacement level since the 1970s. This is argued to be because of different structural conditions and societal response in the USA as compared to Europe (Coale 1986: 97). In spite of some relevance of the theory, therefore, apprehensions are expressed about the possibility of replicating this model in the Third World. This is also because if countries like India or Bangladesh shall have to wait for a normal control of population through 'development', then they would be exposed to a kind of situation, which is no less than catastrophe. As a corollary, birth control and family welfare measures are given utmost importance by the governments of these countries.

Looking at the experience of 'development' particularly the Third World countries during the second half of 20th century, some social scientists have, however, tried to throw a new light on the old debate. Thus, Boserup (1965) has argued that, instead of being a hindrance to development, population growth is actually a prerequisite to such development because it modifies the population-resource ratio by encouraging people to migrate to urban areas. To him, a major stimulus to the 'green revolution' came from the pressure of population on food supply. Similarly, Easterlin (1967) suggested that population pressure can favourably affect individual motivation and lead to changes in production technique to overcome the consequences of human pressure. In Hirschman's (1958) model, population growth was argued to be helpful in increasing the supply of potential decision makers, expanding market base and leading to development via shortage. Simon (1986) also believed that population increase has long-run positive feedback effects which outweigh the short-term negative ones. He, however, agreed that for the developing countries, moderate population growth is more favourable to the growth of living standard than either a stationary population or very fast population growth.

Notwithstanding rather 'positive' evaluation of population-development relationship by some economists, ecologists have expressed serious concern about the common ideologies of 'economic expansion'. First of all, many developing and underdeveloped countries have not just been able to increase their stock of capital and improve the state of art rapidly enough for their growing population despite 'development' here and there. Looking into the experience of Western countries, it can fairly be argued that technological progress does not require population growth and in modern times it depends more on factors like R&D and huge capital investment. Contrarily, population explosion may inhibit the adoption of new technology, if it keeps down the level of human resource investment (Cassen 1999: 53). The economist's version of growth and development is also being criticized for concentrating only on aggregate statistical measures such as Gross National Product or Per Capita Income – numbers that in many instances conceal a multitude of sins. Ecologists are more keenly interested in the components of growth, i.e. the technologies that produce goods, the processes by which these goods are consumed, the cumulative impact of production and consumption on the living systems of the earth. The story of the growth of modern civilization is also marked by excessive use of non-renewable resources like metal ores, coal and oils with corresponding depletion of resources like air, water and land. There are, therefore, some limits to growth and the Club of Rome in their influential study on such limits (Meadows et al. 1972) argued that current trends in population growth, energy demand and resource consumption are pressing hard on the carrying capacity of the earth. The principal defect of the industrial way of life is that it is not sustainable. The 'effluents of affluence' is the major threat of current development paradigm. Changes in the global environment over the past century clearly indicate that more people translate into greater environmental degradation.

Economists, however, sometimes refer to a U-shaped relation between economic growth and environmental reality in this context. Following the concept of Environmental Kuznets Curve (EKC), it has been suggested that after a level of development, damage of environment becomes less. This may be first, due to changes in the economic processes of any country from agriculture to manufacturing and then to services, and second, due to increasing efficiency of such an economy to control population. But the concept of EKC is dependent on the model of economy in which there is no feedback from the reality of the environment to production possibilities and in which trade has a neutral effect on environmental degradation. The actual violation of these assumptions gives rise to fundamental problems in estimating the parameters of an EKC. There are other econometric problems with the estimates of EKC. The idea that further development will reduce environmental degradation is dependent on the assumption that world per capita income is normally distributed (Stern, Common and Barbier 1996). But in reality, the income gap between the rich and the poor in the world has increased several times over the last one century.[3]

It must be kept in mind that the Western model of development and environmental activism cannot provide answers to all problems of developing world. The positive association between development and fertility reduction is often summarized by the slogan 'Development is the best contraceptive'. While there may be some truth in this argument, one should take note of two facts: (a) there are various components of development and (b) the influence of social and cultural factors over and above the economic factors in fertility behaviour need to be explained. Thus, for instance, many of the richest Indian districts like Punjab and Haryana have much higher fertility rates than the Southern states with much lower per capita income but with much higher female literacy and female job opportunities (Sen 2000: 218). Findings of several other researches suggest that social and cultural forces play a vital role in determining the onset of fertility transition (Desai 2004: 73).

The developing countries, therefore, continue to face an uphill task of improving the well-being of a vast majority of poor people, raising their standard of living and providing education, health, housing, drinking water and similar other essential facilities to them. The 'elongated pyramid' age distribution in these countries adds further to their social and economic difficulties. For instance, the percentage of population below the age of 15 is considerably high in poor nations of Africa (41% in all) like Niger (52), Uganda (48), Angola (48), Mali (47), Zambia (46), Burundi (46), Nigeria (44), Senegal (44), Ethiopia (41) and the like. In the Asian continent also countries like Afghanistan (46), Yemen (44), Iraq (43), Jordan (37), Syria (36), Nepal (36), Pakistan (35), Philippines (35), India (31), and Bangladesh (31) have quite sizeable young population (PRB 2012). It has also been estimated that by the year 2050, 53 per cent of the young people (15–24 years) of the world will live in Asia. Managing such 'revolutionary' changes in the age structure of the population of developing countries will become a big challenge for the planners of these countries. This is because, on the one hand, young people are considered 'resources' for the economic development of any country. But, at the same time, they would require adequate infrastructure and facilities to realize essential needs to make a decent or dignified living. Failure to ensure such 'development' on the part of any country would culminate in chaos and disturbances. It is worth mentioning here that presence of a huge segment of young population at present in Indian states like Uttar Pradesh and Bihar did not translate into higher productivity. Conversely, one in every three persons committing suicide in India in 2009 was aged between 15 and 29. On a day-to-day scale, eight Indians commit suicide daily due to poverty, another nine due to bankruptcy and seven because of unemployment (NCRB 2009). One should not, therefore, speak too much about the 'potential' of country's huge young population. The population pressure would, therefore, continue to remain a crucial issue in years to come for developing countries like India.

In the context of globalization of our economies and societies in recent times, a new debate on the possibilities of emergence of Asian powers like China and India has cropped up. It is worth noting here that the tag of 'Asian Tigers' was first assigned to countries like Indonesia, Malaysia and Thailand. Later on, failures on the part of these countries and remarkable economic 'growth' achieved by China and India during early years of the current century, led the World Bank and the International Monetary Fund to shift the focus to these 'new giants'. It was then believed that these countries would be able to utilize their excess man-power and cheap labour to produce goods and services for the world market. The shifting of the base of manufacturing and certain service industries from the North to the South has been cited in this context.

There is some truth in this argument in the sense that economic globalization did open up the scope for the expansion of particularly informal sector industries in the South. But whether such 'informalization' and 'casualization' of our economies having little scope for decent income would be able to develop 'capability and freedom'[4] of average Asians and solve the vicious circle of poverty, unemployment, illiteracy and poor health is a million dollar question. As of now, the current process of globalization of capital or 'globalization from above' is largely proved to be beneficial to the developed countries. In spite of certain movement and mobility of educated Asians, there is hardly any 'globalization of labour' in a world that is rather moving in the directions of protection to national economy while at the same time creating barriers for the competitors.[5] It should also be noted at the same time that the strategies of Human Resource Management are being effectively used globally to restrict the number of workers, contain their human rights and thereby follow unfair labour practices including corporate corruption (Ghosh 2010). The LPG model of development, which has earlier overstressed the role of market over state and society, is now being partially reversed in the context of global recession (and failure of the market) by stressing on the role of the state to protect the poor and the marginalized. Moreover, the technology driven post-Fordist phase of contemporary industrial reconstruction has left very few options for the vast majority of illiterate and unskilled 'reserve army of labour' in India. Concern for the social security of unorganized workers, who constitute 93 per cent of Indian workers at present, has, therefore, been correctly voiced from different sources. It is, therefore, argued that 'market globalism' has been ushered in to increasingly trap the developing countries into an international system of domination and subjugation (Ghosh 2003). As many of them depend on foreign aid to meet even their daily needs, there appears to be no alternative to population control, simply because the road to 'development' is too complex, lengthy, costly, ecologically harmful and often unacceptable as it results in economic and political dependency on a few mercantile power. Hence, controlling the number of resource user and consumer in the Third World is one of the pressing and urgent paths of alternative development. Let me now look into the experience of population explosion in India so as to understand the need for an India-specific model of development.

III

India's Struggle to Grapple with Population Explosion

As per the 2011 Census, India's total population stands at 121 crore which is almost equal to the combined population of the US, Indonesia, Brazil, Pakistan, Bangladesh and Japan. It was earlier estimated that India could overtake China's population by 2035. But, now due to significant success of the Chinese government to cut down birth rate in recent times, India would become the most populous country of the world by 2028. This ever-increasing growth is due to 'population momentum factor' in which the number of women in the reproductive years of 15 to 49 swells the ranks of potential mothers and contributes to such an increase. It is estimated that 60 per cent of our current population is in the age group of below 25 and above 65, requiring additional facilities for a minimum standard of living including safety and security. As a corollary, dependency ratio in India is 3.22 persons per worker. Presence of a huge percentage of young population (50% below 25 years of age in 2011) also means that India's population will continue to grow even after reaching the replacement level. The lower age at marriage[6] of average Indians contribute to such increase rapidly. As per the National Population Policy (NPP) estimates of 2000, India's population would reach replacement level by the year 2045 when its total population strength would become close to 170 crore. But, by 2021–22 alone, we will be adding nearly 138 million to our working age population. This is an astonishing and unbelievable number for any job creation strategy, given our failures during the last two decades. It is beyond anyone's expectations how the country would be able to manage such huge population. How are we to face such a catastrophe? What would be the environmental impact of such an explosion? Have we not already crossed our sustainable limit of population? It is not easy to suggest definite answers to these queries. But we certainly can draw some lessons from an analysis of some major socio-economic and environmental issues facing India today. Let me analyse five such basic issues in order to ascertain the impact of rising population on India's living standard and life pattern.

Concern for Food Security

The relation between the population growth and food supply in South Asia in general, and India in particular, has often been regarded from a neo-Malthusian perspective although these countries could increase food production many more times. Thus, agricultural production has increased from 51 million tonnes in 1950–51 to 246.2 million tonnes in 2011–12 in India. Yet this 'quantum leap' is argued to be inadequate for India's large population. The Green Revolution, which once contributed to the rise of agricultural productivity, has now become fatigued. Due to rising population, declining resource bases, increasing economic and environmental costs and inadequate policy attention to these issues, India could not sustain high agricultural productivity. Rising food prices, spurred by expectations of shortfall in food production, have brought the issues of food security, food stocks management and need for improving food production and productivity to the forefront of national strategy. It is in this context that the Government of India is currently contemplating a Food Security Bill.

In other words, despite some incremental and intermittent gains since 1990s, the trend of expansion of food production in India today is not 'quite clearly upward', as was earlier claimed by Amartya Sen (2000: 206). India has failed to consistently produce foodgrains to meet the per capita nutritional requirement of 750 grams of food a day for an adult person. A significant food crisis is rather impending, as a vast number of India's small and marginal peasants are heavily dependent on the fate of rain god to produce the minimum required quantity of food. A survey by the Economic Research Service of USDA (2009) has reported that food security in 70 developing countries including India is projected to deteriorate over the next decade. The findings of Inter-governmental Panel on Climate Change (IPCC) also suggest that global warming will affect the monsoon patterns in India, causing a significant damage to the health of India's agricultural sector. It has also been suggested that India could experience a 40 per cent decline in agricultural productivity by 2080, as rise in temperature will affect crop yield and particularly Rabi crop (Gautam 2009: 3). I must not remind the readers that agricultural distress during the last two decades has already culminated into the suicidal death of thousands of farmers. According to National Crime Record Bureau's (NCRB 2009) official statistics, in 2009 alone, 17,368 farmers have committed suicide. Researcher and activists have also shown that since 1995, more than 2,53,000 farmers have committed suicide in India, making this the largest wave of suicides in the world (The Sanhati Collective 2012).

One may read the findings of a NCAER study in this context. It argues that despite the existence of a huge Public Distribution System in India, about 80 per cent of the rural population and 70 per cent of the urban households consume less than the recommended levels of 2400 kilocalories and 2100 kilocalories respectively (quoted by Babu 2000). It is not, therefore, surprising that India imported 12 million tonne of grains in 2012. Consequently, India ranked 67 among 81 countries, sliding below Rwanda, according to the 2011 Global Hunger Index.

It is in this context that Indian policy makers today argue for a 'Second Green Revolution' to meet the requirement of nearly 350 million tonnes of food production (as per our nutritional standard) for nearly 133 crore people by 2020. Necessity has so far not proved to be the mother of invention technologically in India. This is in spite of the fact the use of additional inputs has been rising faster than output, suggesting that costs have been rising. Due to limitations of expansion of agricultural land, it is clear by now that the main source of long-term output growth in the Indian context can only be through improvement in yields.[7] But a huge fund is required for such 'gene revolution'. The environmental and social costs of such biotechnology-driven expensive 'revolution' cast serious doubts for the future of majority of India's small and marginal farmers. It is worth mentioning here that in an attempt to increase food production, Indian farmers have also increased fertilizer consumption 250 times from 69,000 tonnes to 17.4 million tonnes during 1951–2002 (Singh 2009: 3). The 'success' of Green Revolution in raising the level of food production should also take into account the environmental costs like depletion of groundwater level, degradation of agricultural land, or decline in Common Property Resources. Moreover, the impact of agricultural modernization programme on socio-economic inequalities, health condition and other social aspects should also be looked into (Sharma and Dak 1989). In a country where nearly 66 per cent of agricultural populace today produce only 14 per cent of GDP (2011–12), one needs to seriously introspect into the basic issue.

Employment Creation

One of the major argument in favour of economic liberalization in India was that it would invite foreign capital to create employment opportunities, as unemployment and poverty are the two most important 'pollutants' in our country. In pre-liberalized India, official efforts through several employment generation programmes could not reduce unemployment backlog and at no point of time India could come closer to the proclaimed target of full employment. This was simply because the rate of population growth had always surpassed the rate of employment creation. Given such a sordid state of affair, globalization of both capital and labour appeared as the only way out. But, the experience of economic reforms during the last twenty years belies many of our expectations.

In fact, starting from early 1990s, jobless growth along with public sector job-cut, compulsory retirement, industrial sickness and closure have made the situation worse. Employment opportunities in the country since 2008 have also been affected by the global financial crisis and economic slowdown in the major markets of India. Consequently, the unemployment rate has risen from 5.99 per cent during 1993–94 to 10.70 per cent in 2009–10. In fact, the logic of 'jobless growth' during 2004–05 to 2009–10 was accepted by the United Progressive Alliance Government after much hesitation. The source of such revelation was the 66th round of NSSO data that made it clear that percentage of self-employed declined from 56.4 per cent of total workforce in 2004–05 to 50.7 per cent in 2009–10. The percentage of regular employees, however, only marginally rose from 15.2 to 16.4, though share of casual employment rose significant from 28.3 per cent to 33 per cent during the same period (*The Times of India* 2013b). All these clearly demonstrate that 'growth' did not lead to substantial increase in employment opportunities in post-reform India. Agricultural stagnation and consequent mass exodus of agricultural labourers to towns and cities have added fuel to the crises. It was in this context that the Government of India had to introduce the largest ever, rural employment programme called Mahatma Gandhi National Rural Employment Guarantee Scheme (MGNREGS) since 2005. But, a review carried out by the Comptroller and Auditor General (CAG) in recent times reveals that due to lack of monitoring, irregularities and widespread corruption, there is decline in rural household employment in the country from 54 days in 2009–10 to 43 days in 2011–12 (*The Times of India* 2013c).

It has been argued that a strong, vigorous and dynamic manufacturing sector has to be central to India's economic growth. If India is to have growth at rates of 8–9 per cent per annum and create jobs for its young population, the manufacturing sector must grow at 14–15 per cent per annum on a sustained basis. But, most of our manufacturing jobs have continued to be in the informal sector despite rapid growth during 2009–2011 (Kant 2012). The mushrooming of low paid temporary/casual jobs for even the skilled workers in the BPO sector only signifies the emergence of a 'risk society' for the common man. Along with informalization and casualization, globalization has also led to escalating displacement of poor and marginalized section of people due to conversion of agricultural land and Common Property Resources for industrial and commercial purpose. It is, therefore, argued that contemporary globalization is governed by free migration of goods and capital. As Daly (2004) argues, the costs and benefits of overpopulation under globalization are now distributed by class more than by nation. Labour bears the cost of reduced wage income; capital enjoys the benefit of reduced wage costs. Under such a situation, how can wages increase, given the nearly unlimited supply of labour? If wages do not increase then what reason is there to expect a fall in the birth rate of the labouring class via the 'demographic transition'? How could we ever expect to have high wages in any country that becomes globally integrated with a globe having a vast oversupply of labour? Global economic integration and growth may not, therefore, bring a halt to population growth.

It should, however, be recognized that labour migration flows from India since 1990s have registered impressive growth (Sasikumar and Hussain 2008). India is the world's largest recipient of migrant workers' remittances. The magnitude of the diasporic Indian community is estimated at 25 million residing in nearly 130 countries (Government of India 2008a). But, most of these migrants are skilled professionals, technicians and knowledge workers. It is very difficult, if not impossible, for a common man to migrate to any foreign country and get employed in any 'descent' job. This is because, one can clearly notice a new trend among Western developed countries to regulate and restrict cross-border migration. An unintended consequence of such migration governance is a steep increase in irregular migrants. Migrant workers encounter various vulnerabilities and the existing institutional and policy structures in place are not adequate to protect the rights of the migrants. Spectacular rise in human trafficking in recent years particularly from developing countries should be noted in this context (Ghosh 2009).

Eradication of Poverty and Inequality

India's achievement in reducing poverty and consequent income inequality has not been very bright. The official definition of the poverty line in the country does not include even such a basic necessity as clothing (even for women) leave aside the need for a roof over one's head or education or medicine. The extent of population living below the poverty line is, however, a contentious issue and recent estimates of even officially constituted committees vary very widely (Dhawale 2010; Chatterjee 2010). Faced with a mounting criticism that government is deliberately keeping poverty estimates low, the Tendulkar Committee was asked to rectify the Planning Commission estimate and the Committee had found poverty to be 37.2 per cent in 2004–05 based on a new methodology.[8] The World Bank's recent estimate, based on the yardstick of earning $1.24 a day, also claims 41.6 per cent of Indians as 'poor' (World Bank 2011). Similarly, the recommendations of the rural development ministry's committee on Below Poverty Line (BPL) surveys (called Saxena Committee) submitted in July 2009, has recommended that 50 per cent of India's population be given below-poverty-line cards. There is, therefore, every reason to believe that economic reforms have widened disparity and distress among people. As poverty exposes poor people to pollution, affects their health, lowers their productive capacity and contrarily increases their family size, it is just not possible to sustain any kind of development in India without solving the problem of rampant poverty and large-scale inequality.

Kelley (1994) had earlier concluded that population growth relates negatively to per capita income. Growing inequalities have also been accompanied by increasing regional and rural-urban disparities. Income inequality in India is so sharp that the richest 20 per cent of the population share more than 46 per cent of total national income, leaving only 8 per cent for the poorest 20 per cent (Human Development in South Asia 2002). Social exclusion of common man and woman and more particularly of people belonging to disadvantaged groups like Scheduled Castes, Scheduled Tribes and Muslims becomes very sharp in such a context. One need not therefore explain the necessity of equitable and holistic development for India and restricting population invariably becomes an important agenda in this context.

Literacy and Human Development

Literacy and human development are key to any qualitative social change and these two factors strongly influence demographic behaviour. In countries like India, efforts to enhance literacy rates and other facilities are strongly affected by an increasing number of human population. The educational system in India, is the second largest in the world (Tilak 2006: 33–34) and the numbers of students in educational institutions in

India outnumber the total population of united Germany, England and Canada taken together. We have been able to increase the enrolment[9] rate many times since Independence. Yet, the pressure to build appropriate educational structure for the burgeoning population is deeply felt in India. India has the world's largest population in the age bracket 5 to 24 years of about 450 million. It has been estimated that India will require 6.0mn more teachers by 2020 to attain the world average in terms of student teacher ratio (Saxena 2012). With nearly 273 million total illiterate in 2011, most of whom are women and people from disadvantaged sections, India tops the rank of largest number of illiterates in the world.

Despite the expansion of the system in recent years, the progress achieved so far has not been satisfactory, both in terms of quantity and quality. A general feeling is that 'education in India is in peril' (Tilak 2006: 34). According to the 'EFA Global Monitoring Report 2010' of UNESCO, India's rank on EDI was 105 among 128 countries. And it continues to figure, along with a bunch of African and one or two Asian countries, such as Pakistan and Bangladesh, in the group of countries with a low EDI. In 2001 also India was ranked 105 among 127 countries (Tilak 2010).

The task of making India a 'full literate nation' in true sense of the term appears daunting because of sheer magnitude of the problem. Apart from questions related to quantity, the question that becomes most pertinent today is how many of the Indians would be able to qualify for jobs available in the market? *Mercer Consulting* has estimated that only a quarter of graduates are 'employable' (Wikipedia 2010). TeamLease India Labour Survey Report 2009 (*The Times of India* 2010) also found that 89 per cent of Indian workers in the age group of 15–49 have had no vocational training; and of the 11 per cent who received such training, only 1.3 per cent was formally trained. This raises serious question about the potentiality of, particularly, our rural youth to develop requisite skill and knowledge necessary for new types of jobs available today. The gainers certainly are the members of particularly English educated, urban or semi-urban classes. It is quite clear by now that there is a large 'class of losers' along with a small 'class of winners' in post-liberalized India.

It must be mentioned here that in the social sector, India could not improve its performance during the last one decade. Thus, as per the Human Development Report 2013, the Inequality-adjusted human development value for India in 2012 was 0.554 on the basis of which India is ranked 136 out of 186 countries of the world. Again, in terms of the Gender Inequality Index, India is ranked 136 with 0.610 score in 2012 (*The Times of India* 2010: 159).

Health and Illness

In the health sector, India's performance is even worse. It has been argued that our medical infrastructure is alarmingly backward to support even the population figures cited by the 1991 Census. In 2009, government expenditure on health per capita (PPP US$) in India is only 21 for the year 2006, and this is much lower than that of China (144), Sri Lanka (104), Bhutan (71), Kenya (51), Ghana (36), and even Bangladesh (26) or Nepal (24) (HDI 2009). In 2012, the Infant Mortality Rate (IMR) in India was 47 as compared to 17 in China and only 2.3 in Japan (PRB 2012). The intensity of the problem of malnutrition and consequent risk of impaired health is known to us. The prevalence of underweight children in India is among the highest in the world, and is nearly double that of Sub-Saharan Africa with dire consequences for mobility, mortality, productivity and economic growth. According to the UNICEF report *The State of the World's Children* (2009), India has the highest rate of child malnutrition in the world. The prevalence of anaemia among children (6–35 months) is as high as 70 per cent during 2005–06 (NFHS 3). More importantly, prevalence of anaemia among pregnant women is found to be 58 per cent during 2005–6 (GOI 2008). In such a context, significant mortality risks to mothers and children in India are often associated with high fertility.

Sorrowfully, after twenty years of immunization, the children of India are not even close to being fully immunized. NFHS-3 (2005–6) data reveal that such coverage is 44 per cent only. The same government report also states that 44 per cent of currently married women in the country do not follow any family planning method. Over and above, there are failures and shortcomings on the part of the government to provide basic health facilities to its expanding population. The Government of India (2008) has also documented that there are acute shortages of doctors, nurses, volunteers, midwives, specialists, technicians and officers to manage health services particularly in rural areas. The Health Survey and Planning Committee (Mudalier Commission) had recommended that there should be one doctor for every 3,500 persons and one hospital with one bed for every 1,000 persons. Keeping in track with global standard, the Medical Council of India has recently proposed the target of 1 doctor for every 1,000 population by 2031. But in reality, India continues to face a poor doctor to patient ratio with only one doctor available for 1,700 people in the country. In comparison, the doctor population ratio globally is 1.5: 1000 (*The Times of India* 2011). It appears that our task in every essential front is quite heavy and with disproportionate governmental spending on health and other welfare measures these days, one may hardly hope for any better life in near future.

Conclusion: Is There a Way Out?

The above discussion leave out many other areas of deprivation and constraints related particularly to population explosion in modern India. After two decades of economic reforms, it can safely be argued now that the current model of 'globalization' has failed to reduce marginalization and deprivation, rather it has given rise to new forms of inequities and exclusion. It is not enough to argue that India has enough 'youth power', it is equally required that our youth are qualified enough for the world market. As the neo-liberal road map to economic growth appears to be sloppy and risky, India cannot have both 'population explosion' and 'Sustainable Society' at the same time. Daly (2004) was right when he argued that our population can be our strength only when we have the power and strength to feed the people, provide them clothing and shelter, good education, health care and jobs and mould them into an asset. Golley and Tyers (2011) have equally argued that in contrast with China, much of India's potential demographic dividend lies in waiting for the decades ahead, with the extent and duration depending critically on a range of policy choices. In other words, we definitely need to strike a healthy balance between quantity and quality. Some optimistic assertions about population growth worldwide did not prevent Amartya Sen to recognize that 'reducing fertility is important not only because of its consequences for economic prosperity, but also because of the impact of high fertility in diminishing the *freedom of people*' (2000: 226 emphasis mine). While population control alone would not suffice for realizing the goal of a sustainable and environment friendly society, such step should be considered as the most urgent and necessary one at the present moment. Interestingly, average Indians also feel such a need. What appears to be debatable is the ways and means of realizing such a goal.

Hence, it is imperative that some bold initiatives are taken before it is too late within the tradition of democracy that India upholds. Some state governments like Andhra Pradesh, Rajasthan, Madhya Pradesh, Uttar Pradesh and Haryana have already come forward with 'disincentives' like withdrawal of government sponsored facilities or debarring a person with more than two children from contesting Panchayat election. It is true that the Chinese coercive method of 'one-child policy' followed since 1979 has certain uncalculated social consequences like neglect of girl child (Sen 2000), and India obviously cannot follow China. Nevertheless, India must take note of the fact that China's significantly greater achievement in social development in combination with a strict population policy have paid dividend in controlling population very rapidly. So far India has failed to come close to Chinese level of development in social sectors. At the same time, the success of any voluntary family planning programme in a country like India depends on huge and therefore, unaffordable government spending for long-term on socio-cultural and health sectors. Moreover, due to socio-structural differences, the much-publicized 'Kerala model' could hardly be replicated in other parts of the country and particularly central and north India. Hence, I would aver that if India cannot opt for coercion to control birth rate (like those followed during 1975–77), it certainly can offer systems of incentives and disincentives[10] to implement the small family norm. And in order to do so, the country needs to be politically committed, administratively efficient and socially responsible for making an India that really shines.

Notes

1. The concept of 'carrying capacity' refers to the maximum (not desirable) number of people that a given environment can sustain over a longer period of time.
2. For 19 countries of Europe along with Japan, the rate of natural increase of population is either zero or less than that.
3. The World Development Report 2000 says that the income gap between the fifth of the world's people living in the richest countries and the fifth in the poorest was 74 to 1 in 1997, up from 60 to 1 in 1970 and 30 to 1 in 1960.
4. Amartya Sen (2000) has argued that development can be seen as a process of expanding capacities and instrumental freedom.
5. At present only five countries officially welcome international migrants. Even among them, USA and Australia face internal pressure to stop immigrants and outsourcing.
6. According to UNICEF (2009), 47 per cent of India's women and 56 per cent of rural women aged 20–24 are married before the legal age of 18. It is shocking to note that 40 per cent of the world's child marriages occur in India.
7. India's wheat and rice yields are 2.6 tonnes and 2.9 tonnes per hectare respectively as compared to 8 tonnes and 10 tonnes being the highest in the world. China produces 40 per cent more food than India on just 60 per cent of the arable land area of India.
8. As compared to the NSSO's estimate based only on needed expenditure of ₹ 368 and ₹ 559 per person per month for rural and urban areas respectively to meet minimum calorie intake, the Tendulkar Committee has modified the estimate of poverty by including spending on food, education, health and clothing along with calorie intake.
9. Data on enrolment in India are subject to serious problem as some government agencies often report fictitious enrolments. Some also fail to distinguish between 'enrolment' and 'attendance' (Tilak 2006: 35).
10. As of now, 'incentives' like scholarship for single girl child or reserving certain kinds of health care services to persons who have sterilized do exist in the country. Similarly, 'disincentives' like forbidding persons with more than two children to contest Panchayat elections or denying maternity leave to mothers of more than two children are used in particular regions.

References

Asian Development Bank. 2011. http://www.iberglobal.com/archivos/ado_sept_2011.pdf, accessed on 5 April 2013.

Babu, S.C. 2000. *The Hindu Survey of Indian Agriculture*, Chennai: Kasture and Sons Limited, 25–30.

Boserup, E. 1965. *Population and Technology*, New Delhi: Oxford University Press.

Cassen, Robert. 1999. 'Population and Development Revisited', in Barbara Harriss-White and S. Subramanian (eds), *Illfare in India*, New Delhi: Sage, 47–72.

Census of India. 2011. Government of India, New Delhi.

Chatterjee, Biswadeb. 2010. 'Doublespeak on Poverty in Bengal', *The Statesman*, Perspective Page, Kolkata Edition, 27 April.

Coale, Ansley (ed.) 1986. *The Decline of Fertility in Europe*, Princeton: Princeton University Press.

Daly, Herman E. 2004. 'Population, Migration, and Globalization', *World Watch Magazine*, 17 (5), www.worldwatch.org/pubs/mag/, accessed on 5 April 2013.

Desai, Sonalde. 2004. 'Population Change', in George Ritzer (ed.), *Handbook of Social Problems: A Comparative International Perspective*, Thousand Oaks, CA: Sage, 69–86.

Dhawale, Mariam. 2010. 'Maharashtra: CPI (M) Organises Convention for Food Security', *People's Democracy*, 34 (13), 28 March.

Easterlin, R.A. 1967. 'The Effects of Population Growth on the Economic Development of Developing Countries', *Analysis of the American Academy of Political and Social Science*, January.

Ehrlich, Paul R. 1968. *The Population Bomb*, New York: Ballantine Books.

Gautam, H. Raj. 2009. 'Effect of Climate Change on Rural India', *Kurukshetra*, 57 (9): 3–5.

Ghosh, Biswajit. 2003. 'Impact of Market Globalism on Indian Society and Culture', *Bharat Vidya*, 3: 73–83.

——. 2005. 'Population and Pollution: Concern for a Sustainable Society in India', in A. Das (ed.), *Environment-Science and Thought*, Vol. I, Kolkata: ACB Publications, 167–180.

——. 2009. 'Human Trafficking in India: Nature, Dimensions and Strategies for Prevention', *The International Journal of Human Rights*, 13 (5): 716–738.

——. 2010. 'How to Govern Corporate Houses? Significance of Industrial Democracy and Social Unionism in the Context of Globalisation', in S.K. Pramanick and R. Ganguli (eds), *Globalisation in India*, New Delhi: PHI Learning, 191–208.

Golley, J. and R. Tyers. 2011. 'Demographic Dividends, Dependencies and Economic Growth in China and India', University of Western Australia Business School Economics Discussion Paper 12.23.

Government of India. 2008. *Eleventh Five Year Plan 2007–2012*, Vol. 1 and 2, New Delhi: Planning Commission.

——. 2008a. *Annual Report 2007–08*, New Delhi: Ministry of Overseas Indian Affairs.

Hirschman, A. 1958. *Strategy of Economic Development*, New Haven: Yale University Press.

Human Development in South Asia. 2002. New Delhi: Oxford University Press.

Human Development Report 1992, 2009 and 2013. United Nations Development Programme, hdr.undp.org/en/reports/global/hdr2009/, accessed on 8 May 2013.

Kant, Amitabh. 2012. 'The Heart of the Matter', *The Times of India*, Kolkata Edition, 29 June.

Kelley, Allen. 1994. 'Population and Income Change: Recent Evidences', World Bank Discussion Paper No. 249, Washington, D.C.

Meadows, D. et al. 1972. *The Limits to Growth: A Report for the Club of Rome's Project on the Predicament of Mankind*, New York: Universe Books.

Mehta, S.R. 1997. 'Poverty, Population and Sustainable Development: Issues and Perspectives', in S.R. Mehta (ed.), *Population and Sustainable Development*, Jaipur: Rawat Publications, 1–37.

National Crime Record Bureau. 2009. *Crime in India 2009*, New Delhi: Government of India.

National Family Health Survey-3. 2007. *India*, Vol. I, Mumbai: International Institute for Population Sciences.

Population Reference Bureau. 2012. *World Population Data Sheet*, Washington, USA.

Roche, Florine. 2012. 'India's Growing Population: Our Strength or Weakness?', published online on 11 July, www.daijiworld.com/news/news_disp.asp?n_id=143174, accessed on 4 April 2013.

Sasikumar, S.K. and Zakir Hussain. 2008. *Managing International Labour Migration from India: Policies and Perspectives*, New Delhi: ILO Sub-regional Office.

Saxena, Aurobindo. 2012. *Indian Education Sector Outlook (Insights on Schooling Segment)*, www.technopak.com, accessed on 5 April 2013.

Sen, Amartya. 2000. *Development as Freedom*, New Delhi: Oxford University Press.

Sharma, M.L. and T.M. Dak. 1989. *Green Revolution and Social Change*, New Delhi: Ajanta Publications.

Sharma, S.L. 1997. 'Towards Sustainable Development in India: Issues and Perspectives', in S.R. Mehta (ed.), *Poverty, Population and Development*, Jaipur: Rawat Publications, 290–303.

Simon, J. 1986. *Theory of Population and Economic Growth*, Oxford: Basil Blackwell.

Singh, Sukhjinder. 2009. 'Boosting Agricultural Productivity', *Kurukshetra*, 57 (7): 3–5.

Stern, D.I., M.S. Common and E.B. Barbier. 1996. 'Economic Growth and Environmental Degradation: The Environmental Kuznets Curve and Sustainable Development', *World Development*, 24 (7): 1151–1160.

Tilak, J.B.G. 2006. 'Education: A Saga of Spectacular Achievement and Conspicuous Failures', in *India Social Development Report 2006*, New Delhi: Council for Social Development, Oxford University Press.

——. 2010. 'Tangible Targets at School', *The Hindu*, 10 February.

The Times of India. 2010. 'Skilled Labourers in High Demand', Kolkata Edition, 7 September.

——. 2011. 'Target: 1 Doctor for Every 1000 by 31', Kolkata Edition, 7 March.

——. 2013a. 'China Performed 330m Abortions since 1971', Kolkata Edition, 17 March, Front page.

——. 2013b. 'CAG Picks Holes in Rural Job Scheme', Kolkata edition, 24 April, 7.

——. 2013c. 'Centre Admits to "Jobless Growth" during UPA-1, Fall in Self–employment', Kolkata Edition, 22 April, Front page.

The Sanhati Collective. 2012. 'Farmer Suicides in India: A Policy-induced Disaster of Epic Proportions', 15 January, sanhati.com/excerpted/4504, accessed on 5 April 2013.

UNICEF. 2009. *The State of the World's Children*, New York.

USDA. 2009. *Food Security Assessment Survey 2008–09: ERS Report Summary*, www.ers.usda.gov/publications/gfa20/gfa20_reportsummary.pdf, accessed on 6 September 2010.

Wikipedia. 2010. 'Tertiary Education in India', en.wikipedia.org/wiki/tertiary_education_in_India, accessed on 6 September 2010.

World Bank. 2011. 'Perspectives on Poverty in India: Stylized Facts from Survey Data', www-wds.worldbank.org, accessed on 28 August 2011.

17

Participatory Approach in Management of Natural Resources

A Reflection of Global Practices at Empirical Level

Sanatan Nayak

Natural Resources Management (NRM) plays an important role for sustainable use of major natural resources, thus leads to sustainable growth and poverty reduction. However, despite considerable emphasis on conservation, natural resources degradation at the national level has accelerated. These were overexploitation of groundwater resources (Dubash 2002; Pant 2004; Nayak 2009), improper use of surface water (Joshi 1987), waterlogging and salinity in soil (Joshi and Jha 1992), unbalanced use and at times excessive dependence on chemical fertilizers (Nayak 2006), deterioration of chemical, physical and biological properties of soil, mining of soil nutrients and depletion of soil organic matter, degradation of perennial biomass in common land and forest department land. Two types of natural resources management programmes were taken up in the country. First, integrated approach under participatory development programmes focusing on simultaneous development of multiple natural resources facilitated by a multidisciplinary team (Government of India 2007). Second, that adopting situation-specific approach dealing with a specific resource. The first approach includes large-scale adoption of people centred participatory approach with direct funding to community-based organizations such as watershed development, irrigation management, and joint forest management. Further, there are three major components under the comprehensive natural resource management programme, viz. Institution Building and Capacity Building; Development and Management of Natural Resources; Diversification and Intensification of Rainfed Farming System.

Looking into degradation of natural resources and the importance of its management, this paper tries to highlight on the importance of knowledge (both traditional and modern) for reclamation and subsequent maintenance of the quality of the soil in various regions of Uttar Pradesh, India. Section I and II analyses conceptual and historical, and various aspects of participatory form of development for sustainable use and management of different types of natural resources at the global level. Section III highlights the relevance of the management of sodic soil through participatory form of development model by the Government of Uttar Pradesh, its methods, and study area and sample size. This is followed by analysis and discussion of results through social categories as various agro-climatic regions are linked with economics, environmental and social characteristics in section IV. Whereas, Section V concludes the relevance of the study by looking into the governance of the participatory development model.

I

Meaning and Issues in Participatory Approach

The concept of community development approach originated in India during 1950s (Guimarães 2009) and subsequently spread to other developing countries during 1960s. The underlying idea of growing knowledge of participatory development had spread from institution building, social welfare through rural development projects such as education, health and employment schemes to management of natural resources for increasing the awareness of the poor and oppressed in 70s and in subsequent periods. There are numerous ways to understand participatory development (Campbell and Vainio-Matilla 2003). First, it can be understood as 'people's participation' or 'community participation' or 'people's own development' (Swantz 1986), or 'community development' and 'self-help' (Verhagen 1987). Second, participation as a means to project implementation and viewing participation as an end that, when achieved, will result in long-term engagement by those involved in the process of solution finding (Arnstein 1969). Third, participation can be seen as a process of empowerment of the deprived and the excluded. This view is based on the recognition of differences in political and economic power among different social groups and classes. Participation in this sense necessitates the creation of organizations of the poor which are democratic, independent and self-reliant (Ghai 1988). Four, participatory development stands for partnership, which is the basis of dialogue among the various actors, during which the agenda is jointly set, and local views and indigenous knowledge are deliberately sought and respected. This implies negotiation rather than the dominance of an externally set project agenda. Thus, people become actors instead of being beneficiaries. Five, participation is a process through which stakeholders influence and share control over development initiatives and the decisions and resources which affect them. Six, if rural development was to realize its potential, the disadvantaged rural people had to be organized and actively involved in designing policies and programmes and in controlling social and economic institutions (Van den Hove 2000). Thus, participatory processes can be understood as both a 'social technology' and a means of achieving efficient policy implementation. There are various issues involved in participatory approach as follows.

Diagnosis, Community Planning, Project Identification and Formulation

This includes all efforts to involve the local population in defining and diagnosing their own problems, setting priorities for their resolution, and identifying and formulating project interventions that may help to solve some of those problems. For example, various rural development projects in Bangladesh, Philippines and India helped rural mass for income-generating activities such as the raising of chickens, ducks and goats through participatory approach (Natpracha 1986; Estrella 2002). However, often lack of political will is identified as the main reason for the tardy progress in irrigation reforms at the state level in Andhra Pradesh, India (Reddy and Reddy 2006). More importantly, devolution of powers to Water Users Associations (WUAs) has still not taken place, as most of the important functions like assessment, collection of water charges, sanctioning of works etc. remain in the hands of the irrigation department.

Research and Extension, Innovation and Knowledge

This part deals with the complexity of indigenous knowledge and its status. This deals with the implicit assumptions concerning the superiority of scientific knowledge that prevailed during the 19th and 20th centuries. These had the advantage of restoring respectability in the form of indigenous or folk, or vernacular, or, in agricultural contexts, farmer knowledge, and to open up the possibility of serious research on this important subject, which tended to reveal situations far more complex, nuanced and dynamic than had been assumed before (Gurung 2003). Another was a substantial increase in the recognition and acceptance of the role that farmers can play in agricultural research, development and extension, particularly after late 1980s onwards (Guimarães 2009).

Natural Resource Management

Natural Resource Management development is a main area of application of participatory approaches. In general terms, the purpose of these interventions is to improve the living conditions of local people, particularly the poor, by helping them manage the natural resources available to them or under their control with greater effectiveness sustainability and equity. Participatory approaches to natural resource management had been implemented in connection with a considerable variety of natural resources, including forests and woodlands, coastal resources (Ira 1997), water and watersheds (Kolavalli and Kerr 2002; Reddy et al. 2004), fisheries in flood-prone ecosystems, soil and water conservation and conservation of nature reserves, to mention only some examples. Participatory approaches are naturally complemented, in some of these cases, by the use of a *sustainable livelihoods framework*, which considers that a livelihood comprises the capabilities, assets (including both material and social resources) and activities required for a means of living. Because it takes into account not only agricultural but other assets and activities as well, the livelihoods approach is particularly suited to natural resource management, where very often the resources being dealt with are not of a strictly agricultural nature. For example, analysing the impact of India's programme of watershed development on the livelihoods of rural communities, by looking at its impact on the five types of capital assets and strategies required for the means of living (Reddy et al. 2004). They also analyse the vulnerability and stability of those capital assets as well as the participation in the programme. They conclude that, even in difficult environmental conditions where the watershed cannot bring direct irrigation benefits on a large scale, watershed development, if supported with other programmes, has potential for sustaining rural livelihoods.

II

Background, Objective and Methods of the Study

Uttar Pradesh Bhumi Sudhar Nigam (UPBSN) has adopted the participatory model sponsored by the World Bank for the sustainability of the quality of land vis-à-vis increasing productivity and reduction of poverty in various regions of the state in various stages. The objective of the model was threefold. First, to develop awareness through various means among the farmers for reclamation of the soil and maintaining the quality of soil for further soil management and support on problems of soil and extension efforts to make farmers aware of technology for both reclamation and prevention. Second, make available inputs such as inputs required for reclamation, fertilizer, irrigation pumps, and electricity connections often through subsidies and credit schemes. Third, strengthen rural institutions and infrastructure in general to make such inputs available to all farmers after the reclamation process is over. The purpose was not only to reclaim the land for reduction of poverty in the present generation but also to maintain the quality of soil for future generations for sustainable agricultural production. However, this model clearly reflects a sustainable development model through reclamation of soil through participatory model.

Keeping in view of the importance of various aspects/themes of the management of natural resources, we have tried to understand some of the important activities, which could reflect the sustainability of the soil quality vis-à-vis increasing productivity and lowering of poverty in the study areas. More specifically, it tries to assess the depth of knowledge on different aspects of the reclamation of different social categories[1] of beneficiaries, which further enable them for managing the fertility of soil vis-à-vis productivity.

We conducted the study in three districts covering one district from each region of Uttar Pradesh. Sultanpur from Eastern Region, Raebareli from Central Region and Mainpuri from Western Region are selected as these districts reflect clearly the nature of reclamation of sodicity in the state. The villages are selected in multistage sampling methods having different environmental, economic, social characteristics. Further, selection is based on the availability of sodic land and land those were reclaimed. We have collected information from 480 respondents covering 12 villages in the three districts. Households were selected based on house's caste, landholding size and religion of the respondent. Data were collected during March to April, 2008 immediately after the rabi crop season.[2]

III

Analysis of the Results

The context of participatory development in conservation and management of natural resources can be broadly looked into four dimensions. First, the relationship between theory and praxis of participatory development in conservation of natural resources (Campbell and Vainio-Matilla 2003). The conservation of natural resources through participatory development leads to economic development and both are two sides in one coin (WCED 1987). Second, issues related to the development of knowledge in the participatory development process. Third, issues pertaining to the work done by the international, national agencies and NGOs supporting for the preservation and protection of natural resources through community based participatory approach (Campbell and Vainio-Matilla 2003). Fourth, the failed development project experience. There are large numbers of evidences that promoters of community-based conservation have not learned from or looked to participatory development. In other words, development organizations themselves have had problems in implementing participatory development in practice (Vainio-Matilla 2000). Considering various dimensions of participatory approach, the present study highlights importance of knowledge on preservation of soil quality in the stated study areas.

Importance of Knowledge for Sodic Reclamation

The stakeholders of training at the village level are categorized into mainly two parts. First, farmers directly involved and benefited in the process of reclamation. This category represents the member of Water User Group including the Group Leaders, Core Team, Mitra Kisan (MK) and Mahila Mitra Kisans (MMK), and Animators. While the second category represent, the institutions developed at the village level for sustaining the level of income and disseminating the knowledge of farm activity, such as the member of Cluster, SHG (both male and female), Micro Enterprises and finally the FFS including the Master Trainers. The development of knowledge to the first category is undertaken through series of interventions at different levels. First, the development of knowledge through various aids/interventions, such as SIC, Bhumitra Patrika, Field Day, Impact Point etc. are undertaken to disseminate to the members of the Water User Groups. Second, the major types of training/exposure visits at the village level to Core Team members, MK/MMK, progressive farmers are provided at District level institute, State level and National Level. Third, some of the trainings are provided to different categories of animators for better employment and income level at the district or state level institutes.

Acquiring knowledge on sodic land reclamation and further maintaining the quality of soil has played a crucial role on improving the productivity of the soil vis-à-vis net income and reduction of poverty in the fragile region of Uttar Pradesh. Therefore, knowledge played a crucial role along with other inputs for improving the productivity of the soil, though it is not the only factor. The knowledge provided by various types of training interventions for sodic reclamation and subsequently maintaining the level of productivity and raising the level of income of the members of Water User Groups (WUGs)[3] and members of Core Team[4] are divided mainly on knowledge on participatory processes, reclamation technology adoption (including knowledge on crop production and horticulture) and village level institutions involved. Some of the important indicators have been considered to assess the level of knowledge on the above-mentioned three aspects of the training at the village level.

Knowledge on Participatory Process

Eight important indicators, viz., the knowledge on the formation of water user group (WUG), selection of leader of WUG, selection of boring site, finalization of alignment of link drain, knowledge on input distribution, knowledge on quantity of inputs, knowledge on site implementation committee (SIC),[5] and awareness campaign have been considered to assess the knowledge on participatory processes. The main purpose of the project was to involve the small and marginal farmers and their families in various activities of the reclamation process to enhance the productivity and maintenance of the quality of the soil. Therefore, providing knowledge on participatory process, technology and various institutions involved for sustainable production and income has been one of the challenging tasks before Uttar Pradesh Bhumi Sudhar Nigam (UPBSN). UPBSN has provided various types of training, viz., technical training, participatory training, refresher training and training on attitudinal behaviour for leadership development to its various stakeholders. The impact evaluation of knowledge on participatory process, technology and institutions by the beneficiary farmers including group leaders can provide some idea about intensity and efficient involvement of the beneficiaries in the reclamation process as a whole.

It is observed that more than 70 per cent of the total respondents are aware about all the important elements of the participatory process except finalizations of link drains and distribution of quantity of inputs distributed (Table1). However, finalization of link drains is one of the important factors for clearance of sodic water, where only 63 per cent of the farmers are involved. Another important factor is that nearly 99 per cent of the beneficiaries are aware about distribution of inputs, i.e. gypsum; whereas only 50 per cent of the beneficiaries were well aware about the quantity of gypsum distributed per acre or among the farmers. There was no formal well-communicated mechanism to inform all the beneficiaries about the distribution of gypsum.[6] Another 50 per cent of the beneficiaries were not informed about the quantity of distribution of gypsum. Therefore, many of the small and marginal farmers were denied to have gypsum proportionately according to their size of landholding.

So far as caste-based development of knowledge on different aspects of participatory process, least percentage of farmers belong to SCs category are aware about most of the indicators and highest percentage of farmers belong to OBCs category are aware about most of the indicators followed by general category (Table 1). The awareness on the quantity of gypsum is one of the important factors for the reclamation of the soil.[7] Most importantly, nearly 25 per cent of SCs beneficiaries are aware about the quantity of gypsum to be distributed for the reclamation of sodic land, whereas it is 28 percentage from OBCs category and 50 per cent from the general category. It may reveal here that the marginal and small land-holders belong to the SCs community are least aware about the participatory process as well as the quantity of gypsum required for reclamation.

Knowledge on Technology of Reclamation

The knowledge on application of technology during the reclamation and post-reclamation period constitutes an important aspect for controlling the quality of soil. Training has been provided on technology, On Farm Development (OFD), crop production, horticulture, drainage network to different category of stake-holders such as members of core team, MK/MMK[8] and progressive farmers. Subsequently, further dissemination has been done by these people to other beneficiaries for effective and efficient reclamation and sustainable farm production. In addition, village level training at SIC and various aids such as cultural parties, field days, Bhumitra Magazine etc. are also used. In this regard, selected six important indicators, viz., knowledge on the number of days water is standing in the farm after mixing gypsum with soil (leaching process), type of irrigation to be used for reclamation, type of preventive measures for disease management, type of horticultural plants to be developed in the sodic condition, benefits of application of organic manure, number of days paddy

Table 1

Caste-wise Level of Knowledge on Participatory Process

Caste	*Formation of WUG*	*Selection of Group leader*	*Selection of the Boring Sites*	*Finalization of Link Drain*	*Awareness on Distribution of Input*	*Awareness on amount of Distribution*	*General Meetings of the SIC*	*Awareness Campaign*
General	73 (76.04)	70 (72.92)	71 (73.96)	63 (65.63)	95 (98.96)	48 (50)	79 (82.29)	80 (83.33)
OBC	141 (82.46)	135 (78.95)	128 (74.85)	115 (67.25)	169 (98.83)	48 (28.07)	142 (83.04)	149 (87.13)
SC	69 (74.19)	61 (65.59)	54 (58.06)	52 (55.91)	92 (98.92)	23 (24.73)	70 (75.27)	78 (83.87)
Total	283 (78.61)	266 (73.89)	253 (70.28)	230 (63.89)	356 (98.89)	119 (49.59)	291 (80.83)	307 (85.28)

Source: Field Survey

nursery to be developed for further plantation[9] in the sodic condition are taken into consideration to assess the knowledge on technology aspect of sodic reclamation.

It is observed that most of the factors are highly acknowledged by substantial high proportion of respondents at the village level on technology development (Table 2). More than 75 per cent of the beneficiaries are aware about all of the factors except type of horticultural plants and number of days of paddy nursery to be planted in the sodic condition, which also plays crucial role for agricultural development. Caste-based knowledge on technological aspects of the reclamation and further activity clearly reveals that least percentage belong to SCs category in the selected villages (Table 2). Highest percentage of people are aware, who belong to the OBCs category followed by general category, so far as all the six indicators are concerned. Knowledge on benefits of organic manures is having highest per cent, whereas knowledge on type of horticultural plants to planted are the lowest.

Table 2

Caste-wise Level of Knowledge on Technological Aspects

Caste	*Leaching process*	*Ground water irrigation*	*Disease Management*	*Type of horticulture*	*Benefits of Organic Manure*	*No. of days Paddy nursery to be planted*
General	82 (85.42)	85 (88.54)	93 (96.88)	63 (65.63)	88 (91.67)	69 (71.88)
OBC	134 (78.36)	134 (78.36)	150 (87.72)	113 (66.08)	165 (96.49)	121 (70.76)
SC	67 (72.04)	70 (75.27)	80 (86.02)	46 (49.46)	80 (86.02)	56 (60.22)
Total	283 (78.61)	289 (80.28)	323 (89.72)	222 (61.67)	333 (92.5)	246 (68.33)

Source: Field Survey

Knowledge on Individuals and Institutions Involved in the Reclamation

Some of the individuals, such as MK/MMK, Animators[10] play crucial role for developing awareness and helping people for enhancing their income. The purpose of incorporating such indicators and their activities in a knowledge testing process reflects the level of involvement of these people for the development of awareness. They are not only considered as beneficiaries of the project but also key individuals or facilitators for other farmers at the village level. MK/MMK used to act as channel between government officers and village level beneficiary for disseminating and updating the information relating to the farm protection and production. Moreover, different types of animators have been developed by providing various trainings with the background of their respective interest and involvement for additional income and better life as well as disseminating the technique on agriculture. Similarly, development of institutions at the village level plays a crucial role for the success of the project. Farmers Field School (FFS) and Micro Enterprisers or Self Help Groups (SHGs) were developed by providing various training for updating the knowledge on farm production and income generating activities under the project. The major challenge is the development of FFS, which is an innovation to address the key issues affecting the project sustainability, like timely availability of agricultural inputs, credit, technical know-how, drainage maintenance. The development of SHGs/Cluster at the village level for generation of income is one of the important aspects for sustainability of the project. Keeping in mind the important role of the above-mentioned individual/institutions for the overall development of the villagers, we have identified seven important indicators to assess the knowledge and involvement of the beneficiaries with above-mentioned individuals/institutions. There are seven important indicators, viz. formation of core team, selection of Mitra Kisan and Mahila Mitra Kisan, function of MK/MMK, selections of Animators, functions of Animators, location and benefits of FFS and formation of SHG are included to capture the knowledge on institutions involved at the village level.

It is observed that substantially lower percentage of beneficiaries are aware about some of the important institutions and their activity under this category. Only 14 per cent of the beneficiaries are aware about formation of core team, which clearly reveals that no beneficiaries are involved in the selection process. Therefore, leader of the core team have been elected on the basis of their share of land in specific WUG or by their influence in the village or by the government officials. Similarly, selection and functions of different types of animators portrays that hardly 8 to 9 per cent of beneficiaries are aware about them. It is reflected here that the selection of animators are purely on individual basis without the knowledge beneficiaries. Moreover, their functions are hardly known to village beneficiaries (Table 3). Similarly, the location and function of FFS shows only 11 per cent of the beneficiaries are aware about the FFS, which clearly reveals that these institutions have very limited influence on the beneficiaries. Caste based awareness on the involvement of individuals and institutions also reflecting the same trend as the earlier two aspects are reflected. The awareness of SCs on different individuals and their respective functions are quite lower compared to general and OBCs categories (Table 3).

Table 3

Caste-wise Level of Knowledge on Institutions

Caste	*Formation of Core Team*	*Selection of MK/MMK*	*Function of MK/MMK*	*Selection of Animators*	*Functions of Animators*	*Location and utility of FFS*	*Formation of SHG*
General	10 (10.42)	72 (75)	71 (73.96)	6 (6.25)	6 (6.25)	14 (14.58)	59 (61.46)
OBC	32 (18.71)	113 (66.08)	120 (70.18)	22 (12.87)	21 (12.28)	22 (12.87)	108 (63.16)
SC	7 (7.53)	58 (62.37)	60 (64.52)	4 (4.3)	2 (2.15)	5 (5.38)	35 (37.63)
Total	49 (13.61)	243 (67.5)	251 (69.72)	32 (8.89)	29 (8.06)	41 (11.39)	202 (56.11)

Source: Field Survey

Summary and Conclusion

Participatory approach for management of natural resources plays an important role for its safeguards, with different social, ecological and economic background of farmers (Purushothaman et al. 2013; Sekher 2001). The success of natural resources management depend on the strategies of collective action of various attributes, such as attributes of goods or resources, attributes of user groups and attributes institutional arrangements (Sekher 2001). This paper highlights the awareness of users on the resources and institutions for its sustainable use. The knowledge on reclamation of natural resources (sodic land) is classified into participatory process, knowledge on input use and technology, and knowledge on institutions and individuals. First, it reveals that knowledge on input use and technology was substantial, whereas knowledge on participatory process and institutions and individuals are moderate and some indicators provide poor performance. For example, in the participatory process, awareness on distribution of inputs was substantial, whereas the knowledge on distribution of quantity of inputs, it is quite low (only 50 per cent of the beneficiaries). It is further investigated that marginal and small landholder belonging to SCs category have least awareness on the participatory process. Knowledge on individuals and institutions was rather poor in most of the cases except few. Most of respondents have viewed that they were unaware about selection and role of animators, existence of FFS, formation of core group etc. However, these are some of the important institutions and individuals, which could have played a serious role for the development of project areas. Though, institutional efforts in promoting sustainable agricultural practices were relatively more efficient and beneficial for small farmers and less developed districts as observed by Purushothaman et al. (2013), yet our study does not confirm the same very directly. Some of the most recommended practices of participatory structure of Sekher (2001), such as active involvement of users, executive leadership, user groups obligations to the local needs and conditions, regular consultation/interactions with appropriate needs are not strongly present in the present study. The participatory model does not address much to the people, to whom it was meant for.

Notes

1. Various social categories are general, OBCs, SC and ST. However, ST category is not available in this study.
2. This is part of the study conducted earlier in 2008 titled, 'Socio-economic Impact Assessment of Some Sodic Affected Villages in Uttar Pradesh', under the sponsorship of ICSSR, New Delhi.
3. Around a shallow tube well, the beneficiaries themselves formed Water User Group (WUG). All the project activities were carried out through these WUGs. In subsequent years, WUG evolved as an effective implementation unit. It took over all the responsibilities, i.e. from planning, OFD works, boring development, input distribution, crop planning and most importantly payment to its members. The responsibility of these groups initially was to ensure proper and timely availability of irrigation water to their members. The beneficiaries were asked to select about 4 ha command area for formation of WUGs.
4. The Core Team was constituted for effective monitoring and review of project activities carried out by the WUGs. It had leaders of WUGs, Mitra Kisan and Mahila Mitra Kisans and chairperson of WSHGs. Selection of core team was done at the SIC meeting.
5. The SIC has been an independent informal body under the chairmanship of *Gram Pradhan*. It was made a legal sub-committee of Gram Sabha, through a Government order which was later amended and at present, it continues to be an informal and most effective village committee. It is the real platform, where all the people get together (both male and female) for any kind of decision-making. SIC meeting has been tested by providing options of satisfactory, less and very less members. However, we have taken satisfactory as the answer in our analysis, which would show the result of other two variables automatically.
6. It means no meeting was held at the village level in advance. However, this question was asked only to 240 respondents in last 8 villages.
7. The percentage of the people aware about the quantity of the distribution per unit of land is mostly same with the percentage of recipients of gypsum in the participatory model.
8. MK stands for Mitra Kisan and MMK Stands for Mahila Mitra Kisan. One Mitra Kisan and one Mahila Mitra Kisan were selected from among the beneficiaries of 50 ha sodic land. While selecting them, care was taken that they represented the majority and had leadership qualities. Those having

some formal education were preferred. Those who were cooperative in nature and willing to work for the benefit of others were given priority. Mitra Kisan was essentially acceptable to all the beneficiaries. They are the real co-coordinator between government officials and village people. The purpose of developing MK/MMK by providing various training under this project is to develop such categories of beneficiaries, who can act as bridge between UPBSN and village level farmers for further dissemination of agricultural related information and income generating activity.

9. The knowledge on sustainability of reclamation or prevention of sodic patches has been drawn by receiving various important answers, such as continuation of cultivation, use of groundwater, use of organic and green manure, decline of chemical fertilizer etc.
10. The process of the development of animators was basically started during the reclamation period of phase-I. There was need of many boring and pump set mechanics at the time of reclamation in each village. Hence, UPBSN has provided training to some of the interested farmers with the background of same work during reclamation period and subsequently they were enabled to earn for their livelihood. Later on, the development of other categories of animators was encouraged. Four types of animators, such as boring mechanic, pump set mechanic, animal husbandry, and health were developed. The major purpose is to train some of the beneficiaries to increase their level of employment and income in addition to their primary activity and subsequently they would be able to help and disseminate the knowledge to other farmers at the village.

References

Arnstein, S. 1969. 'A Ladder of Citizen Participation', *Journal of the American Institute of Planners*, 35 (4), July: 216–224.

Campbell, L.M. and A. Vainio-Matilla. 2003. 'Participatory Development and Community-based Conservation: Opportunities Missed for Lessons Learned?', *Human Ecology*, 31 (3): 417–437.

Dubash, Navroz K. 2002. *Tubewell Capitalism: Groundwater Development and Agrarian Change in Gujarat*, Delhi: Oxford University Press.

Estrella, M. 2002. 'Rural Development: Mobilizing the People', *ADB Review*, 34 (2): 18–19.

Ghai, D. 1988. 'Participatory Development: Some Perspectives from Grassroots Experiences', Discussion Paper No. 5, UNRISD, Geneva.

Government of India. 2007. *Report of the Working Group on Natural Resources Management, Vol. I: Synthesis*, Planning Commission, Chapter III.

Guimarães, J.P. de Campos. 2009. 'Participatory Approaches to Rural Development and Rural Poverty Alleviation', Working Paper, Economic and Social Commission for Asia and the Pacific, United Nations.

Gurung, A.B. 2003. 'Insects – A Mistake in God's Creation? Tharu Farmers' Perception and Knowledge of Insects: A Case Study of Gobardiha Village

cent in 2015 and 3.5 per cent in 2016. The highest growth rates in developing economies, particularly in India and China, need explanation on various sources of growth and the final supremacy in the world lower growth rates hovering around 2.5 per cent to 3.5 per cent. So the double growth rates of India and China may motivate other world economies to share the experience of higher growth order. Nevertheless, it is also being observed that developing nations are suffering from malnutrition and chronic poverty.

I

Malnutrition and Hunger

Protein deficiency is the main source of malnutrition/hunger. It is basically a lack of calories and protein. Food is converted into energy by humans, and the energy contained in food is measured by calories. Protein is necessary for key body functions including provision of essential amino acids and development and maintenance of muscles.

No one really knows how many people are malnourished. The statistic most frequently cited is that of the United Nations Food and Agriculture Organization, which measures 'undernutrition'. The most recent estimate released in October 2010 by FAO, says that 925 million people are undernourished. As the figure below shows, the number of hungry people has increased since 1995–97, though the number is down from last year. The increase has been due to three factors: (1) neglect of agriculture relevant to very poor people by governments and international agencies; (2) the current worldwide economic crisis, and (3) the significant increase of food prices in the last several years which have been devastating to those

Chart 1

925 million hungry people in 2010

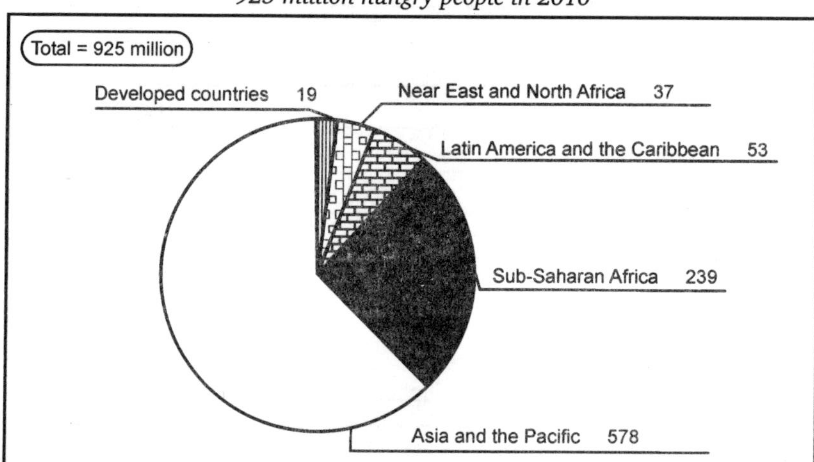

with only a few dollars a day to spend. 925 million people is 13.6 per cent of the estimated world population of 6.8 billion. Nearly all of the undernourished are in developing countries.

In round numbers there are 7 billion people in the world. Thus, with an estimated 925 million hungry people in the world, 13.1 per cent, or almost 1 in 7 people are hungry. The FAO estimate is based on statistical aggregates. It looks at a country's income level and income distribution and uses this information to estimate how many people receive such a low level of income that they are malnourished. It is not an estimate based on seeing to what extent actual people are malnourished and projecting from there as would be done by survey sampling. Undernutrition is a relatively new concept, but is increasingly used. It should be taken as basically equivalent to malnutrition (FAO 2010; Shah 2011).

The issue of corruption is very much interrelated with other issues. At a global level, the 'international' (Washington Consensus-influenced) economic system that has shaped the current form of globalization in the past decades requires further scrutiny for it has also created conditions whereby corruption can flourish and exacerbate the conditions of people around the world who already have little say about their own destiny. A difficult thing to measure or compare, however, is the impact of corruption on poverty versus the effects of inequalities that are *structured into law*, such as unequal trade agreements, structural adjustment policies, so-called 'free' trade agreements and so on. It is easier to see corruption. It is harder to see these other more formal, even legal forms of 'corruption.' It is easy to assume that these are not even issues because they are part of the laws and institutions that govern national and

Chart 2

Number of Hungry People, 1969–2010

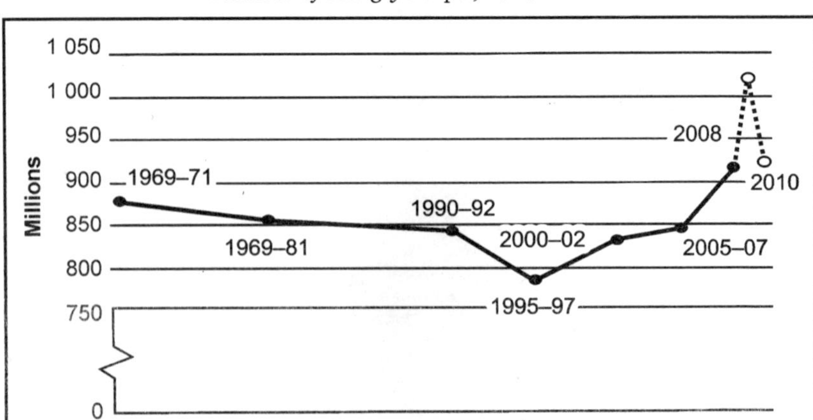

Source: FAO

international communities and many of us will be accustomed to it – it is how it works, so to speak.

The geography of poverty and social deprivation has changed dramatically over the last two decades. More than 70 per cent of the world's poor now live in middle-income countries. This pattern, likely to continue into the next decade, raises important questions. Have poverty reduction and human development kept up with income growth? Is growth incomplete without social progress and gender-inclusiveness? Consider South Asia, where the poverty rate fell from 60 per cent in 1981 to 40 per cent in 2005 – not fast enough, given population growth, to reduce the total number of poor people. In fact, the number of poor people (defined as those living on less than $1.25 per capita per day at 2005 purchasing power parity) in South Asia increased from 549 million in 1981 to 595 million in 2005, and from 420 million to 455 million in India, where almost three-quarters of the region's poor reside. In other words, while South Asia's economies have not underperformed on poverty reduction, merely matching global trends may not be enough for the region with the world's largest concentration of poor people. India has experienced slower income growth than has China, which partly explains its higher poverty rate. But a country's poverty rate also depends on the degree of income inequality – a reduction in which makes growth more pro-poor – and inequality in China has, in fact, increased *more* rapidly than in India. So a rising tide really *can* lift all boats, with growth trumping inequality when it comes to poverty reduction. Moreover, income growth has contributed to improved education. Adult literacy rates in South Asia match the global norm. But education outcomes lag behind when it comes to secondary and tertiary education, which are becoming increasingly necessary to thrive in today's world. Nor have health indicators kept up with income growth. South Asia has the world's highest rates of malnutrition and the largest number of undernourished children, who have higher mortality rates, lower cognitive performance, and a greater likelihood of dropping out of school. More than 200,000 people in India die annually from malaria, mainly in poor regions. And, while much of the existing international healthcare assistance is focused on sub-Saharan Africa, India, along with Bangladesh, Bhutan, Nepal, Pakistan, and Sri Lanka, are just as devastated by neglected tropical diseases. Indeed, India alone accounts for one-quarter of intestinal worm infections worldwide and more than one-half of all cases of elephantiasis, leprosy, and visceral leishmaniasis. Although gender parity in primary education has improved, dropout rates for girls are higher than for boys. The dowry tradition puts pressure on girls' families to marry them early, leading to a preference for sons – and thus to sex-specific abortions targeting female fetuses. Legislation, courts, and law-enforcement mechanisms have failed to address the high incidence of violence against women. Death rates for young girls are much higher than for boys. These indicators are symptomatic of a general pattern of discrimination. The expectation that girls will grow up to do little other than serve their husbands reduces parents' incentives to invest in their daughters' education. Uneducated women then have few alternatives, and the expectation becomes self-fulfilling, leaving women in a continuous cycle of powerlessness that has had significant adverse long-term effects. In 2008, India's female labour-force participation rate averaged 35 per cent, while low-income countries averaged 58 per cent. Moreover, a large proportion of women in the region are employed in the informal sector.

The paradox of South Asia is that growth has been instrumental in reducing poverty and improving social outcomes, but poverty rates and social outcomes have not improved fast enough to reduce the total number of people living in misery. As a result, policymakers should begin to consider direct policy interventions to accelerate social progress, with a particular focus on human development and gender inclusiveness. In today's uncertain world, social turmoil, gender deprivation, and rising conflict have tested countries' abilities to create jobs, promote gender equity, equip young people with skills, and design effective social protection programmes. Tackling these challenges requires a clear understanding of how economic opportunities can be broadened to ensure faster poverty reduction, promote human development, and stimulate gender-inclusive growth. Greater gender equality can contribute to economic growth and development, and major initiatives to increase opportunities for women can transform society. If more girls had gone to school a generation ago, millions of infant deaths could have been averted each year, and tens of millions of families could have been more educated, healthier, and happier. Deeper social disparities should never be viewed as the inevitable price of rapid growth, and more egalitarian outcomes in education, health, and gender should not be considered 'second-stage' reforms. A development strategy that promotes growth first, and only then deals with human misery, is not sustainable. Policies designed to make redistribution more efficient need not hamper growth itself.

II

Persistence of Poverty

Poverty is not simply a deficit, for that is simply to adopt the commodity based perspective of the North. Poverty can also be a resource. It may also be understood that poverty elimination does not mean equalization to Western culture and quality of life (Pieterse 2010). The agents of poverty reduction can benefit by poverty control activities such as market expansion and income escalation. That channel needs to work by visualizing the size of poverty and exclusion. Almost half the world – over three billion people – live on less than $2.50 a day. Approximately 1 billion people were living in slum conditions. The poorest 40 per cent of the world's population accounts for 5 per cent of global income. The richest 20 per cent accounts for three-quarters of world income. If current trends continue, the Millennium Development Goals target of halving the proportion of underweight children will be missed by 30 million children, largely because of slow progress in Southern Asia and sub-Saharan Africa. Number of children in poverty is 1 billion that is 50 per cent of total children (every second child) of the world. Rural areas account for three in every four people living on less than US$1 a day and a similar share of the world population suffering from malnutrition. However, urbanization is not synonymous with human progress. Urban slum growth is outpacing urban growth by a wide margin. Approximately half the world's population now lives in cities and towns (Shah found 2011). In 2005, the wealthiest 20 per cent of the world accounted for 76.6 per cent of total private consumption. The poorest fifth having only 1.5 per cent. The poorest 10 per cent accounted for just 0.5 per cent and the wealthiest 10 per cent accounted for 59 per cent of all the consumption. For every $1 in aid a developing country receives, over $25 is spent on debt repayment. A mere 12 per cent of world's population uses 85 per cent of its water, and these 12 per cent do not live in the Third World (Aghion and Aghion 2006; Shah 2011).

There are two basic types of malnutrition. The first and most important is protein-energy malnutrition – the lack of enough protein which all of the basic food groups provide. This is the type of malnutrition that is referred to when world hunger is discussed. The second type of malnutrition, also very important, is micronutrient (vitamin and mineral) deficiency. This is not the type of malnutrition that is referred to when world hunger is discussed, though it is certainly very important. Protein-energy malnutrition is the most lethal form of malnutrition/hunger. It is basically a lack of calories and protein. Food is converted into energy by humans, and the energy contained in food is measured by calories. Protein is necessary for key body functions including provision of essential amino acids and development and maintenance of muscles.

The geography of poverty and social deprivation has changed dramatically over the last two decades. More than 70 per cent of the world's poor now lives in middle-income countries. This pattern, likely to continue into the next decade, raises important questions. Have poverty reduction and human development kept up with income growth? Is growth incomplete without social progress and gender-inclusiveness (Ghani 2010)? Consider South Asia, where the poverty rate fell from 60 per cent in 1981 to 40 per cent in 2005 – not fast enough, given population growth, to reduce the total number of poor people. In fact, the number of poor people (defined as those living on less than $1.25 per capita per day at 2005 purchasing power parity) in South Asia *increased* from 549 million in 1981 to 595 million in 2005, and from 420 million to 455 million in India, where almost three-quarters of the region's poor reside. While emerging South Asia's economies have not underperformed on poverty reduction, merely matching global trends may not be enough for the region with the world's largest concentration of poor people … . Moreover, income growth has contributed to improved education. Adult literacy rates in South Asia match the global norm. But education outcomes lag behind when it comes to secondary and tertiary education, which are becoming increasingly necessary to thrive in today's world. South Asia has the world's highest rates of malnutrition and the largest number of undernourished children, who have higher mortality rates, lower cognitive performance, and a greater likelihood of dropping out of school. More than 200,000 people in India die annually from malaria, mainly in poor regions (Ghani 2010).

The paradox of South Asia is that growth has been instrumental in reducing poverty and improving social outcomes, but poverty rates and social outcomes have not improved fast enough to reduce the total number of people living in misery. As a result, policymakers should begin to consider *direct policy interventions* to accelerate social progress, with a particular focus on human development and gender inclusiveness (Ghani 2010). Asia could reduce poverty in late 20th century because of focus on decentralization such as Philippines (Lyons and Hasan 2002). India also reduced poverty through decentralization and constitutional reform (1993–94). As a result, many states of India such as Karnataka, Kerala and West Bengal developed. Decentralization was done in Brazil also to control feudal structure, in Chile to control over local government and for more development in Bolivia. The experience of evolution of local government and welfare states in Scandinavia convinces regarding positive results of decentralization (Heller 2001). Central and Latin America has shown special interest in decentralization and local government across party lines, perhaps more so than Asia and Sub-Africa. Sub-Saharan African countries are generally much less formally decentralized (for example, in terms of proportion of public expenditure or public employment of local government compared) to Central Government (Olowu 2003).

III

Employability and Empowerment

The concept of empowerment confines to the following issues:

1. Government employment strategies with thrust on skills generation.
2. Lifetime friendly job guarantee without discrimination in the labour market.

Employability linked empowerment of an individual is an absolute employability. It depends on assets and deployment mechanism. An individual's 'employability assets' includes knowledge, skills and attitudes. Knowledge means how much they know, skill means what they do with given knowledge and an attitude means how they do it. The talent transferability for these skills is also an issue for sustainability of employability. People also need the capability to use their knowledge and skill assets for efficient and optimum marketing. Deployment is the second issue. There is an interrelationship between assets and deployment. The significance of employment opportunities available may affect their willingness to undertake human development measures for further prosperity. Participation is an additional issue of employability. It shows the ability to demonstrate 'employability' assets and expose before the relevant market. Relative employability is 'the relative chances of getting and maintaining employment'. Employability not only depends on whether one is in a position to satisfy the conditions of specific jobs, but also on how a person compete relative to others. Taking the supply and demand of labour into account, challenges, credentials, knowledge and social structure may guarantee a good position in the job market. A situation of obtaining skill asset for getting good employment has led surplus competitors chasing the same top assignment. Further, there is a clear difference between individuals' expectations and the realities visualized in the labour market. Thus, employability has the dual sides of supply and demand of labour in order to exhibit the forwarding one's claim in the labour market.

Given the present quantum of poverty and social exclusion, the question that apparently hints is, after all, what generates it. In economic literature we have three economic systems – Market based, Public based and Mixed based. An influential policy approach emphasizes the need to promote market forces at the expense of the state-led development in poor and middle income countries commonly known as 'Washington Consensus' and advocated by IMF, World Bank with the backing of US Treasury. The underlying view is that the only effective way to reduce poverty is to promote growth via 'Trickle Down'. This consensus weakened recently because of less success in Latin American countries that followed this approach since 1980s (Aghion and Aghion 2006). This consensus was challenged by orthodox tradition such as Nancy Birdsall and Augusto de la Torre. It is known as 'Washington Contentious'. They argue for a package of policy measures that directly confronts poverty: better social safety nets, schools for the poor progressive taxes, promotion of small businesses, protecting workers rights, anti-discrimination policies, land reforms, improvement in public services and reducing protectionism in rich countries (Banerjee et al. 2006). Kuznet opined that inequality would follow an inverted U with the level of development. Rising inequality is an inevitable by-product of early stages of economic development and one just has to wait for enough development to occur for inequality to fall (Piketty 2006).

Many developing economies have adopted development strategies with an urban bias that have prevented rapid growth in agricultural productivity by imposing controls on prices and trades of agricultural commodities. Therefore, deregulation and trade liberalization will reduce poverty in developing countries (Eswaran and Kotwal 2006). In addition, traditional societies tend to be characterized by relatively low level of corruption owing to smallness, simplicity and high levels of monitoring of leaders by society's members. Developing societies, characterized by an increasing size and complexity of public decision making, tend to involve more delegation of authority by political leaders to bureaucrats, which leads to the emergence of corruption. At later stage of development, political leaders have greater resources available while bureaucrats become wealthier Corruption across countries with differing per capita incomes is supportive of the idea of corruption first going up then going down with development. The openness to trade and foreign investment also reduce corruption (Laffont 2010). Thus, the reasons of poverty are redistributional, colonial, geographic, lack of self-control on population growth, beneficiary unfriendly planning, public resource plundering by government and bureaucrats, market imperfections, undergrowth of social capital and infrastructure. One can even say a lack of discipline in a democratic federation. Poverty reduction depends first and foremost on the capacity of the poor to gain access to productive capacity and to demand a higher share of total surplus – through organized civil action. This requires sufficient representation in political parties and the state to ensure government transformation of power, institution mobilization and promotion of poverty reduction (Vedeld 2003). Poverty reduction is here understood as a complex and multifaceted concept, with the following dimension (Crook 2003).

(a) Empowerment of the poor understood as effective participation and voice – met by some degree of responsiveness on the government part (hence, it complements 'democracy').
(b) Social or human capacity development through better access to health, education, water, infrastructure (Sen 1999).
(c) Economic gains by the poor through pro-poor growth or improved economic opportunities.
(d) Social inequality reduction through income redistribution hereunder reduced inters-regional disparities.

Empowerment can be done through various service deliveries, microfinance, and guaranteed sources of employment and income. Better education and health can always prove better capacity to produce and earn for individual and national purposes. Several pro-poor strategies may lead to poverty reduction and escalation of economic activities. Redistribution is a type of equity-efficiency trade-off in which society's redistribution goals must be weighted against the supply side distortion that taxes and transfers create in labour supply, savings and other economic decisions. The canonical formalization of this trade-off is the theory of Mirrlees (1971) which has three main implications for the structure of optimal redistribution policy: (1) redistribution should be in cash rather than in kind; (2) it should combine a relatively high level of guaranteed income with a steep phase-out rate (the rate at which benefits are reduced as earnings increase: and (3) marginal rates at the top should be very low (i.e. the highest marginal incomes should not be taxed (Banerjee et al. 2006). In USA, Earned Tax Income Credit (EITC) is a tax transfer programme for the poor. Although it is not possible to give basic income support but complement it with subsidy to poor people based on the hours of work which may decline and fall in a phase out region. Several welfare payments including nominal unemployment benefits can also be planned (Saez 2006). Market imperfections lead to unemployment and loss of efficiency. The manifestations of market failures take the forms of poverty traps. Certain segments of the population are simply unable to accumulate the critical level of assets that would allow them or their offspring to achieve their productive potential, participate efficiently in economic activity, and ultimately escape poverty. However, even when it is sometimes possible to escape poverty, the same mechanism can lead to serious misallocations, resulting in large efficiency losses and making poverty a very persistent condition. Imperfections in credit market impeded efficient investment by the poor in human capital and education. Such investment for poor bring higher rate of return than the rich. Poor children do not possess better environment for education. This is also a kind of market failure. A poverty perpetuating mechanism arises when resources fall short of a basic nutrition – and health related efficiency wage level below which workers cannot effectively function or children properly develop. In South Asia and Africa malnutrition triggered because of drought (Ravallion 2006). Therefore, investment in education of children at secondary stage and skill formation are quite essential. This would also reduce child labour practices in poor countries of ASSACLA. Microfinance can create entrepreneurship and this is a good mechanism of self-made poverty reduction. However, this may be complemented by efficient service delivery system of the society and the government. Lastly, political leaders should have the requisite will to combat corruption; corruption can thereafter decline with further development (Laffont 2006). Capability enhancement is a very crucial factor of employability. It depends on quantity and quality of skill formation and up gradation.

Employment is a legal agreement between two individuals or groups or capitalists and labours. Here one individual or group serve as recruiter and the other are being recruited. An employee is recruited. As per dictionary they may be defined as: ‘A person in the job of another under any agreement, express or implied, oral or written, where the employer has the power or right to control and direct the employee in the material details of how the work is to be performed’ (Black’s Law Dictionary 5th edition 1979: 471). Employment depends on skill of the employee or worker who happens to be a poor person in our analysis. A skill is the learned capacity to carry out predetermined results often with the minimum outlay of time, energy, or both. Skills can often be divided into domain-general and domain-specific skills. For example, in the domain of work, some general skills would include time management, teamwork and leadership, self-motivation and others, whereas domain-specific skills would be useful only for a certain job. Skill usually requires certain environmental stimuli and situations to assess the level of expertise being shown and used. Many people need a large range of skills in order to contribute to an economy.

Employability also refers to a person's capability of gaining entry level employment, maintaining employment, and obtaining innovative and enjoyable employment. Employability is about being capable of obtaining and keeping fulfilling work. Employability is thus the capability to move self-respectfully within the labour market to realize potential through sustainable employment. For individuals, employability depends on the knowledge, skills and abilities (KSAs) they possess, the way they use those assets and present them to employers, and the context (e.g. personal circumstances and labour market environment) within which they seek work.

Employability is about work and the ability to be employed as:

1. the ability to gain initial employment; hence the interest in ensuring that 'key skills', careers advice and an understanding about the world of work are embedded in the education system.
2. the ability to maintain employment and make 'transitions' between jobs and roles within the same organization to meet new job requirements, and
3. the ability to obtain new employment if required, i.e. to be independent in the labour market by being willing and able to manage their own employment transitions between and within organizations.

Employability is a two-sided equation. Many individuals need various forms of support to overcome the physical and mental barriers to learning and personal development (i.e. updating their assets). Employability is not just about vocational and academic skills. Individuals need relevant and usable labour market information to help them make informed decisions about the labour market options available to them. They may also need support to realize when such information would be useful, and to interpret that information and turn it into intelligence. Finally, people also need the opportunities to do things differently, to access relevant training and, most crucially, employment. Both the supply and demand of labour need to be taken into account when defining employability, which is often dependent on factors outside of an individual's control. For individuals, employability depends on:

1. Their assets in terms of the knowledge, skills and attitudes they possess;
2. The way they use and deploy those assets;
3. The way they present them to employers;

IV

Governance

The government policy is aimed at:

1. more at the development and accreditation of knowledge and vocational skills than at the soft skills and attitudes
2. more on the demonstration of assets than their deployment – particularly for adults (e.g. lack of provision of a careers education and guidance service for adults)
3. more at individuals looking to enter the labour market (e.g. from education or unemployment) than within
4. more on the individual and the supply side, than on employers and the demand side (i.e. the labour market contextual factors).

- For the state, as well as raising the skill profile of the existing workforce, especially at lower levels to boost flexibility and competitiveness, there are a number of potential priority groups including:
- labour market entrants
- labour market re-entrants
- disadvantaged groups
- insecure or underutilized employees
- where different policies may need to be targeted according to different circumstances.

For employers the priorities might be to help key groups of staff to develop both those assets which have explicit, immediate value to the organization as well as those transferable ones which have a wider, longer term currency, thereby engendering a sense of security, encouraging commitment, risk-taking and flexibility among employees.

A brief review of government initiatives in this area suggests that policy is aimed:

1. more at the development and accreditation of knowledge and vocational skills than at the 'softer' skills and attitudes
2. more on the demonstration of assets than their deployment – particularly for adults (e.g. lack of provision of a careers education and guidance service for adults)
3. more at individuals looking to enter the labour market (e.g. from education or unemployment) than within

This policy orientation may reflect a variety of factors such as difficulties in defining, assessing and verifying 'soft skills', and difficulties identifying and accessing specific groups of employees at which to target being limited.

Thus, some key questions for future policy interventions include:

- who are the priority groups
- where the most serious gaps are for such groups be they related to, e.g. which assets, dimensions of deployment or presentational skills

- how these gaps might best be remedied and
- which of the arms of public policy are best placed to add such value and how through interventions.

Finally, whatever the interventions, they need to be evaluated so that lessons can be fed back into further improvements and to the decision to continue with, change or stop such interventions. Potential measures include those relating to input measures, e.g. possession of vocational qualifications; the views of management and the labour, and outcome measures; the speed at which people are able to get jobs or 'measurements of failure'; the numbers or proportion of people with difficulty finding or keeping work, or the number of job changes, however defined. Obviously, there is room for some combination of all three. Whatever route is chosen, it is important to take account of the overall state of the job market.

From the beginning of the current century there has been a discourse on poor but lousy governance in developing economies. The issue of corruption in service deliveries is frequently reported by media and those working on parallel economy. This is further followed by inefficiency in public offices. Virtually, corruption and inefficiency are going together. It has also been observed that there is gross absenteeism in public offices. This costs heavy losses to the exchequer. The terms of appointments in developing economies are such that the jobbers are unsackable and unaccountable. The long-term staffs are more corrupt, inefficient and unaccountable. The short-term and contractual staff are more responsive to society, more sincere and honest. The politicians are corrupt but at the time of fighting re-election they start distributing freebies such as enough subsidies, loan waivers, reservations in admission and appointments based on caste and distribution of physical goods. In recent socio-political movement, Aam Admi Party emerged and is becoming popular day by day. The upsurge of this party is basically its fighting against corruption and indecency at public places. Thus, because of poor and unresponsive governance, skewed income distribution has taken place. As a consequence malnutrition and poverty has persisted in developing nations. The time of global integration must also create an environment of social responsibilities of all segments of society such as public bureaucrats, corporate CEOs and giants, politicians and other influential capable persons. This process can also contain other social evils.

V

Conceptual Observations

Conceptualization usually depends on assumptions. These assumptions are often false. That is why it is called a concept. Theorization also requires a few assumptions which are lucid and necessary. A key and crucial assumption is one on which conclusions depend. It is also necessary that crucial assumptions are slightly realistic. This is because dubious assumptions are misleading and culminate to suspected results (Solow 1956). We wish to assume a fair and transparent economy. Either it is democratically planned and mixed one or market based does not matter. We also note it that the planning for which various programmes are made, proper execution and enforcement of these programmes are also perfectly made. It is also crucial to assume that government is having good governance values. This indicates an existence of little corruption while executions of public programmes and other interventions. It is also assumed that beneficiaries are having sound ethics and nationalism and they are hardly dragged into any kind of malpractices and are honest and sincere. Neither are they lured by corrupt practices nor do they encourage it. The issue of corruption is very much interrelated with other issues. At a global level, the 'international' (Washington Consensus-influenced) economic system that has shaped the current form of globalization in the past decades requires further scrutiny for it has also created conditions whereby corruption can flourish and exacerbate the conditions of people around the world who already have little say about their own destiny. A difficult thing to measure or compare, however, is the impact of corruption on poverty versus the effects of inequalities that are structured into law, such as unequal trade agreements, structural adjustment policies, so-called 'free' trade agreements and so on. It is easier to see corruption. It is harder to see these other more formal, even legal forms of 'corruption.' It is easy to assume that these are not even issues because they are part of the laws and institutions that govern national and international communities and many of us will be accustomed to it – it is how it works, so to speak (Shah 2011; Laffont 2006).

We also assume that crucial aggregate variables such as rational consumer and investment behaviour, multiplier and accelerator quickly work in the economy with the motive to reduce poverty. Central banking institutions facilitate to the beneficiaries in crediting production loans. The economy is tending to be monetized. Lastly, it is assumed that since the purpose is sustainable end of poverty, therefore budgetary allocations are made on projects which are efficient and viable and having long lasting results. Market is perfect. It is being observed that relative poverty persists in all economies with varying degrees and the percentage varies in Asian, Sub-Saharan African, Central and Latin American (ASSACLA) and European economies. Of course the percentage is higher in many economics of Asia, Africa and Latin America. The percentage is meagre in European economy. This is virtually least in several other OPEC economies. Since the definition of poverty in different economies are not uniform therefore, a poor of a rich country can be treated as rich for a poor country, if definitions are changing. However, it is true that the global scenario of poverty can be also visualized on the basis of per capita USD 2.5 a day. Accordingly, there are approximately three billion poor in the world. Population grows in Malthusian framework, so is also the growth of poor population in developing economies. It is being observed that the rate of growth of population is rising by an annual growth rate of 2 per cent. The number of poor population is rising even if the percentage of population is reduced slightly with government interventions. Therefore, it is assumed to control population up to a reasonable limit of 1 per cent annually.

Conclusion

Global growth data shows that in recent years developing country giants such as China and India have performed better than developed countries. At the same time, malnutrition and poverty is also high in developing country giants. This situation can be explained through rising population, inefficiency in offices, unaccoutable and unsackable bureaucracy and unjust governance. Poverty is a historical commodity deficit. Its reduction is possible, partly by birth and health control. At the same time decent decentralization, micro management, transparent working at the government level can lead to its alleviation. In the present fluctuating developing economy, social turmoil, gender deprivation, and rising conflict have tested countries' abilities to create jobs, promote gender equity, equip young people with skills, and design effective social protection programmes. Tackling these challenges requires a clear understanding of how economic opportunities can be broadened to ensure faster poverty reduction, promote human development, and stimulate gender-inclusive growth. Greater gender equality can contribute to economic growth and development, and major initiatives to increase opportunities for women can transform society. Deeper social disparities should never be viewed as the inevitable price of rapid growth, and more egalitarian outcomes in education, health and gender should not be considered 'second-stage' reforms. A development strategy that promotes growth first, and only then deals with human misery, is not sustainable. Policies designed to make redistribution more efficient need not hamper growth itself. Decentralization and democracy are also in their early stages (Crook 2003). Given if this review found connection between democratic detraction and poverty reduction at this stage in the process, it is likely that while short economic growth can be achieved with authoritarian systems, long-run economic growth that also address inequalities, requires a rule-based and responsive government that continuously adapt to people's rights and demands (North 1990). For sustainable poverty alleviation, however, focus should be more on long-term strategies rather than short-term strategies. Unfortunately, the ruling government comes into power for a defined and even fluctuating period and that is why they do not plan for the long-term basis. When planning for poverty reduction is done for a short-term then poverty quantum would re-occur and it can be never-ending. Thus, it is essential to theorize poverty alleviation and act accordingly. Else, poverty will remain the property of a few powerful persons. Lastly, agents of governance should be culturally skilled and specialists of the concerned area. Unfortunately, higher education has not been linked to top government machineries. This needs to be looked into for sustainable poverty alleviation

References

Aghion, Philippe and Beatriz Armendariz de Aghion. 2006. 'New Growth Approach to Poverty Alleviation', in Abhijit Vinayak Banerjee, Roland Benabou and Dilip Mookherjee (eds), *Understanding Poverty*, New York: Oxford University Press, 73–84.

Banerjee, Abhijit Vinayak, Roland Benabou and Dilip Mookherjee (eds) 2006. *Understanding Poverty*, New York: Oxford University Press, 161–168.

Black, Henry Campbell. 1979. *'Black's Law Dictionary'*, 5th Edition, West Publishing Company.

Bossuyt, J. and J. Gould. 2000. 'Decentralization and Poverty Reduction: Elaborating the Linkages', IDS Policy Management Brief 12, Helsinki: IDS.

Bratton, M., M. Coulibali and F. Machado. 2002. 'Popular View of the Legitimacy of the State in Mali', *Canadian Journal of African Studies*, 36 (2): 197–238.

Brown, P. and A. Hesketh. 2004. *The Mismanagement of Talent: Employability and Jobs in the Knowledge Economy*, Oxford: Oxford University Press, retrieved from http://en.wikipedia.org/wiki/employability.

Crook, R. and A.S. Sverrission. 2001. 'Decentralization and Poverty Alleviation in Developing Countries: A Comparative Analysis', IDS Working Paper 130, Brighton: IDS.

Crook, R.C. 2003. 'Decentralization and Poverty Reduction in Africa: The Politics of Local Central Relations', *Public Administration and Development*, 23 (1): 77–88.

Eswasan, M and Kotwall, Ashok. 2006. 'The Role of Agriculture in Development' in Abhijit Banerjee et al. (ed.), *Understanding Poverty*, New York: OUP.

FAO. 2010. *World Hunger Report*.

Ghani, Ejaz (ed.) 2010. *The Poor Half Billion in South Asia: What is Holding Back Lagging Regions?*, New Delhi: Oxford University Press.

Heller, P. 2001. 'Moving the State: The Politics of Democratic Decentralization in Kerala, South Africa and Porto Alegre', *Politics and Society*, 29 (1): 131–163.

Jenkins, Stephen P. and John Micklewright. 2007. *Inequality and Poverty Re-examined*, New Delhi: Oxford University Press, 3–36.

Johnson, C. 2001. 'Decentralization in India: Poverty, Policy and Panchayati Raj', Draft Working Paper 199, London: ODI, September.

Laffont, Jean Jacques. 2006. 'Corruption and Development', in Abhijit Vinayak Banerjee, Roland Benabou and Dilip Mookherjee (eds), *Understanding Poverty*, New York: Oxford University Press.

Lyons, M. and S. Hasan. 2002. 'Researching Asia's Third Sector', *Voluntas: International Journal of Voluntary and Non-Profit Organization*, 13 (2), June: 107–112.

Mirrlees, James A. 1971. *Optimal Taxation in Theory and Practice*, Harvard USA.

Narayan, Deepa (ed.) 2002. *Empowerment and Poverty Reduction: A Source Book*, Washington, D.C.: World Bank.

North, C.D. 1990. *Institutions, Institutional Change and Economic Performance*, Cambridge: Cambridge University Press.

Olowu, D. 2003. 'Local, Institutional and Political Structures and Processes: Recent Experience from Africa', *Public Administration and Development*, 23 (1): 41–52.

Pieterse, Jan Nederveen. 2010. *Development Theory*, 2nd edition, New Delhi: Sage.

Piketty, Thomas. 2006. 'The Kuznets Curve: Yesterday and Tomorrow', in Abhijit Vinayak Banerjee, Roland Benabou and Dilip Mookherjee (eds), *Understanding Poverty*, New York: Oxford University Press.

Rahnema, M. 1992. 'Poverty', in W. Sachs (ed.), *The Development Dictionary: A Guide to Knowledge as Power*, London: Zed Publications, 158–176.

Ravallion, Martin. 2006. 'Transfer and Safety Nets in Poor Countries: Revisiting the Trade Offs and Policy Options', in Abhijit Vinayak Banerjee, Roland Benabou and Dilip Mookherjee (eds), *Understanding Poverty*, New York: Oxford University Press, 203–230.

Saez, Emmanuel. 2006. 'Redistribution towards Low Incomes in Richer Countries', in Abhijit Vinayak Banerjee, Roland Benabou and Dilip Mookherjee (eds), *Understanding Poverty*, New York: Oxford University Press, 187–202.

Schneider, K. and H.U. Otto. 2009. *From Employability towards Capability*, Luxembourg: Inter-Actions.

Sen, A. 1999. *Development as Freedom*, Oxford: Oxford University Press.

Shah, Anup. 2011. Global Issues, www.global issues.org/issue/2/causes of poverty.

Solow, R.M. 1956. 'A Contribution to the Theory of Economic Growth', *Quarterly Journal of Economics*, 70 (1): 65–94.

Vedeld, Trond. 2003. *Democratic Decentralization and Poverty Reduction: Exploring the Linkages*, Forum for Development Studies No. 2, Bergen, Norway: Norwegian Association for Development Research, 159–204.

World Bank. 2000. *Reforming Public Institution and Strengthening Governance: A World Bank Strategy*, Washington, D.C.

——. 2008 and 2013. *World Bank Development Indicators*, Washington, D.C.

Index